THE FIRST PASSIVE SOLAR HOME AWARDS

by
Franklin Research Center
Philadelphia, PA

for
The U.S. Department of Housing and Urban Development
Office of Policy Development and Research

University Press of the Pacific
Honolulu, Hawaii

The First Passive Solar Home Awards

by
National Solar Heating and Cooling Information Center

for
U.S. Department of Housing and Urban Development
U.S. Department of Energy

ISBN: 1-4102-2496-1

Reprinted from the 1979 edition

University Press of the Pacific
Honolulu, Hawaii
http://www.universitypressofthepacific.com

EDITOR'S NOTE

The Passive Design Competition is a step forward in the refinement of solar design and building. The grant application introduced a new calculation procedure which allowed for the comparison of numerous designs. Heretofore, passive solar designers were free to develop their own methods of estimating their building's **thermal load, auxiliary energy** needs, and **solar contribution**. In the competition, over 550 applicants completed the same procedure (see Chapter 9—Energy Calculations).

These 550 submissions were then subjected to a technical evaluation by a team of some 50 solar experts. During this review process, two related concerns emerged: 1.) some of the numbers in the applicants' energy calculations were questionable, although the designs seemed workable; and 2.) a number of commonly-made thermal mistakes appeared.

1. While the technical reviewers were able to re-calculate and double check each questionable number before they made a final determination on the reliability of the performance estimate, in general it was decided to present the original designers' calculations rather than these new values.

In a few cases however, where a recheck of the original numbers showed significant problems, we have corrected them so as not to confuse the reader. We have included a chapter on the calculation procedure. We urge the reader to study this section and to carefully check the numbers given for **thermal load, auxiliary energy** and **yearly solar fraction** before attempting to apply them to a different building. Remember, each passive solar building will have a unique set of thermal requirements based on its particular size, location, climate and system performance.

2. Secondly, the concerns over commonly-made thermal mistakes lead us to include the material in Chapters 7 and 8. These sections address the common mistakes in design and the importance of passive solar construction details. Here again, the reader will benefit from the findings of the technical reviewers.

HUD, DOE, SERI and the contractors who prepared this text assume no responsibility for the accuracy or reliability of the applicants' calculations for use in construction.

NOTE: SECOND PRINTING

The use of the words **Auxiliary Energy** in the project summaries throughout this book has been a source of some confusion. The number given for **Auxiliary Energy** should more properly be designated **Heating Energy**. **Heating Energy** is defined on page 208 as

$$\left[\begin{array}{c} \text{Heating Energy} \\ \text{BTU/DD-ft}^2 \end{array} \right] = \frac{\text{(Internal Sources)} + \text{(Auxiliary Heating)}}{\text{(Degree-Days) (Floor Area)}}$$

In addition, the method of calculating the passive solar performance has been improved since the publication of the first edition of the **First Passive Solar Home Awards**. The newer Solar Load Ratio Method was used to evaluate all the projects submitted under HUD's Cycle 5 solicitation. At this time, all the projects listed in this book are being recalculated using the newer method of Cycle 5. A table containing updated values for the **gross thermal load, auxiliary energy** and **yearly solar fraction** is being prepared. This table will be available from the National Solar Heating and Cooling Information Center.

It has also been noted that it is not possible to derive the **yearly solar fraction** from the values given in each project summary. Values of the **gross thermal load** included in the above mentioned table may be combined with the information given in the project summaries to derive the **yearly solar fraction** independently.

Finally, it should be noted that the abbreviation **MBTU** stands for **Million British Thermal Units**. The M does not mean 1,000 as in some other scientific notation.

TO THE READER:

The National Solar Heating and Cooling of Buildings Program, managed by the U.S. Department of Energy, has now been underway for over four years. As part of this program, the U.S. Department of Housing and Urban Development administers the residential demonstration, residential market development, standards, and information dissemination activities. In the demonstration portion of the program, HUD has funded solar systems in over 12,000 dwelling units through more than 400 grants.

As more people became aware of the potential of solar energy during this period, more also came to appreciate the potential value of utilizing solar energy through better design and materials selection in our housing. While such "passive" uses of solar energy have been known for centuries, only recently have design and analysis techniques been developed which permit the designer, the builder, and the homeowner to estimate the portion of a home's energy demands which can be met by passive solar methods.

To identify promising new passive solar designs and to publicize this information widely throughout the housing industry, HUD, with the cooperation of the Department of Energy and the Solar Energy Research Institute, conducted a Residential Passive Solar Design Competition and Demonstration during the summer of 1978. In response to the call for designs for passive homes, over 550 applications were submitted; 162 of these designs were selected for awards, 145 for new homes and 17 for "retrofit" installations of passive solar elements on existing homes. An especially promising sign is that 80 of the new home designs are being built for sale on the open market by builders who believe that passive solar will sell.

This report of the results of the design competition is intended to provide examples of the best passive architecture now available, to discuss in some depth the particular design decisions which need to be made in creating a passive solar home, and to point out design mistakes to avoid. Sad to say, while almost one third of the designs were worth awards, about as many exhibited a very limited understanding of how energy flows within a building structure.

Finally, while these designs are for passive solar homes which make little use of conventional energy sources except for backup, the design principles involved can be used in any home design. It may not always be possible, due to local market conditions or other factors, to build and sell a completely passive solar home design. It is, though, always possible to design a home to make significant use of solar energy through passive methods.

The homes in this report are a first step. We in the Federal solar program will continue to work with the design professions and the home building industry to improve design procedures, to develop new materials and building components, and to take other steps to encourage the use of solar energy as one answer to our growing need for alternative ways to obtain the energy we need.

Sincerely,

Donna E. Shalala

TABLE OF CONTENTS

INTRODUCTION

The passive solar home makes use of the materials from which the dwelling is constructed to directly capture, store, and distribute solar heat to its occupants. Sunlight enters the building through south-facing glass or plastic. Dense material such as brick, block, concrete, stone, adobe, or water absorbs this solar energy as heat as it enters the house. Living spaces of the home are carefully arranged so that they are in direct thermal contact with this storage mass allowing these spaces to be heated directly without the expense of special plumbing or forced hot air distribution systems.

Passive designs require considerations of solar and heat flow considerations in every detail and component. Floor plan layout, circulation patterns, window location, and the selection of wall and floor materials all affect how well a passive design will work. The entire house *is* a solar energy system with many of its components now having dual functions; both the traditional function of providing a quality enclosure and the solar function of collecting, storing, and distributing heat. Windows not only let light in and allow a view, but collect heat as well. Walls which subdivide and enclose space can also store and radiate heat. Components whose functions were primarily structural, spatial, or aesthetic may double as solar heating components. Often a traditional building component can be replaced by one with a dual solar function and built-in nature of passive designs, the builder is able to realize new economies in solar heating at minimum risk.

Passive is not a special trade or industry per se; rather it is an approach to building which allows the sun to enter and be used in the home naturally. The key to the passive approach rests in the proper arrangement of common materials for residential construction. The passive approach to solar homes cannot be subcontracted to plumbing or solar equipment installers. It is not an add-on item. Only the home builder can produce the passive solar home.

The passive approach to solar heating is not new. Deeply rooted in climatically-responsive building techniques, many cultures throughout the world have used passive features for centuries in traditional home construction. Before white settlers arrived on this continent, the Pueblo Indians of the southwest built whole communities which were passively solar heated.

In the 1933 Chicago World's Fair, the "Crystal House" successfully demonstrated a modern direct gain passive heating system. Built with ample concrete masonry and south-facing glass, this house proved that a residential living space can, without mechanical assistance, collect, store, and distribute solar heat comfortably to its occupants. The designers of this house, George and William Keck, saw a number of their direct gain designs successfully built during the 1940's and 1950's. However, at that time, the costs of conventional fuels was still very low and solar heating by passive or other means received little attention.

As the cost of fuels doubled in the mid 1970's, solar energy became attractive as a source of heat to a growing number of Americans. However, the widespread marketing of solar homes was still hampered by high initial costs and by maintenance requirements.

By the simplicity of its construction, the passive solar approach can effectively surpass these market hurdles. With few or no moving parts, operating and maintenance costs are minimal. The initial costs for passive solar are not fixed by the costs of equipment, but are determined largely by the design and materials which the builder selects. The passive approach has produced some of the most cost-effective solar homes in existence today.

HEATING COMPARISON
VARIOUS HOME TYPES

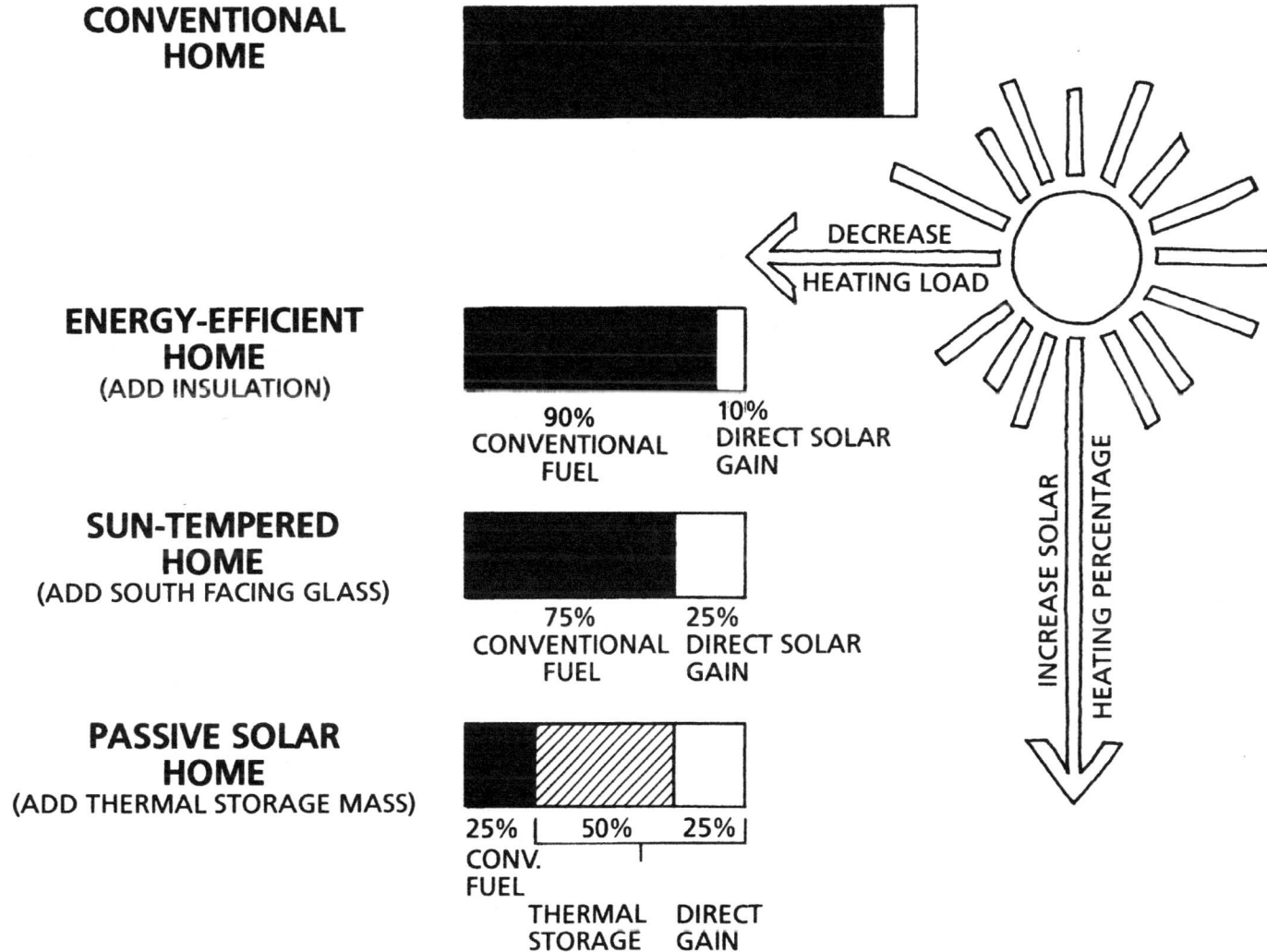

CONVENTIONAL HOME

ENERGY-EFFICIENT HOME
(ADD INSULATION)

90% CONVENTIONAL FUEL   10% DIRECT SOLAR GAIN

SUN-TEMPERED HOME
(ADD SOUTH FACING GLASS)

75% CONVENTIONAL FUEL   25% DIRECT SOLAR GAIN

PASSIVE SOLAR HOME
(ADD THERMAL STORAGE MASS)

25% CONV. FUEL   50% THERMAL STORAGE   25% DIRECT GAIN

DECREASE HEATING LOAD

INCREASE SOLAR HEATING PERCENTAGE

3

Due to the recent surge of interest in passive buildings and the results of recent demonstration homes, the U.S. Department of Housing and Urban Development, with the cooperation of the U.S. Department of Energy and the Solar Energy Research Institute, conducted a design competition for passive solar residential designs to stimulate the development of a passive solar builder's market. Submissions for single family home designs using solar heating and cooling systems were received from over 550 architects, engineers, builders and home designers, during the summer of 1978.

A passive solar home is defined in the HUD competition guidelines as "one in which the thermal energy flow is by natural means". Low horsepower fans and pumps were permitted in the competition to assist this natural energy flow, but the major use of mechanical support systems was prohibited. All successful entries in the competition include three common elements. Each passive home has: *1) a very effective thermal envelope; 2) transparent south-facing surfaces for collection; 3) a building mass which serves to absorb and store thermal energy; and 4) distribution.*

The first element, *a tight and thoroughly insulated building enclosure* to contain and regulate the transfer of heat between inside and outside, is not limited to passive solar homes. Many builders have been marketing for several years what are commonly labeled "energy-efficient" homes. Measures taken to reduce the heat losses in a home have generated the greatest savings to the homeowner for each dollar spent. Saving a BTU of heat is generally easier and cheaper than producing one from any source. This applies especially to the passively heated home. Many of those builders who entered this "energy-efficient" home market several years ago are moving into the passive solar market today.

The second element, *transparent surfaces for solar collection or "sun tempering,"* will be found to some degree in any home which has exposed south-facing windows, which are efficient solar energy collectors. The average home receives 10% of its heat entirely by accident from ordinary windows on its south side! This winter heat is a free bonus to the homeowner.

Some homes do not receive this heat supplement because of carports or solid walls on their south sides. Others with large south-facing windows or sun tempering receive twice the solar gain of the average home. In conventional light-frame construction, the heat gain from these windows is useful only during sunny daylight hours and in the early evening which immediately follows. This "sun-tempered" approach, without additional thermal mass for heat storage, can only provide about 25% of a home's heating

needs. Still, sun tempering is a very practical approach to provide supplementary heat to a home at very little extra cost.

The third element which distinguishes the "passive solar" home from sun-tempered and energy-efficient homes is *the ability of the home to store excess heat during the daytime in the mass of its construction for use at night.* The location of the heat storage mass varies greatly from one design to the next.

Some designs have dense floor materials such as concrete, brick, or tile to absorb heat. Others include masonry walls, ceilings, free-standing mass forms, and water-filled containers to provide additional heat storage. In no case should this mass for heat storage be directly exposed to the outside. The insulated envelope of the dwelling must enclose this mass to hold its heat. Glass areas have little insulation value and should have double glazing with moveable insulation, if possible.

The masonry chimney which rises through light frame construction is a traditional form which has provided supplemental heat storage when placed on an inside wall. The limited surface area of this chimney, however, does not allow it to absorb a very significant portion of a home's heating needs. Not only must a *sufficient mass be inside the home to store heat, but it must be in a configuration with sufficient surface area* to allow the transfer of this heat into the mass during the daytime and into the heated space at night. Although many passive homes are very simple to build, they all require a very comprehensive thermal design to allow a natural, balanced flow of heat.

The home designs presented in this book use passive solar as their primary heat source. The passive approach can be used incrementally for small percentages of passive solar at minimal costs. However, to avoid confusing the home-buying public, we must maintain a clear distinction between "passive solar" homes and their look-alike cousin, the "sun-tempered" home. While both have south-facing glazed surfaces for heat collection, the passive home contains thermal mass for heat storage while the sun-tempered house does not. Thermal storage is what allows a home to supply over 50% of its heat by solar means. In fig. 1-1 you can see a graphic comparison between the house types discussed here. The energy payoffs for marketing homes which heat passively are very significant. To integrate a working thermal storage mass into the home in a low-cost manner which is both pleasing and tasteful to the home buyer is the true challenge facing the passive home designer and builder today.

CHAPTER 1
MARKETING THE SOLAR PASSIVE HOME

To successfully enter the passive solar home market, you must completely understand the home type which you are building. Although simple to construct, the passive home is a very complex thermal design which must be tuned for peak performance before the first brick is laid. Builders who lack experience with passive should begin by reading all that they can about the design of passive homes. (See bibliography on page 223.) Visiting passive homes in your region will also be very helpful. A discussion with the owner or occupant of a passive home will reveal some of the inherent and unique characteristics of living in a passive solar home.

Local conferences and educational programs on passive solar design and construction are now becoming more common. Although these may be time consuming, they can put you, the home builder, in touch with passive home designers from your area, other builders with previous experience, and passive solar enthusiasts who are potential clients. If passive solar building is a new area for you, a professional liaison with a home designer who has had passive solar experience will be invaluable. The costs for a thermally-balanced home design tailored to your local climate can be recovered from building several homes, each only modified slightly from the original design. It is of paramount importance that your first passive solar home project successfully demonstrate the cost-effective elegance of the direct solar heating approach. The credibility of passive solar homebuilding and the future expansion of the market are contingent on the selection of thermally-balanced designs which are comfortable to live in and which use a minimum of supplementary fuels.

The passive solar market began from a small number of unorthodox designers and builders, many of them owner-builders who were not concerned with the wide-scale marketability of their final products. By trial and error, they explored numerous ways to build homes of common and often indigenous materials which were heated directly by the sun. As these early prototypes proved very successful, additional builders who were searching for a low-cost and direct approach to solar moved into this arena. To support this growing industry, reliable testing and simulation methods were developed by prestigious engineering research centers, most notably the Los Alamos Scientific Laboratories in New Mexico. Their findings further documented the positive results of early prototypes.

With the groundwork laid for an expanding passive home building industry, the challenge now rests with the homebuilder to adapt these trial configurations to passive residential construction in ways that are attractive to the market at large. Passive solar homes have far more masonry in their interiors than do conventional ones. This offers a more durable quality to the interior than typical light-frame construction, but has less flexibility for rewiring and modifications. To ease the transition to masonry walls from a home market which is used to monolithic gypsum board, plaster or stucco can be applied to some of these surfaces to soften their appearance.

Masonry floors are more of a challenge. Carpeting and hardwood floors are currently popular because of their soft and resilient qualities, but are not nearly as thermally effective as masonry. Where masonry floors are planned for thermal mass, brick pavers, tile, slate, and terrazzo can be added to give texture and character to a concrete slab. Padded resilient tile or carpet can be added in high traffic areas to cushion the floor surfaces and area rugs can be added to central living areas. Carpeting, however, should be avoided over masonry floors which are intended to store heat, particularly in areas where the sunlight strikes the floor directly. Carpeting acts as insulation over the floor mass, making it virtually ineffective as heat storage.

When building passive solar homes, trade-offs are encountered between thermal and user requirements. The means of heat storage must be compatible with the home buyer's expectations. Where conflicts arise, substitutions often can be made. Where hard floor surfaces are unacceptable to the home market, an equivalent wall mass may be substituted. If this wall mass receives sunlight directly, it will store four times as much heat as a wall or floor which receives heat indirectly. The passive solar home must cater to the needs of the user and not only to thermal efficiencies. Studies have shown that the American public is interested in quality solar homes, but their prime interest is in good quality homes for their investment; the importance of solar features is secondary.

The greatest marketing bonus comes to the builder who constructs a passive solar residence for himself. Living in a passive home offers first-hand information on the essential qualities of this home type, in addition to revealing any hidden weaknesses of a particular design. In this home, the builder can also experiment with options which require user involvement such as moveable insulation for protecting glass areas. The knowledge gained from this expe-

rience is only part of the reward. A unique and genuine confidence in passive homes and building techniques is shown by the builder who lives in one.

Building a passive solar house for speculation is another excellent way to enter into passive construction. Unlike the custom-built home, the homebuilder retains control over the final product and directly experiences the kinds of trade-offs encountered in a passive solar project. To make the most of this speculative venture, the builder should consult a passive solar architect or designer.

A drawback to speculative houses is that capital risks are assumed by the builder. However, builders who can include passive design features at little extra cost will find a great marketing advantage from a small additional risk.

Some custom builders enter the passive solar arena based on requests for passive homes from their clients. People knowledgeable or interested in passive solar heating are seeking qualified builders to construct their dream homes. This is an exciting opportunity to a builder who lacks passive solar experience. Traditionally, the builder has made most of the structural and material decisions in home construction prior to the selection of finish surfaces. However, the passive solar client will probably want more control over all phases of construction. Any naive assumptions that a client or builder may have on passive designs must be corrected before construction begins. A design consultant with a passive solar background can alleviate many potential problems here, and make this situation more promising to both the builder and client.

Some homes, particularly those which are passive, have a distinct appearance which can be a tremendous asset in the right location.

In a passive solar village or community, the solar features reinforce each other and tie together a visual fabric for the community. The marketability of a solar village goes well beyond mere visual advantages. Street layouts in a solar development can be carefully planned so that every lot and home site has a good solar orientation. New environmental concerns in land use can be addressed in a community of environmentally-oriented housing. Water conservation, preservation of existing vegetation, and erosion control can be included in the community guidelines. Solar rights can be effectively solved by a simple covenant in the deed, or a homeowner's association document. In these communities, the use of solar energy can be a unifying element in homes, in gardens, and in people's daily lives.

Passive solar construction is an open market at this stage. Those who can get in first, and establish their reputation for sound, workable homes will find their services in great demand as the market for energy-conscious, passive solar homes continues to expand.

One marketing strategy for passive solar homes which cannot be overemphasized is to let the product sell itself. Do not oversell the home and create expectations from the buyer which exceed its performance. A recent study indicated that home buyers who expected 80% solar heating and got only 60% were very disappointed, while those who expected 40% and got 60% were enthralled at their home's performance. Try to be on the low side of the anticipated performance when you present a home to clients, and when they discover how well the system actually works, they may even sell your next two homes for you! As Wayne Nichols, a pioneer in building and marketing passive solar homes, once said, "There is something magic to the American public about solar heating." The builder only needs to provide the passive solar home. The buyer will discover its magic.

CHAPTER 2
HOW TO USE THIS BOOK

The information presented in the individual project descriptions was selected from the applications of the 162 projects selected to receive awards. To make the report most useful to project designers and builders, several of the best projects in each of the passive categories—Direct Gain, Indirect Gain, and Solarium—were selected for in-depth discussions and analyses of the design approach being used. These project write-ups were prepared by some of the best-qualified passive solar experts in the country, and reflect the leading edge of the state of the art.

Following the in-depth discussions, all of the other design awards in the same category are listed with brief illustrations and descriptions. Space and time considerations precluded providing extensive discussions of all of the 162 designs receiving awards. It should be noted, also, that some of these designs may have been revised before construction to meet specific client requirements. Such revisions have not been included in this report.

For additional information on any project listed in the report, the name, address, and telephone number of the designer and the builder can be obtained from the National Solar Heating and Cooling Information Center, P.O. Box 1607, Rockville, MD 20850.

In addition to the project descriptions, the report includes general material on selecting the best type of solar project for a given area, issues involved in marketing passive solar homes, and calculating the solar gain from passive systems. In addition, there is a bibliography listing some of the most important publications on passive solar design and construction.

INDEX
PROJECT CLASSIFICATION
All of the 162 projects in this book are divided into three general categories; DIRECT SOLAR GAIN, INDIRECT SOLAR GAIN, and SOLARIUM. Many, if not most of the designs considered in this competition are combinations of two or three of these general system types. The fact that a particular design appears in only one section is a result of examining the design closely and deciding which one of the many passive features was primary. The key criteria used to identify the primary passive feature (or system) was: "What part of the passive system is in contact with the occupants and their primary living spaces?"

PAGE	DESIGNER	LOCATION	NO. OF UNITS	DEGREE DAYS	% SOLAR	SYSTEM TYPE
44	William L. Burns, The Burns/Peters Group	Tsaile, AZ	1*	7,467	59	DIRECT
45	Jim Raney, Sun System Engineering	Cottonwood, AZ	1	2,548	94	DIRECT
45	Paul Fellers	Davis, CA	2	2,374	82	DIRECT
46	Peter Calthorpe, Calthorpe/Fernau/Wilcox	Dillon Beach, CA	1	3,413	89	DIRECT
46	Larry and Jacqueline Morgan	Georgetown, CA	1	4,450	71	DIRECT
47	Astrin, Jonathan Allan Stoumen, Architect	Miranda, CA	1	3,270	92	DIRECT
30	Peter Calthorpe, Calthorpe/Fernau/Wilcox	Occidental, CA	1	3,019	82	DIRECT
47	Lynn S. Nelson, The Habitat Center	Pacheco, CA	1	2,697	92	DIRECT
48	David Elfring	Aurora, CO	2	5,936	40	DIRECT
48	F. R. Rutz, Fenton A. Bain & Frederick Rutz	Boulder, CO	1	6,202	45	DIRECT
49	Jeffrey Ellis, Superstructures	Boulder, CO	1	5,446	52	DIRECT^R
49	Stephen Sparn	Boulder, CO	1	6,173	59	DIRECT
50	Jim Logan and Paul Nicely, Logan Construction	Boulder, CO	1	5,803	77	DIRECT
50	Rick Cowlishaw, Natural Systems, Inc.	Colorado Springs, CO	5	6,423	55	DIRECT
40	David Wagner, David G. Wagner and Associates	Ft. Collins, CO	1	6,599	83	DIRECT
22	Joe Migani, Teamworks	New Haven, CT	2*	5,840	44	DIRECT
51	Stephen Lasar	New Milford, CT	1	6,673	51	DIRECT
51	Thomas Lee Goodson, Goodson and Ahlquist Assoc.	Columbus, GA	1	2,383	82	DIRECT
52	W. Jeff Floyd, Jr.	Norcross, GA	1	2,940	51	DIRECT
52	Robert L. Beecher II	Indianapolis, IN	1	5,699	48	DIRECT
53	C. F. Abercrombie	Salina, KS	1	5,628	76	DIRECT
53	John J. Rossini, Architects, Inc.	Amherst, MA	1	6,575	26	DIRECT
54	Mollie Moran, Massdesign	Andover, MA	1	5,750	64	DIRECT
54	Ole Hammarlund, Solsearch Architects	Sharon, MA	1	5,750	82	DIRECT
55	John P. Ross, John Ross/Trellis and Watkins, Inc.	Gaithersburg, MD	1	4,142	54	DIRECT
55	Conrad Heeschen, Sunsystems	Wilton, ME	1	8,505	75	DIRECT
56	Glen Langley, Philip S. Tambling/Glen Langley	York, ME	1	7,446	47	DIRECT

*Attached home   R—Retrofit system

PAGE	DESIGNER	LOCATION	NO. OF UNITS	DEGREE DAYS	% SOLAR	SYSTEM TYPE
26	Tom Ellison/John Carmody	Burnsville, MN	1	8,382	93	DIRECT
56	Peter Pfister	Minneapolis, MN	1	8,248	46	DIRECT[R]
57	L. W. Berg, Berg and Associates	Maple Grove, MN	1	8,062	87	DIRECT
57	Michael Cox	Minneapolis, MN	1	8,382	80	DIRECT
58	John Hueser	Kansas City, MO	1	4,711	71	DIRECT
58	William R. Watkins, Jr., Sunshelter Design	Wake Forest, NC	1	3,352	55	DIRECT
59	Bob Chase, Maxwell Starkman and Associates	Las Vegas, NV	1	2,625	60	DIRECT
59	Steven J. Strong, Solar Design Associates	North Conway, NH	1	7,613	72	DIRECT
60	Carl Fike	Somersworth, NH	1	7,252	79	DIRECT
36	Doug Kelbaugh, Environmint Partnership	Princeton, NJ	1	4,911	77	DIRECT[R]
60	Robert Nicolais/Quentin C. Wilson	El Rito, NM	1	5,605	71	DIRECT
61	Monika Lumsdaine, E & M Lumsdaine, Solar Consult.	Las Cruces, NM	1	3,260	65	DIRECT
61	William L. Burns, The Burns/Peters Group	Moriarty, NM	1	5,559	71	DIRECT
62	Ken Brooks, Rocky Mountain Sun Power	Santa Fe, NM	1	5,913	54	DIRECT
62	Jim Hays, Jim and Laura Hays/Bob Slattery	Santa Fe, NM	1	5,720	NA	DIRECT[R]
63	Thomas C. Regino, Energy Resources Group/Albin Assoc.	Bronx, NY	1*	4,778	34	DIRECT[R]
63	Anne Hersh, Connell and Hersh Architects	Woodhull, NY	4	6,532	73	DIRECT
64	Alfred De Vido, Alfred De Vido Associates, Architects	East Hampton, NY	2	5,400	100	DIRECT
64	Richard D. Strayer	Fredericktown, OH	1	5,543	95	DIRECT
65	Lee Kersh	Eugene, OR	1	4,599	87	DIRECT
65	Thomas A. Meados	West Salem, OR	1	4,599	73	DIRECT
18	Jim Morgan	Bedminster Twp., PA	1	3,911	47	DIRECT
66	Michael V. Arnold and TEA	Carlisle, PA	1	5,300	55	DIRECT
66	David Smith, Design Services	Bovina, TX	1	5,176	35	DIRECT
67	Richard A. West, Spartan Technologies, Inc.	Spring, TX	1	1,434	63	DIRECT
67	Sam Cravotta, Star Tannery Design Studio	Winchester, VA	1	5,662	82	DIRECT
68	James Guy, Barbara Guy and Christopher Umberger	Wytheville, VA	1	4,675	49	DIRECT

*Attached home   R—Retrofit system

PAGE	DESIGNER	LOCATION	NO. OF UNITS	DEGREE DAYS	% SOLAR	SYSTEM TYPE
68	Douglas C. Taff, Parallax, Inc.	Shelburne, VT	2	7,727	63	DIRECT
69	Peter Pfister	Amery, WI	1	8,248	54	DIRECT
69	James Cardwell	Beloit, WI	5	7,863	38	DIRECT
70	Douglas Steege	Chilton, WI	1	7,852	46	DIRECT
110	Jack Cohen, Goldblatt, Cohen and Aros	Tucson, AZ	6	1,855	94	INDIRECT
111	Dick Munday	Devonshire, CA	1	8,208	34	INDIRECT
111	Tom Carver, Sierra Engineering	Fair Oaks, CA	1	2,374	88	INDIRECT
112	Paul Shipee, Colorado Sunworks	Boulder, CO	4	5,540	92	INDIRECT
112	Deidre McCrystal, McCrystal Design	Boulder, CO	1	6,051	72	INDIRECT
113	Doug Graybeal, J. Welch and D. Graybeal	Boulder, CO	1	5,275	80	INDIRECT[R]
113	Craig Christensen, Rohde & Christensen	Boulder, CO	5	5,368	61	INDIRECT
100	Ron Shore, Thermal Technology Corp.	Carbondale, CO	1	7,339	88	INDIRECT
114	Doug Davis, Sunshine Design	Carbondale, CO	2	7,340	69	INDIRECT
114	Darrel Smith	Loveland, CO	1	6,202	61	INDIRECT
115	Bob King, Allen-King Builders	Lyons, CO	1	5,446	81	INDIRECT
115	Carl Mezoff, Wormser Scientific Corporation	Shelton, CT	1	5,102	85	INDIRECT
116	Preston Stevens, Jr., Stevens and Wilkinson	Shenandoah, GA	1	2,800	61	INDIRECT
116	C. Eugene Moeller	Boner Springs, KS	1*	4,711	43	INDIRECT[R]
117	Gifford Pierce, Beckman, Blydenburgh & Associates	Groton, MA	1	6,424	76	INDIRECT
117	Richard Zamore, Suntech Homes, Inc.	Georgetown, ME	1	7,246	75	INDIRECT
118	Thomas J. and Jill E. Newhouse	Grand Rapids, MI	1	7,279	69	INDIRECT
118	Bruce Monroe, B. Monroe/R. Pryor	Three Rivers, MI	1	6,782	48	INDIRECT
119	Jeffrey Barger, Sol-Terra Design	Columbia, MO	1	5,046	55	INDIRECT
119	Nicholas Peckham and Bradley Wright	Columbia, MO	1	5,083	86	INDIRECT
120	Terry A. Hoffman, Hunter Hunter Associates-Archt.	Glencoe, MO	1	4,705	87	INDIRECT
120	G. Herbert Gill	Joplin, MO	1	4,088	44	INDIRECT[R]
84	Ted Hoskins and Daniel Koenigshofer, Integrated Energy Systems, Inc.	Chapel Hill, NC	1	3,454	50	INDIRECT

*Attached home   R—Retrofit system

PAGE	DESIGNER	LOCATION	NO. OF UNITS	DEGREE DAYS	% SOLAR	SYSTEM TYPE
121	Charles Pearson, B. V. Pearson Associates	Chester, NH	1	7,246	82	INDIRECT
121	Peter B. Olney	Hampton, NH	1	7,246	66	INDIRECT
122	Richard Holt, Evog Associates, Inc.	Plymouth, NH	1	8,177	37	INDIRECT
122	Doug Kelbaugh, Environmint Partnership	Andover, NJ	1	5,696	83	INDIRECT
123	Tom Wilson, Star Route Studios	Flemington, NJ	1	5,733	57	INDIRECT
123	William Collins	Stillwater, NJ	1	5,810	100	INDIRECT
124	Doug Kelbaugh, Environmint Partnership	Upper Freehold, NJ	1	4,911	85	INDIRECT
96	Ken Brooks, Arch. Res. Cons./Rocky Mt. Sun Power	White Rock, NM	1	5,853	57	INDIRECT
124	W. A. Scott	Pojoaque, NM	1	6,170	99	INDIRECT[R]
125	Buck Rogers, B. T. Rogers/J. Iowa/M. Johnson	Ramah, NM	1	6,576	75	INDIRECT
125	Robert Peters, Addy Associates	Santa Fe, NM	1	5,586	68	INDIRECT
92	Susan Nichols, Communico	Santa Fe, NM	4	5,586	91	INDIRECT
126	Mark Jones	Santa Fe, NM	1	6,016	78	INDIRECT
126	Susan Nichols, Communico	Santa Fe, NM	4	5,586	99	INDIRECT
127	William Kolar, SOA Energy Consortium	Bedford Heights, OH	1	5,993	49	INDIRECT
127	Gregory Goss and Terry Sefchick	Cleveland, OH	1	6,351	90	INDIRECT
128	Donn D. Knokey, T. Kuntzman/D. Knokey	Burns, OR	1	6,871	63	INDIRECT[R]
106	David Noferi, D. Noferi/R. Shafer	Eugene, OR	1	4,599	87	INDIRECT
128	James Emerson, Phase I	Sisters, OR	1	6,598	71	INDIRECT
129	Charles Skowronski	Lakeville, PA	1	6,202	71	INDIRECT
129	J. Wylie Bradley, deVitry, Gilbert and Bradley	Lititz, PA	1	5,251	32	INDIRECT
130	Mike Ondra, Shelter Design	New Tripoli, PA	1	5,827	73	INDIRECT
130	Don Prowler, South Street Design	Unionville, PA	1	5,101	76	INDIRECT
131	Ralph F. McCay, Solar Engineering	Summerville, SC	1	2,146	79	INDIRECT
131	Bill Barth, Barth and Ransbottom	Knoxville, TN	1	3,494	45	INDIRECT
132	Mack Caldwell, Phillip Mack Caldwell	El Paso, TX	1	2,700	97	INDIRECT
132	John W. Stewart, Solarama	Cedar City, UT	1	5,200	99	INDIRECT

*Attached home   R—Retrofit system

PAGE	DESIGNER	LOCATION	NO. OF UNITS	DEGREE DAYS	% SOLAR	SYSTEM TYPE
133	John W. Stewart, Solarama	Kanab, UT	1	5,200	100	INDIRECT
133	Billy Born, Architectural Design Branch, TVA	Duffield, VA	1	4,121	79	INDIRECT
75	Danny Brewer, Architectural Design Branch, TVA	Duffield, VA	1	4,121	74	INDIRECT
134	Richard Fitts	Norfolk, VA	1	3,384	56	INDIRECT
134	Walter Roberts, Jr.	Reston, VA	1	4,962	97	INDIRECT
88	Victor Habib, One Design, Inc.	Strasburg, VA	1	5,978	76	INDIRECT
75	Adolphus Chester, Architectural Design Branch, TVA	Duffield, VA	1	4,121	81	INDIRECT
135	Harris Hyman, R. Finkle/H. Hyman	Middlebury, VT	1	7,988	56	INDIRECT
135	Don Schramm, Prado	Madison, WI	1	7,596	50	INDIRECT
136	David Reynolds	Harrisville, WV	1	4,688	35	INDIRECT
136	Michael Framson, Framson General Construction	Riverton, WY	1	8,433	72	INDIRECT
164	Michael Frerking, Environmental Architecture	Flagstaff, AZ	1	7,177	85	SOLARIUM
165	James J. Hoffman, James Hoffman Design Group	Tempe, AZ	1	4,929	90	SOLARIUM
165	Jonathan Allan Stoumen, Architect	Anderson, CA	1	2,415	96	SOLARIUM
166	Rob Anglin	Pleasanton, CA	1	2,582	63	SOLARIUM[R]
166	Peter Calthorpe, Calthorpe/Fernau/Wilcox	Santa Cruz, CA	1	2,863	81	SOLARIUM
160	Richard Fernau, R. Fernau and Berkeley Solar Group	Santa Rosa, CA	1	2,912	58	SOLARIUM
167	Jim Plumb	Woodland, CA	1	2,600	76	SOLARIUM
167	Bruce Downing, Downing/Leach Associates	Boulder, CO	4*	6,283	81	SOLARIUM
168	James Moore	Buena Vista, CO	1	7,812	73	SOLARIUM
168	Lawrence Atkinson	Denver, CO	1	7,432	96	SOLARIUM
169	Peter O'Connor	Golden, CO	1	6,016	55	SOLARIUM[R]
169	Barry Sulam, B. Sulam/L. Deutsch	Longmont, CO	1	6,360	59	SOLARIUM[R]
170	Carl Mezoff, Sunborne Designs	Fairfield, CT	1	5,102	80	SOLARIUM
170	David Block	Ames, IA	1	6,774	86	SOLARIUM
171	James Rosenbarger/Terry White	New Albany, IN	1	4,605	48	SOLARIUM
171	L. Bradley Cutler, Associated Architects	Beverly Farms, MA	1	5,627	44	SOLARIUM[R]

*Attached home   R—Retrofit system

PAGE	DESIGNER	LOCATION	NO. OF UNITS	DEGREE DAYS	% SOLAR	SYSTEM TYPE
172	DeFrancesco & Baker Associates	Manchester, MA	1	5,529	49	SOLARIUM
172	Seigfried Porth, S. Porth/L. O'Connor	Southampton, MA	1	6,851	61	SOLARIUM
146	Gary Cook	Lima Twp., MI	1	6,267	73	SOLARIUM
173	Richard McMath, Sunstructures, Inc.	Dexter, MI	1	6,267	57	SOLARIUM
173	Charles Williams	Duluth, MN	2	9,930	59	SOLARIUM
174	Warren L. Cargel, Interface Design Group	St. Louis, MO	1*	4,900	57	SOLARIUM^R
174	Steven Fisher, Graphicon	Carrboro, NC	2*	3,338	50	SOLARIUM^R
175	Donald Barnes, Jr.	Morrisville, NC	1	3,338	50	SOLARIUM
175	John Meachem, Sunshelter Design	Raleigh, NC	1	3,352	92	SOLARIUM
156	Mike Funderburk, Sunshelter Design	Raleigh, NC	1	3,352	81	SOLARIUM
176	John Alt	Randleman, NC	1	3,731	79	SOLARIUM
150	Vinton Lawrence, Harrison Fraker Architect	Hopewell, NJ	1	4,911	79	SOLARIUM
176	Ted Bickford and Charly Lowery, Harrison Fraker Archt	Penington, NJ	1	4,911	92	SOLARIUM
177	Robert Richardson	Los Lunas, NM	1	4,292	77	SOLARIUM
177	Stephen Merdler, Soltec Associates	Santa Fe, NM	1	5,913	72	SOLARIUM
178	Stephen Yaussi, Moran and Yaussi, Architects	Henderson, NY	1	7,273	80	SOLARIUM
140	Fuller Moore	Oxford, OH	5	5,280	74	SOLARIUM
178	Arden Handshy	Rogue River, OR	1	5,008	47	SOLARIUM
179	Steve Nearhoof, Ecol·lection	Chicova, PA	1	5,905	37	SOLARIUM
179	Peter-Paul d'Entremont	Feasterville, PA	1	5,364	79	SOLARIUM
180	Randy Granger, Helio Thermics, Inc.	Greenville, SC	1	2,955	100	SOLARIUM
180	Dan Fenyn, Land Systems, Inc.	Oak Ridge, TN	1	3,507	54	SOLARIUM
181	Jamie M. Rohe, Concept Consultants, Inc. Magnus Magnusson, The Ehrenkrantz Group	Arlington, TX	1	2,209	43	SOLARIUM
181	L. R. Bachman, Design Technology, Inc.	Hockley, TX	1	1,354	75	SOLARIUM
182	Ray Boothe, Boothe and Associates, Architects	Parker Co., TX	1	2,234	100	SOLARIUM
182	R. Ashelman, Natural Sun Homes	Berkeley Springs, WV	1	5,428	73	SOLARIUM
183	Bruce K. Kieffer, Northland Country Homes, Inc.	Middleton, WI	3	7,605	51	SOLARIUM

*Attached home   R—Retrofit system

CHAPTER 3
DIRECT SOLAR GAIN

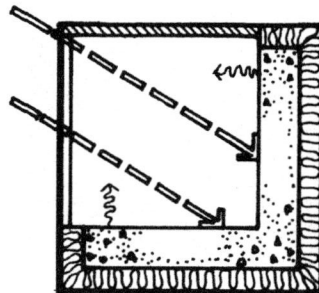

DIRECT SOLAR GAIN

DEFINITION
The DIRECT SOLAR GAIN concept is the most common passive solar building solution and has many historic precedents. Simply diagrammed as <u>sun to living space to storage mass</u>, the solar radiation is collected in the living space and then stored in a thermal storage mass. Thus, the actual living space is directly heated by the sun and serves as a "live-in" collector.

REQUIREMENTS
The basic requirements for the Direct Gain building are: a large south-facing glazed (collector) area, with the living space exposed directly behind; a floor and/or wall storage mass of significant dimension for exposure and solar heat storage; and a method of isolating the storage from exterior climatic conditions. For the first requirement, a large expanse of collector glazing, often double glazed to minimize heat loss, is oriented south to admit the maximum useful radiation (while facilitating the prevention of solar gain in summer). Secondly, a considerable amount of thermal storage mass in the walls, floors, ceilings, or in free standing mass is incorporated in the building to store solar heat and provide longer term heating. The absence of thermal storage mass in most conventional homes is what limits the possibility of storing the heat gained through large picture windows. Thirdly, the distribution of heat is controlled by the relationship between the storage mass and the living space. Proper insulation between the storage mass and the outdoors, or ground, is critical in preventing unnecessary heat loss.

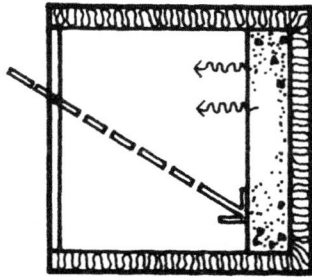

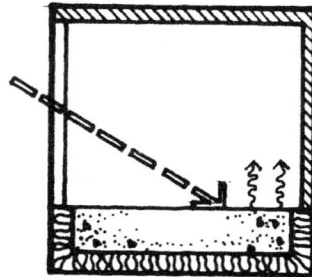

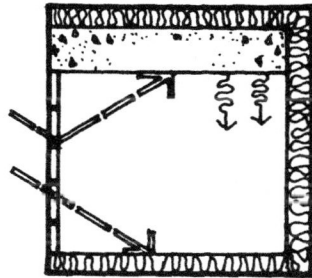

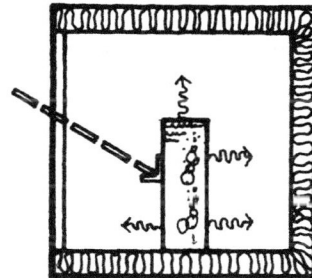

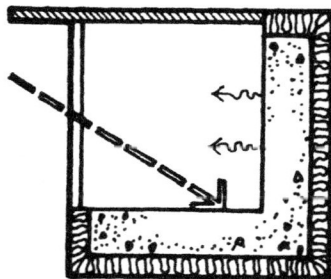

VARIATIONS

Beyond these basic requirements, there are a series of variations and controls that demonstrate alternatives in passive solar heating by Direct Gain. The most common variations are found in the location and materials of the thermal storage mass. The best location of the storage mass is often decided by the physical laws governing natural heat flow. For optimizing the absorption and storage of solar energy, it should be understood that mass which directly receives sunlight will store up to 4 times as much heat as mass which is not in direct sunlight. The use of glazings which diffuse sunlight could help to shower light on all wall and floor masses for maximum solar storage and distribution. For optimizing the radiant distribution of solar energy, physical proximity of the occupant to the radiant storage mass is important. Typical location alternatives include: a) the external building walls, b) the floor surface, c) the ceiling surface and d) internal walls and freestanding masses. In addition to storage location, there are significant variations in storage materials, and the massing of those materials, which provide different heat capacities and different time lag properties. Storage materials vary from concrete, brick, and ceramics, to water and other liquids, either singly or in various combinations, all radiating heat to the living space.

CONTROLS

To add to the efficiency and the usefulness of Direct Gain and other passive systems, several controls must be considered. To prevent unwanted heat gain, sunshading is required for the large expanse of south-facing glass. Due to the high location of the southern summer sun, overhangs can provide adequate protection for vertical southern glazing, but other solutions must be found for tilted glazing, or those with east and west orientations (faced with low sun angles). Exhausts and vents will also help cool interior spaces when summer temperatures are high. To prevent unwanted heat loss, insulation for the glazed collector area is necessary to improve the low R-value of glass. Moveable insulation panels, curtains, shutters, Skylids™, or Beadwall™ all work effectively to prevent unwanted heat losses on sunless winter days and nights and also prevent thermal heat gain on hot summer days. Without these control considerations, a Direct Gain system could cause tremendous discomfort due to winter losses and summer overheating.

The important issue to understand with DIRECT GAIN passive solar homes is that the occupant will be in direct contact with the collection, storage, and distribution components of solar heating.

Bedminster Township, PA

Jim Morgan, AIA
New York, NY

HEATED AREA: 996 FT^2

NUMBER OF DEGREE DAYS: 5,300

NET THERMAL LOAD: 30.2 10^{6}BTU/YR

AUXILIARY ENERGY: 2.5 BTU/DD/FT^2

YEARLY SOLAR FRACTION: 42%

Open market sale.

The solar heating fraction for this weekend retreat is expected to exceed 95%
during its occupancy.

CONTEXT
Bucks County, Pennsylvania, experiences approximately 5,300 heating degree days and 1,000 cooling degree days. The emphasis of
climate control in this region should be primarily on winter heating and to a lesser degree, summer cooling. The average daily
solar radiation (on a horizontal surface) during the winter months
is approximately 600-700 BTU/sq. ft., indicating frequent partial
cloudiness and a significant portion of diffuse sky radiation.

The house has two bedrooms and two baths with an additional
sleeping loft. The structure is conventional wood frame, well insulated, enclosing an interior masonry floor and partial wall which
provide for solar absorption and heat storage in winter. The
masonry is insulated at the perimeter edge to prevent heat loss by
conduction to the exterior. Summertime cooling is accomplished
by cross ventilation and by shading of the south glazing. Movable
insulating shutters improve the net heat gain of the glazing in
winter and also double as shading devices to keep the sun out in
summer.

CONSERVATION
The house is situated at the edge of a gentle south-facing hill.
Terracing the building into the hill minimizes heat loss by reducing
the exposure of the north wall to outdoor air. By elongating the
building in the east-west direction, its southern exposure is
maximized for winter solar collection. The floor plan is arranged

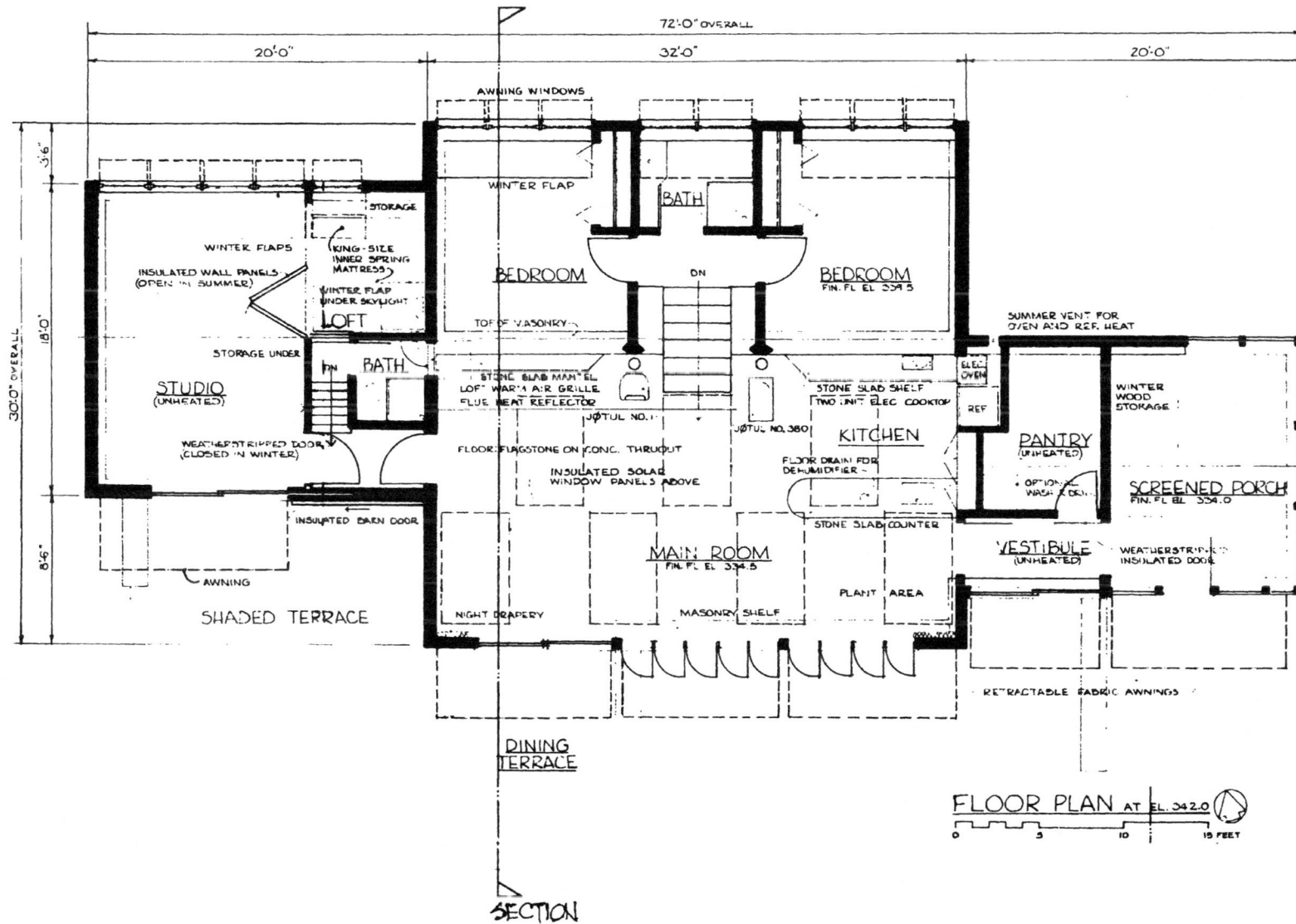

FLOOR PLAN at EL. 342.0

SECTION

72'-0" OVERALL

20'-0"   32'-0"   20'-0"

3'-6"
18'-0"
30'-0" OVERALL
8'-6"

AWNING WINDOWS

WINTER FLAP

BATH

BEDROOM

BEDROOM
FIN. FL EL 334.5

SUMMER VENT FOR
OVEN AND REF. HEAT

WINTER FLAPS
INSULATED WALL PANELS
(OPEN IN SUMMER)

KING-SIZE
INNER SPRING
MATTRESS

WINTER FLAP
UNDER SKYLIGHT

LOFT

STORAGE

STORAGE UNDER

DN

TOP OF MASONRY

STUDIO
(UNHEATED)

BATH

STONE SLAB MANTEL
LOFT WARM AIR GRILLE
FLUE HEAT REFLECTOR

JØTUL NO. 1

DN

STONE SLAB SHELF
TWO UNIT ELEC COOKTOP

ELEC
OVEN

REF

WINTER
WOOD
STORAGE

WEATHERSTRIPPED DOOR
(CLOSED IN WINTER)

JØTUL NO. 380

KITCHEN

PANTRY
(UNHEATED)

SCREENED PORCH
FIN. FL EL 334.0

FLOOR: FLAGSTONE ON CONC. THRUOUT

FLOOR DRAIN FOR
DEHUMIDIFIER

INSULATED SOLAR
WINDOW PANELS ABOVE

OPTION:
WASH & DRY

INSULATED BARN DOOR

STONE SLAB COUNTER

VESTIBULE
(UNHEATED)

WEATHERSTRIPPED
INSULATED DOOR

MAIN ROOM
FIN. FL EL 334.5

PLANT AREA

AWNING

NIGHT DRAPERY

MASONRY SHELF

SHADED TERRACE

RETRACTABLE FABRIC AWNINGS

DINING
TERRACE

FLOOR PLAN at EL. 342.0

0   5   10   15 FEET

SECTION

19

so that the living spaces are located along the south face of the building while bedrooms and service spaces, with small heating and lighting requirements, are located along the colder north wall. The studio and pantry, also infrequently used spaces, are placed to the northeast and northwest to further buffer the living spaces from the exterior. An entrance vestibule provides an air-lock to reduce the large infiltration heat losses often associated with the entry.

All south glazing is insulated at night by drapes and sliding panels. Windows on the north, east, and west sides are kept small since they result in a net heat loss during winter.

HEATING
Sunlight is admitted into the building through south-facing windows and translucent skylight panels. The skylight panels diffuse direct sunlight over the surface area of the masonry interior. During the daytime hours, sunlight is absorbed and partially stored in the masonry as heat. By spreading direct sunlight over a large surface area of masonry, a greater percentage of solar heat is stored in the mass at the end of the day. This reduces the temperature fluctuation in the space over the day, providing for greater thermal comfort. In cloudy climates, any additional masonry thicker than 4" to 8" is not useful in storing heat. It is more effective to spread mass throughout the space than put it in one place, for example as in a two foot thick masonry floor. Also, in cloudy winter climates with large amounts of diffuse solar radiation, south-facing skylights are more effective collectors than south vertical glazing.

COOLING
The house is kept cool in the summer by shading the major glass areas and providing for natural ventilation. Adjustable awnings are located in a pocket above the south glazing. These are a potentially better solution to shading than a fixed overhang. A fixed shading device will provide the same shading on September 21st and March 21st since the sun's movement across the skydome is the same on these days. These awnings however, can be regulated seasonally to partially shade the glazing in September, when the weather is warm, and then adjusted to admit full sunlight in March, when it is cold. The insulating panels over the skylights are closed in the daytime to double as a shading device. The outside face of the panels should be painted white or made of a reflective material to prevent excessive heat build-up in the skylight.

The open floor plan of the house facilitates cross ventilation. Operable windows on the south and north wall allow for the natural flow through the building of the prevailing summer breezes (southwest).

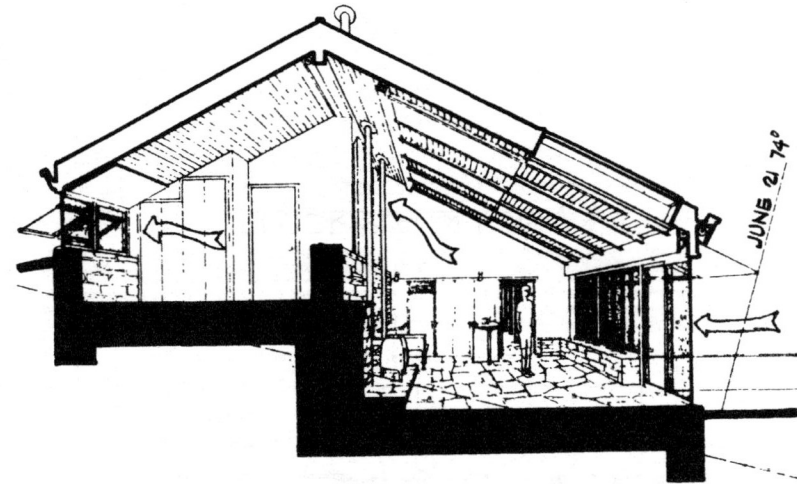

SUMMER MODE

CONCLUSION
The home makes excellent use of a direct solar gain approach through its many light diffusing apertures. The checkerboard pattern of skylights on the roof of this home takes advantage of the diffuse sky radiation in this climate. With over 40% of its heat from the sun, this home does well for this hazy and sometimes overcast region.

Although generous in total square footage, this home zones only the main living area and two main bedrooms for active backup heating. This arrangement should be very workable for its intended purpose, as a second home.

To market this as a regular home, some modifications would be needed. The bedrooms would require some acoustical separation from the main living areas. Masonry walls here would allow some additional thermal storage mass and probably free some of the floor area for rugs or soft furnishings. In general, this design has many features of a passive country home which can be applied to other designs.

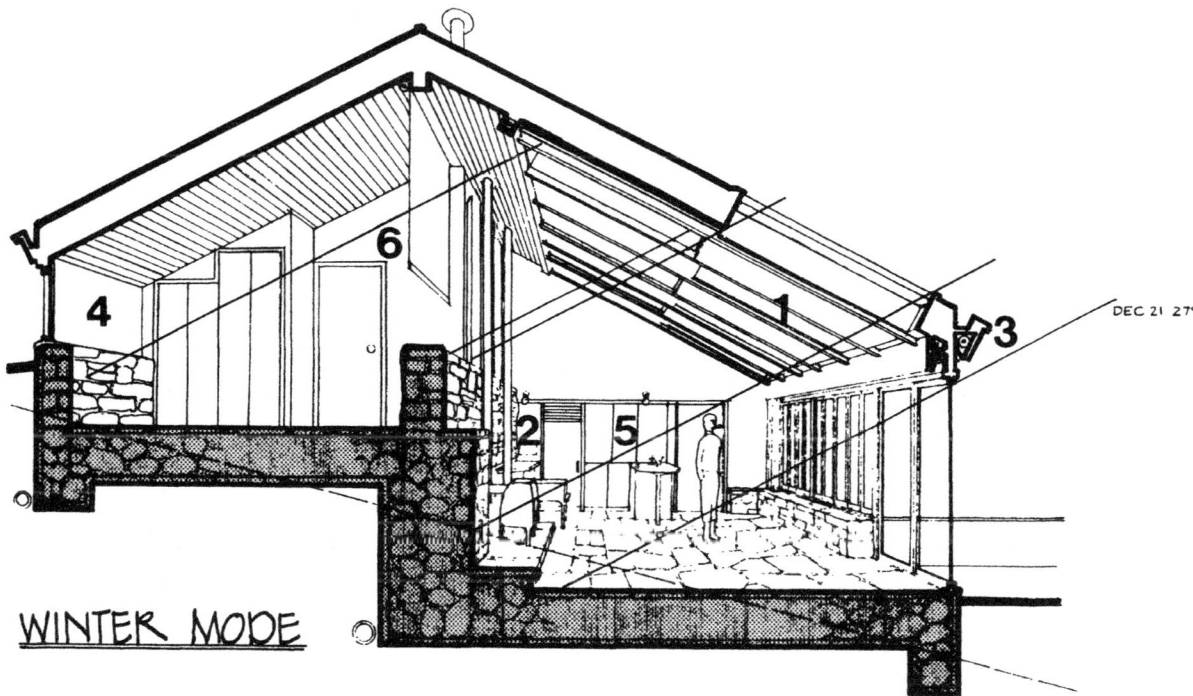

WINTER MODE

1. CEILING HATCHES OPEN ON SUNNY DAYS, WARMING THERMAL MASS.

2. STOVE AND/OR FIREPLACE HEAT OCCUPANTS AND THERMAL MASS; HEAT FROM FLUE GASES REFLECTED INTO ROOM.

3. SOUTH WALL HAS INSULATED SHADES AND NIGHT DRAPERY TO RESIST HEAT LOSS.

4. NORTH WALL AWNING WINDOWS COVERED BY INSULATED WINTER FLAPS.

5. WEATHERSTRIPPED DOORS ISOLATE THE UNHEATED AREAS OF THE HOUSE.

6. HEAVY CURTAIN ISOLATES BEDROOMS FROM MAIN ROOM AT NIGHT.

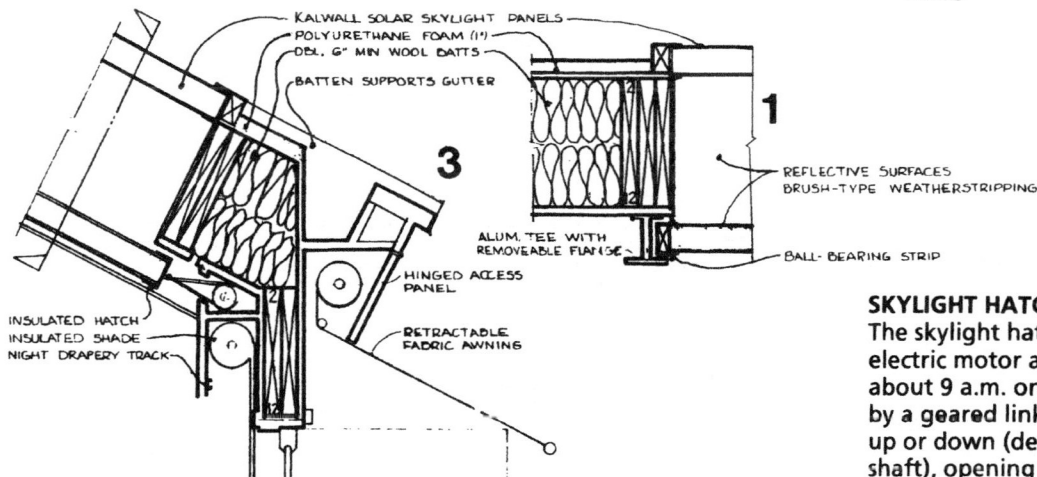

DEC 21 27°

DETAILS

KALWALL SOLAR SKYLIGHT PANELS
POLYURETHANE FOAM (1")
DBL. 6" MIN WOOL BATTS
BATTEN SUPPORTS GUTTER
REFLECTIVE SURFACES
BRUSH-TYPE WEATHERSTRIPPING
ALUM. TEE WITH REMOVEABLE FLANGE
BALL-BEARING STRIP
HINGED ACCESS PANEL
INSULATED HATCH
INSULATED SHADE
NIGHT DRAPERY TRACK
RETRACTABLE FABRIC AWNING

SKYLIGHT HATCHES

The skylight hatches work on the principle of a scroll: a small electric motor at the top is activated by a photoelectric cell at about 9 a.m. on a sunny winter day. It turns the upper shaft (and by a geared linkage the lower shaft) clockwise and the hatches roll up or down (depending on how the cables are wound around the shaft), opening the skylights. At about 3 p.m. a timer activates another motor on the lower shaft which rotates counterclockwise until the hatches again cover the skylights. A manually operated over-ride is provided.

MAY STREET

Ansonia, CT

Joe Migani, Teamworks
Cambridge, MA

HEATED AREA: 1,032 FT²

NUMBER OF DEGREE DAYS: 5,840

NET THERMAL LOAD: 92.6 10⁶BTU/YR

AUXILIARY ENERGY: 8.98 BTU/DD/FT²

YEARLY SOLAR FRACTION: 44%

Open market sale.

CONTEXT
Designed for an urban neighborhood in Ansonia, CT, this two-story duplex demonstrates a simple and effective direct gain solar system. The compact site has excellent orientation, providing for north entries from the street and southern exposure for the private backyards. The neighborhood is a mix of colonial and Victorian residences, having two-story houses with front porches and street setbacks. The duplex repeats these features, integrating the new aspects of passive solar with the more traditional quality of the street. The winter climate is quite demanding, with 6,000 degree days heating loads and mostly cloudy skys. The building uses concrete block exterior and party walls, with foam insulation and stucco applied to the exterior. The floors and roof are precast concrete planks, providing additional heat storage mass. Aluminum blinds are used to reflect direct gain sunlight onto the ceiling mass in these two-bedroom units. The size, construction, siting, and solar system all combine to make this design an efficient and low-cost solution for urban neighborhoods.

CONSERVATION
Central to energy conservation in this scheme is the party wall construction. By choosing to attach these dwellings, the designers have reduced the exterior suface area by 20%. In addition, the overall shape and size of the building is tight, with few projections or jogs, further reducing the surface area and resultant heating loads. The interior layout places the bathrooms, stairs, kitchen, and entry on the north side, creating a buffer from the more heavily used livingroom and bedrooms. The entry employs an airlock vestibule to reduce infiltration, and the entry yard has a row of evergreens which serve as both a windbreak and privacy screen. Several deciduous trees are located to the south to offer summer shade while allowing penetration of the winter sun.

The building structure is well-insulated, with one exception. The roof has an R-factor of 21, the equivalent of 6" of fiberglass in a frame roof, which is generally considered inadequate for this climate. Moreover, the underside of the roof is used for heat storage, making the insulation in this location particularly critical. All other areas of the building are well-insulated: R-21 walls are achieved by 3½" of urethane insulation over the block walls; 3 feet of earth berming surround the east, west, and north sides; all windows are double pane and have insulating shutters for night use. In summer, southern overhangs shade the majority of glass, with the deciduous trees creating a cooler general environment.

HEATING
The passive heating system in this building resolves some of the pitfalls of many direct gain designs. One common design approach

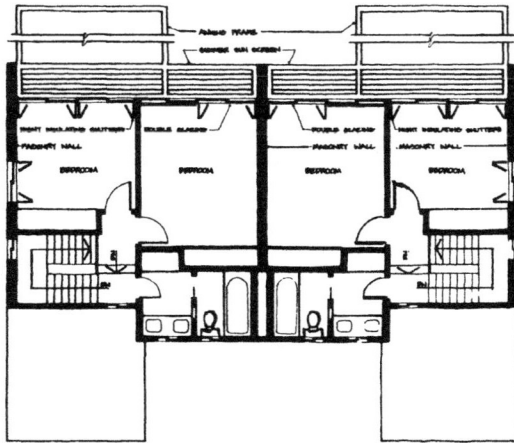

SECOND
FLOOR

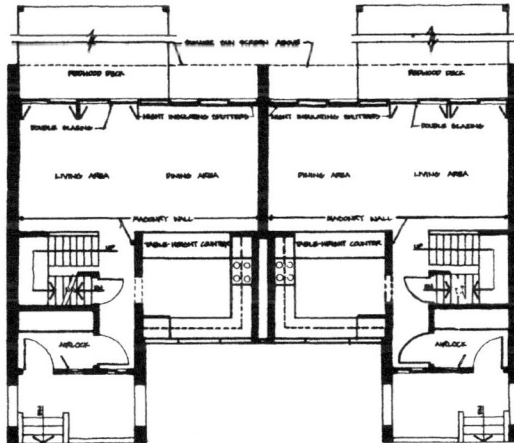

FIRST FLOOR

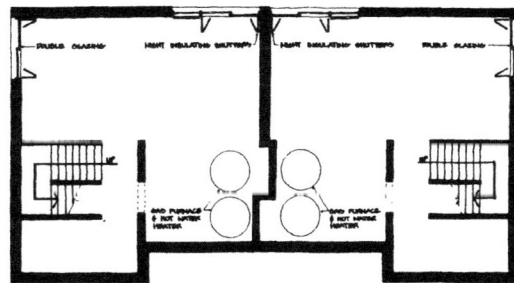

BASEMENT

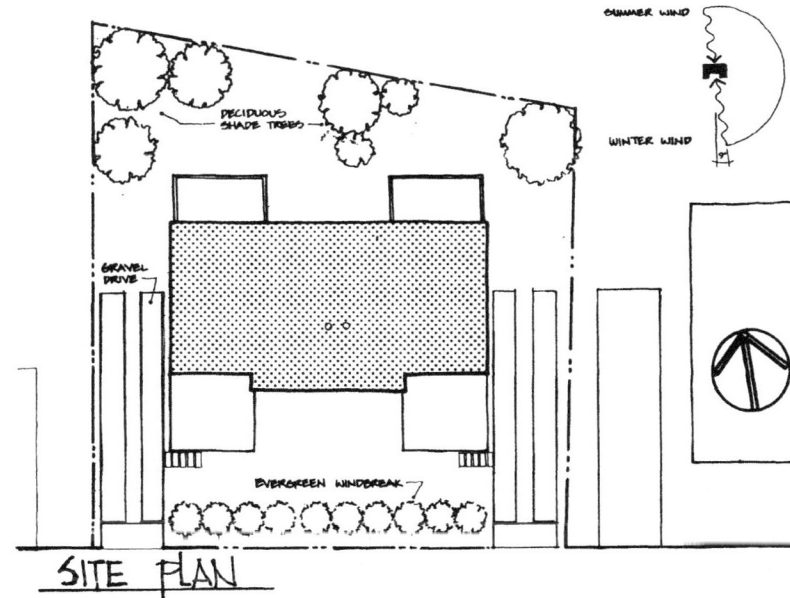

SITE PLAN

places a concrete slab inside of south-facing windows to absorb store the entering solar energy. Unfortunately, floors are often occupied, their surface being obstructed by furniture, rugs, and people. Rather than heating the mass floor, the sunlight strikes these low mass objects and overheats the space. In addition, the furniture and rugs, not to mention the people, often suffer from this exposure. These designers have elected to place the storage mass on the ceiling in the form of precast concrete planks, "Spancrete", and redirect the sunlight to the ceiling by reflective louvers in the south windows. This scheme provides a clear, unobstructed surface for the sunlight to strike and does not limit the use of floor space in any way. Also, it allows the floors to be carpeted, the clear preference of many homeowners.

The reflective louvers are located between two layers of glass to eliminate any need for cleaning. The louvers, installed with the concave surface upward to focus the light, and their angles are adjusted (on a monthly basis) by a thumb-operated dial. The concrete ceiling is plastered and in this design is painted with a selective surface coating. These selective surface coatings were developed for use in flat-plate collectors to absorb as much incoming

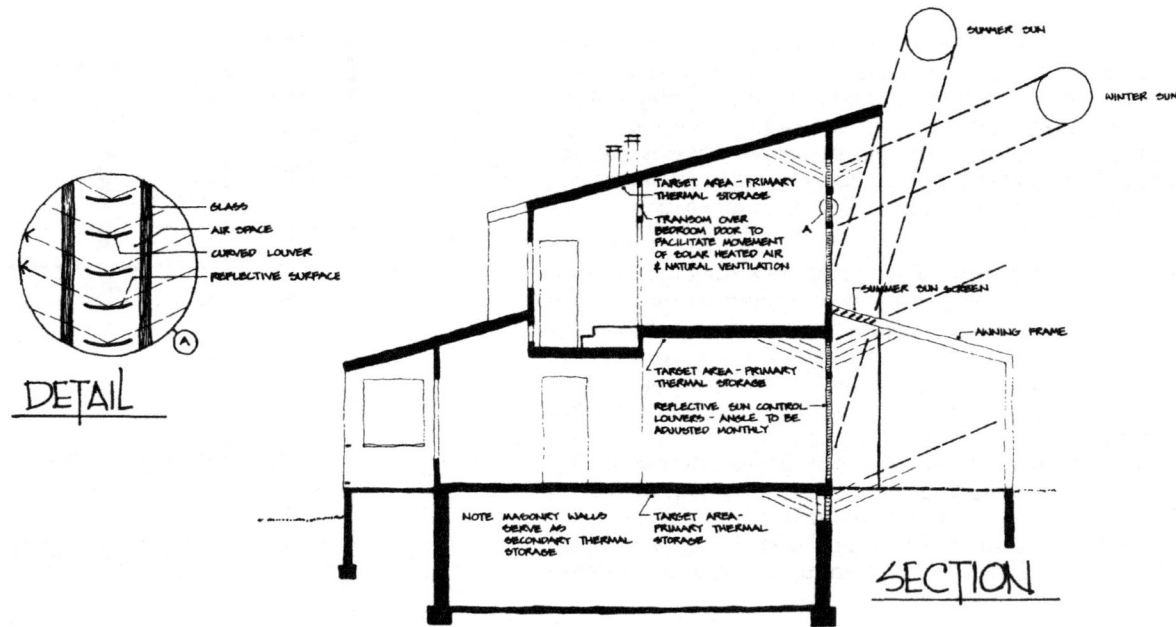

DETAIL

SECTION

(Diagram labels — Detail: GLASS / AIR SPACE / CURVED LOUVER / REFLECTIVE SURFACE / A)

(Section labels: SUMMER SUN / WINTER SUN / TARGET AREA - PRIMARY THERMAL STORAGE / TRANSOM OVER BEDROOM DOOR TO FACILITATE MOVEMENT OF SOLAR HEATED AIR & NATURAL VENTILATION / SUMMER SUN SCREEN / AWNING FRAME / TARGET AREA - PRIMARY THERMAL STORAGE / REFLECTIVE SUN CONTROL LOUVERS - ANGLE TO BE ADJUSTED MONTHLY / NOTE MASONRY WALLS SERVE AS SECONDARY THERMAL STORAGE / TARGET AREA - PRIMARY THERMAL STORAGE)

light as possible without re-radiating any light or heat back out. In this case, it is probably an inappropriate application because the ability of the ceiling mass to give up heat at night is severely limited by this selective surface. In addition, high absorption may not be desirable because it reduces the amount of light reflected to and stored in the masonry walls.

In direct gain systems such as this, the total surface area of the heat storage mass is as important as the thermal capacity of the storage material. In concentrated sunlight, the surface layers of solid mass materials heat up quickly, reducing the ability of the mass to absorb more heat. These hot surfaces also begin to immediately give off heat at a rate which can overheat the space. By contrast, in the Trombe systems heat must migrate through the wall before space heating will occur, creating a time delay and lower surface temperatures. To avoid overheating, the solar gain should be distributed over as large a surface area as possible. In this way the heat build-up in the surface layers is reduced and the energy is more effectively stored.

To achieve this spreading of the heat, the total surface area of the mass should be maximized and the light allowed to partially re-

flect from one surface to the others. In this design, a lighter ceiling will absorb a percentage of the sunlight and reflect the rest to the storage mass walls. If all the mass in the ceiling and walls is used in this way, the temperature fluctuations in the space will be reduced.

One key element for direct gain systems is the ratio of surface area of storage mass to area of window. If only the ceiling is used in this design, the ratio is about 1:1.5; if the light is spread to the walls and rear ceiling, the ratio is 1:5. This distribution not only employs more mass more effectively, but also creates a lighter, more pleasant atmosphere. It should be noted that these issues are not as critical in the bedrooms because daytime overheating in these unoccupied areas will cause little discomfort.

These townhouses employ the maximum south glass possible given the building's shape. This results in approximately 400 sq. ft. per unit of south window area, providing for approximately 66% of the heating. The overall thermal mass, including the walls, totals 12,000 BTU/°F. or around 30 BTU/°F. per sq. ft. of window area. This figure is considered, as a rule of thumb, adequate for Trombe and water wall systems, but because of the smaller temperature swings in the mass of direct gain systems, the thermal capacitance may be too small for this building. For backup heat, the duplex employs a gas furnace with hot water radiant fin tubes.

COOLING
Cross ventilation, shading, and massive building elements combine to temper the summer climate in this building. The south side, which contains 95% of the glass in the building, has overhangs and sun screens which will fully shade this facade in mid-summer. In addition, the decks have awning frames to which canvas is attached to create covered porches. The porches, along with the deciduous trees, shelter the south side of the building and help to cool the summer breezes. The windows are well-positioned to take full advantage of the typically southerly winds and provide cross ventilation to the north side. Unfortunately, the bedroom doors must be opened to allow this cross ventilation. The living areas are located on the lowest floor, which is cooler during the day. Finally, the massive walls and ceiling, dampen the temperature swings in the space by absorbing heat during the day and releasing it the following night. The combination of these features significantly increases the building's comfort levels through most summer conditions.

CONCLUSION
This design demonstrates an extremely efficient and low-cost solar system. The small size, party walls, and site orientation combine to reduce the loads and provide optimum solar access. The passive system has an elegant simplicity which resolves many of the characteristic problems of direct gain systems. The louvers direct the light away from the living areas to the ceiling, providing light and views without the glare, furniture fading, and discomfort of some direct gain systems. By allowing the sunlight to partially reflect from the ceiling to other massive components in the space, more heat storage is employed at lower temperatures, preventing any overheating of the spaces. The economy of plan and solar system should appeal to a broad range of urban housing needs.

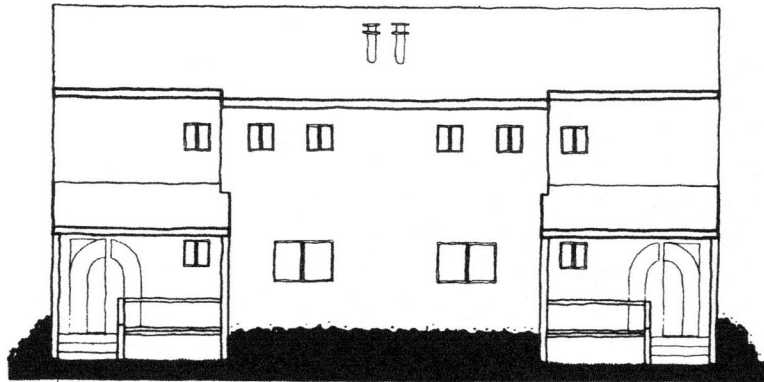

NORTH ELEVATION

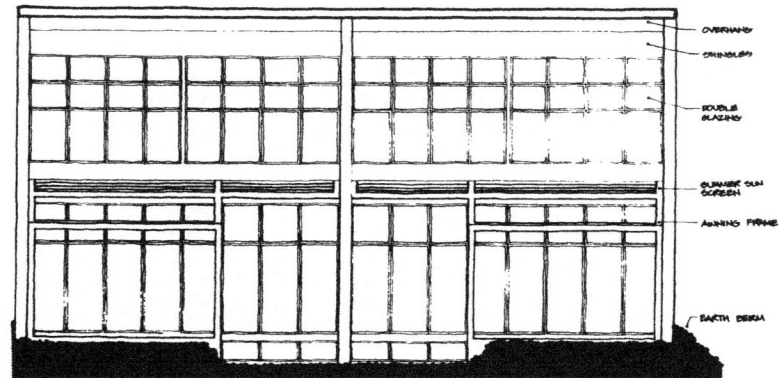

SOUTH ELEVATION

Burnsville, MN

Tom Ellison/John Carmody
St. Paul, MN

HEATED AREA: 2,000 FT²

NUMBER OF DEGREE DAYS: 8,382

NET THERMAL LOAD: 41.9 10⁶BTU/YR

AUXILIARY ENERGY: 2.33 BTU/DD/FT²

YEARLY SOLAR FRACTION: 93%

Private client.

CONTEXT

This underground or earth-sheltered passive solar home is located 15 miles outside of Minneapolis, Minnesota. The region experiences 8,400 heating degree days with average wind speeds of 10 miles per hour. Although the house is earth-covered, it is positioned on top of a ridge with land falling off in all directions for excellent drainage. There is a newly-emerging interest in underground homes in this region and this design attempts to respond to the expectations of local homebuyers. The size and arrangement of the living spaces are typical for an average size family. There are three bedrooms, a living room, family room and two-car garage. The construction is primarily concrete retaining/bearing walls with wood frame on the south wall. The unit is estimated to cost $40 to $50 per square foot.

CONSERVATION

The two-level configuration is compact to minimize exterior surface area. Large oak trees provide shade in summer but not in winter. The slope of the land, the surrounding vegetation and the garage location break the prevailing winter winds from the northwest. The sloping roof allows a higher ratio of south-facing window area to volume of space heated. The east and west walls are buried below grade, and have no windows, so that ninety-five percent of the window area faces south. The north slope opens up to allow an entry door and window. Because the house is basically one room deep, all rooms except bath, laundry and storage rooms receive natural light. The sloping roof helps to reflect daylight deeper into the interior of the second floor.

The primary conservation feature is the earth cover over the walls and roof. In addition to the earth, there are 4" of styrofoam to achieve an R-factor of 24 in the walls and in the roof. The designer was careful to insulate the outside edge of the concrete roof planks where they are exposed to the cold air. The south wall is made of 2 x 6 wood studs with glass fiber batt insulation and 1" styrofoam sheathing for an R-factor of 22. All windows are double glazed and the doors have insulation cores. Tight-fitting insulated curtains are closed over the south-facing window area at night.

The vertical organization of living spaces with bedrooms on the first floor and living spaces on the second floor is the reverse of the conventional home. This configuration minimizes the discomfort associated with heat stratification between floors, one of the most common problems of 2-story passive solar buildings. In this way, living spaces are located on the warmer level with cooler sleeping spaces below.

HEATING

Direct gain supplies the solar heat to this house. South-facing window walls and clerestory windows total approximately 300 sq. ft. of "collector". This collector area supplies roughly one-half the heat needed to keep this 2,000 sq. ft. house warm over an average winter. The solar heat gain is stored in the concrete walls and floors. For every square foot of collector there is about one cubic foot of concrete in direct sunlight which stores 30 BTU's of heat for each degree F rise in temperature. The concrete plank floor is covered with dark floor tiles to increase absorptivity. To protect the house from heat loss through the large window area, insulated drapes are manually closed at night. Insulating shutters can be substituted if a higher R-factor is desired.

A 50,000 BTUH electric forced hot-air furnace located on the second floor with a single distribution duct that runs the length of the south wall, supplies auxiliary heat to both levels. There is a high return air register on the second floor and a low return on the first floor. These register locations will allow the central furnace to redistribute air from the most extreme thermal zones of the house to reduce temperature differences between the two floors. Overheating can be prevented by closing drapes or cor-

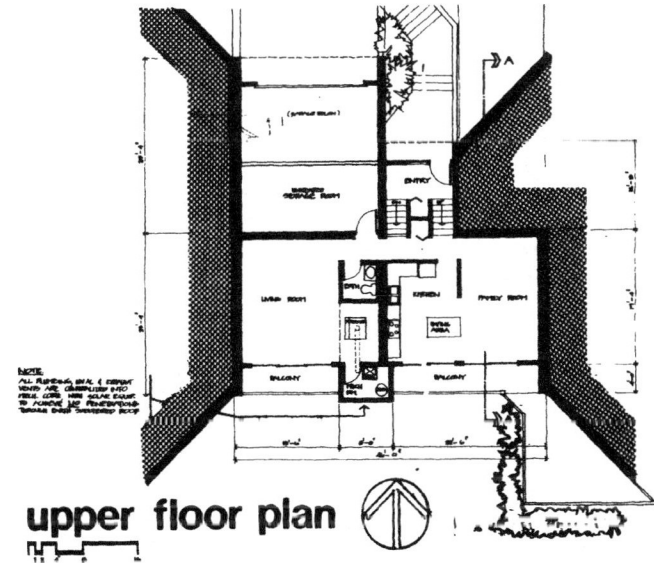

upper floor plan

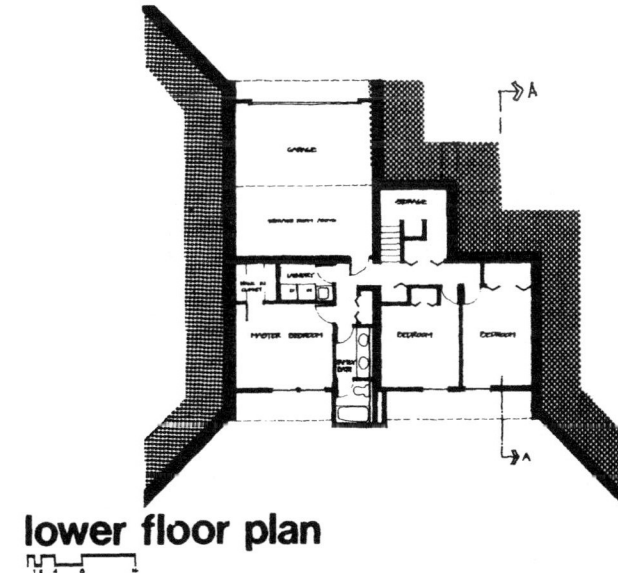

lower floor plan

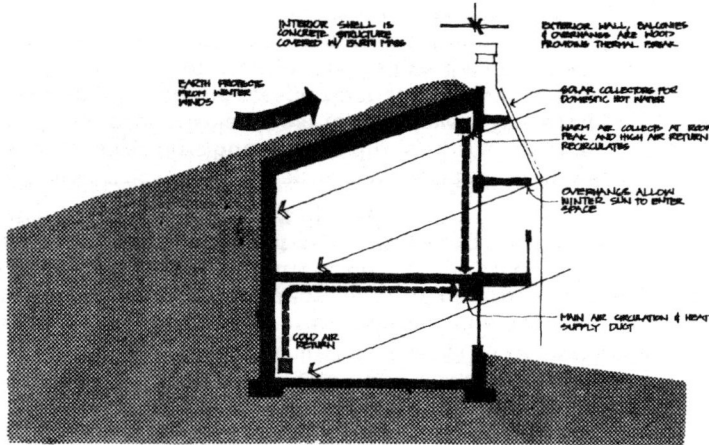

winter heating

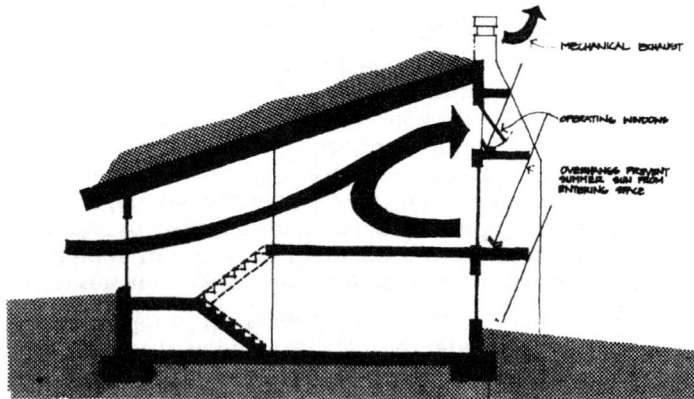

summer cooling

rected by opening windows. The considerable amount of concrete mass that is not in direct sunlight will help to dampen daily temperature fluctuations. Two flat-plate active solar collectors form a south-sloping roof on the mechanical room for domestic hot water preheat.

COOLING

In summer, the extremely heavy mass of the earth that surrounds this building severely reduces daily room temperature swings. The house is cut deep enough into the ridge so that the first floor is cooled by the conduction of heat to the surrounding earth, since ground temperatures average about 45°F. Four-foot overhangs (sized for 44°N. sun positions) and the balcony shade the collector area from the high summer sun. Wind and temperature induced ventilation from the front door on the north to the operable windows placed high on the south side also provides cooling.

CONCLUSION

Generic advantages of building underground include the thermal and acoustic insulation of earth cover, protection from sun and wind, low environmental impact, low air infiltration, low building skin maintenance, and summer cooling. Obvious disadvantages include the need for water proofing, constraints on natural lighting and cross ventilation, additional roof and wall structure to carry heavier dead loads and marketability restraints.

This house could be further improved by a few additions. An air-lock vestibule would cut air infiltration. Natural ventilation and lighting could both be increased by the introduction of operable skylights. The ability to adjust the width or angle of the overhangs in front of the south-facing glass or "collector" area would enhance solar collection in late winter and early spring when the mid-day sun is high in the sky, but heating needs are still significant. However, the design has solved many of the obvious disadvantages associated with underground homes. The use of 1/16" butyl rubber sheets and standard concrete floor planks addresses the waterproofing and structural problems. The front door and operable windows on the north side allow cross ventilation and daylighting as well as creating an inviting yet more conventional north elevation. The design also allows for various locations of the garage to suit different site and client needs. By grouping the heating, ventilation and plumbing stacks into a single core on the south wall, the designer has avoided any penetrations through the earth-covered roof. All considered, this home demonstrates the possibilities of underground construction and direct solar gain heating to provide for the thermal needs of residents in this cold region.

section a-a

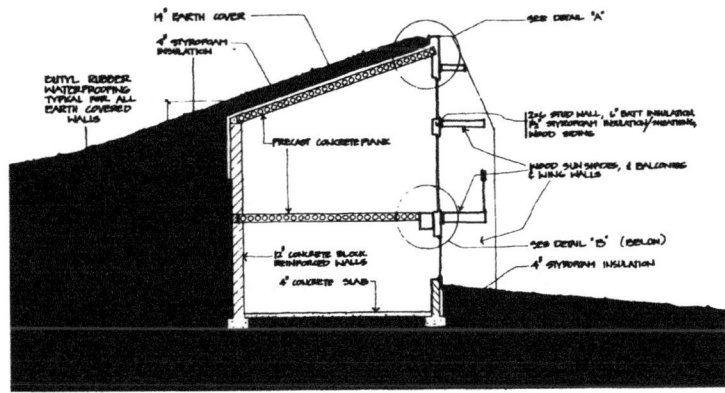

south elevation

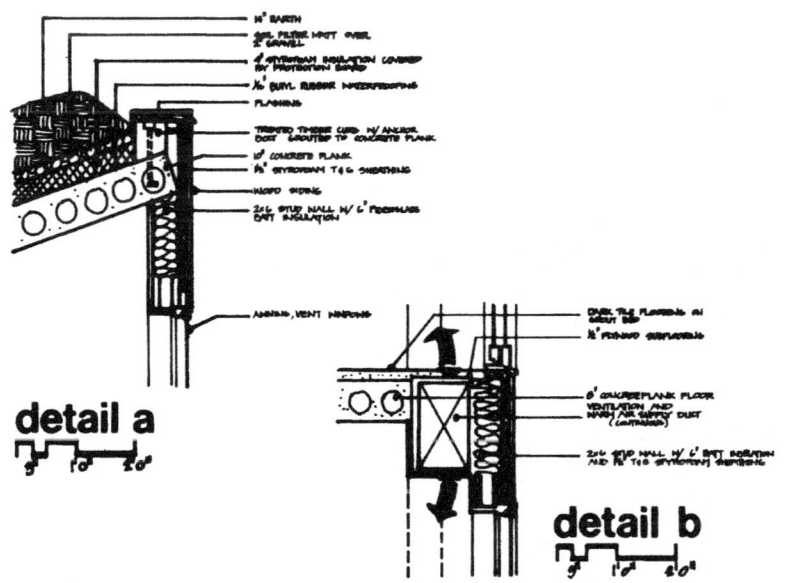

detail a

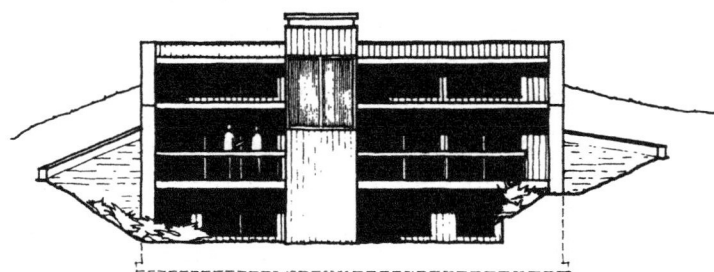

detail b

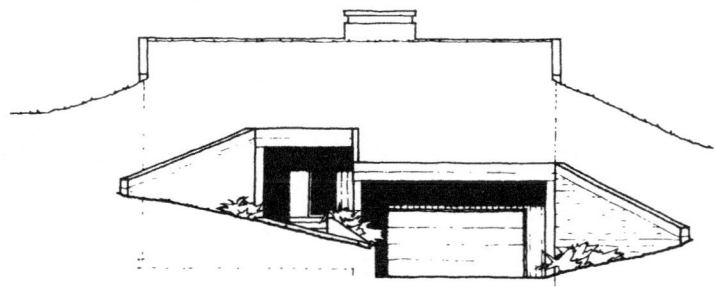

north elevation

Occidental, CA

Peter Calthorpe, Calthorpe/Fernau/Wilcox
San Francisco, CA

HEATED AREA: 1,111 FT²

NUMBER OF DEGREE DAYS: 3,019

NET THERMAL LOAD: 30.5 10^6BTU/YR

AUXILIARY ENERGY: 4.0 BTU/DD/FT²

YEARLY SOLAR FRACTION: 82%

Private client.

CONTEXT

This small two-bedroom house is located in Northern California and employs a hybrid solar heating system. The climate is moderate with approximately 3,000 degree days of heating loads with partly cloudy winters and hot summer periods. Like most of the construction in this area, this building is a two-story wood frame structure on continuous foundation with a 2' crawl space. The hybrid system employs south windows with Venetian blind "louvers" to absorb the sun's heat. This heat is blown to a rock bed in the crawl space where it is stored. The site has good solar exposure with an open meadow to the south and winter wind protection from a redwood grove to the north. The design uses the local standard building practices while providing a controllable solar heating system.

CONSERVATION

The basic two-story rectangular shape, with the long side facing south, reduced the building's surface area while accommodating solar gains. The north side is protected by a stand of redwood trees, and the east and west sides are small. The mild climate allows a greater degree of freedom with window placement and two-story spaces. The building has double glazing throughout,

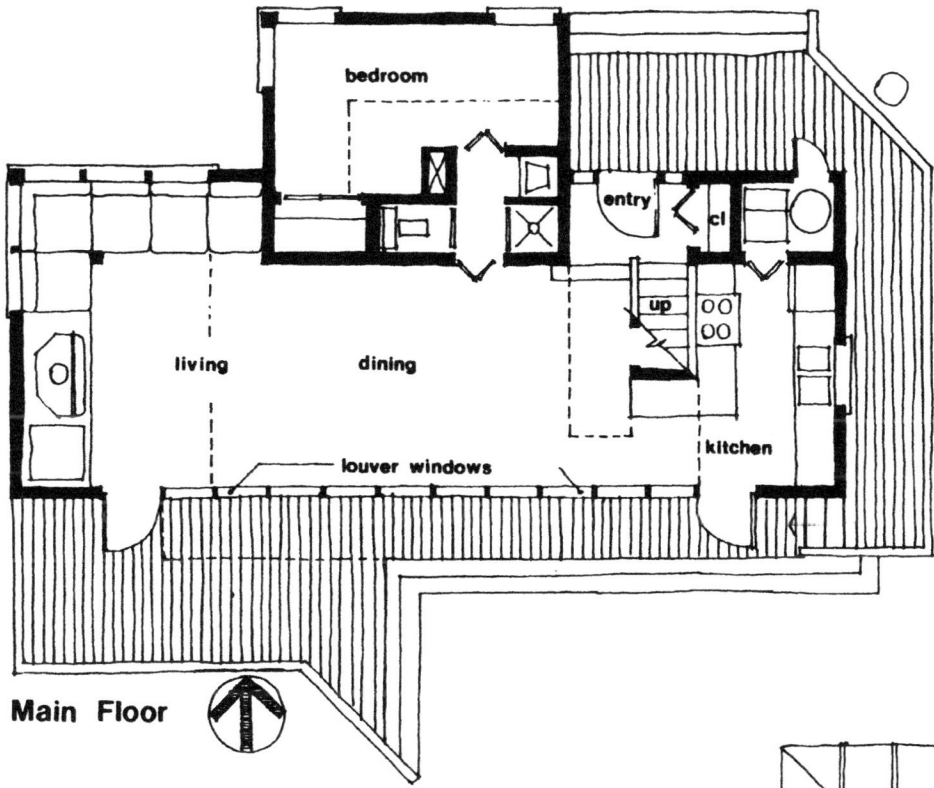

Main Floor

bedroom

entry

cl

living

dining

louver windows

up

kitchen

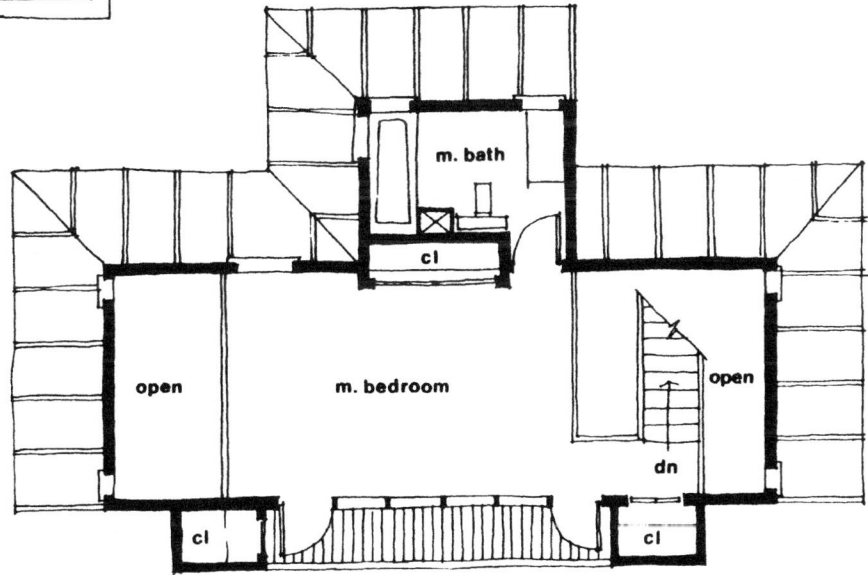

m. bath

cl

open

m. bedroom

open

dn

cl

cl

Second Floor

31

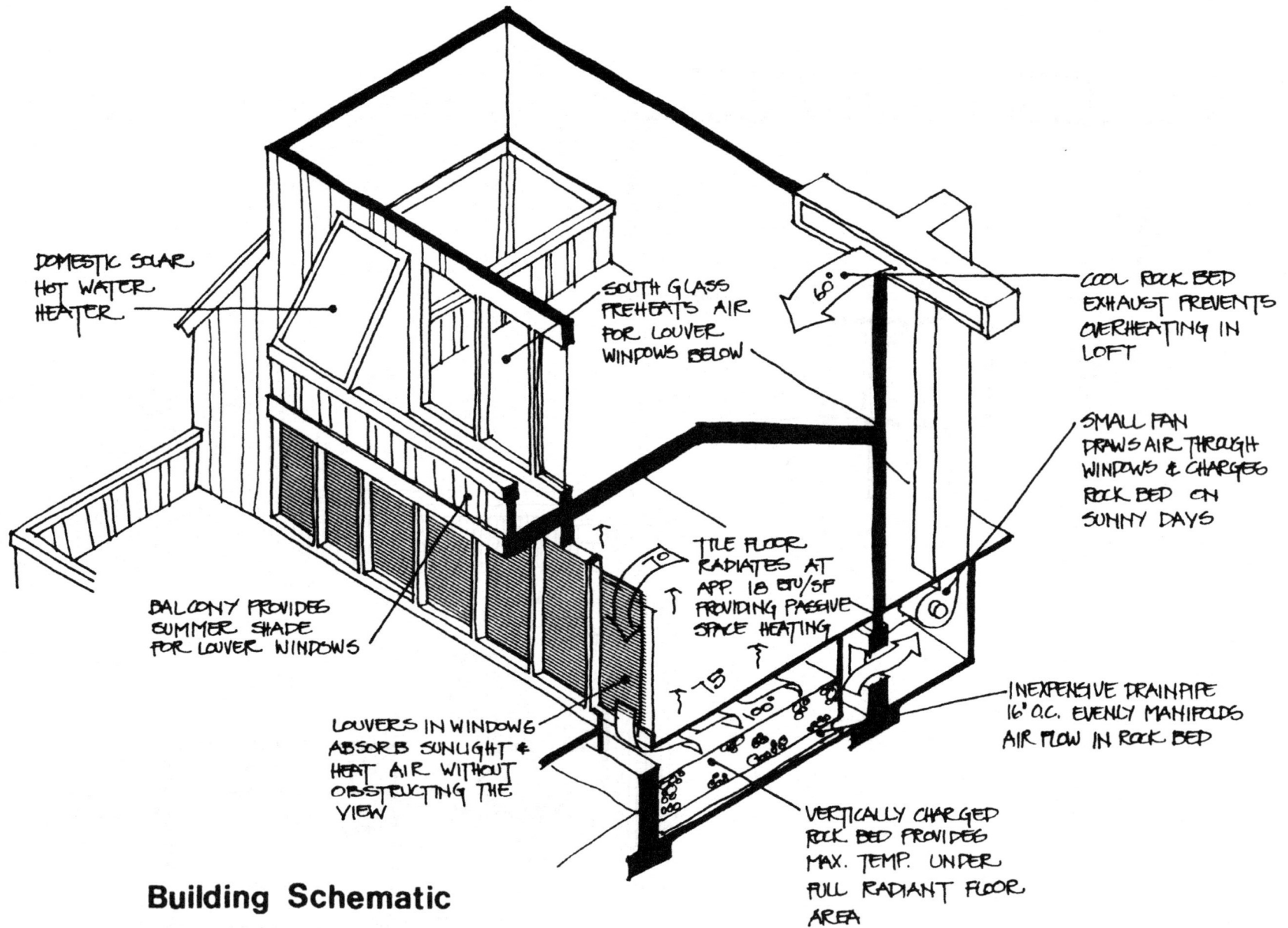

DOMESTIC SOLAR
HOT WATER
HEATER

SOUTH GLASS
PREHEATS AIR
FOR LOUVER
WINDOWS BELOW

60°

COOL ROCK BED
EXHAUST PREVENTS
OVERHEATING IN
LOFT

SMALL FAN
DRAWS AIR THROUGH
WINDOWS & CHARGES
ROCK BED ON
SUNNY DAYS

TILE FLOOR
RADIATES AT
APP. 18 BTU/SF
PROVIDING PASSIVE
SPACE HEATING

70

75

100°

BALCONY PROVIDES
SUMMER SHADE
FOR LOUVER WINDOWS

LOUVERS IN WINDOWS
ABSORB SUNLIGHT &
HEAT AIR WITHOUT
OBSTRUCTING THE
VIEW

INEXPENSIVE DRAINPIPE
16' O.C. EVENLY MANIFOLDS
AIR FLOW IN ROCK BED

VERTICALLY CHARGED
ROCK BED PROVIDES
MAX. TEMP. UNDER
FULL RADIANT FLOOR
AREA

Building Schematic

with triple glazing on the south-side collector windows. Its overall small size is also a conservation feature not to be overlooked.

HEATING

The solar heating system in this house uses a fan to transfer heat into the rock bed heat storage mass and allows natural energy flows, radiation and convection, to distribute the heat back to the living space when needed. Because the system uses a fan to actively "change" the storage while allowing for a natural or passive "discharge", it is considered a hybrid. Specifically, a dark-colored Venetian blind is placed behind the double glazing on the south side to absorb the sunlight. A second layer of glass is placed inside the blind to contain the heat while a small, 1500 CFM fan draws air from the window cavity and down into the rock bed. The heat from the window is deposited in a "vertically charged" rock bed below the floor, returning cooler air to a diffuser high in the second story. This cool air, approximately 60°F, prevents any overheating that may be caused by the sunlight entering the second-story south window. While dependent on a fan to operate, this collection system offers several distinct advantages. The glare, discomfort, furniture fading, and large temperature swings associated with some direct gain solar systems are eliminated by the louvers, while the views and light from the windows are maintained—the best of both worlds. On overcast days, the blinds can be raised to allow a direct heat gain from the diffuse light outside. (During the summer the windows are shaded so that the blinds can also be raised.) At night the blinds can be closed, providing an added insulation layer in the window and privacy.

The rock bed is placed in an essentially free container, the standard foundation crawl space. A 2" draining slab seals the bed from any ground moisture and 4" perforated plastic drain pipe is laid out 16" on center to create a return plenum with uniform airflows throughout the bottom of the bed. These pipes run into a larger, 8" x 24" manifold on the north side which is connected to the fan and exhaust distributors on the second floor. Three quarter inch river rock, 24" deep, is placed over the drainpipe, leaving a 6" airspace below the plywood floor, which acts as the intake plenum from the windows. The hot air from the windows enters this top plenum and, encountering equal resistance everywhere over the bed, distributes evenly over the entire plan area before moving downward. As it moves downward, the air deposits its heat in horizontal layers with the warmer layers at the top and coolest at the bottom. These warm upper layers heat the floor above, which in turn heats the space. After depositing its heat, the air passes into the lower drainpipe plenum and is returned high on the second floor.

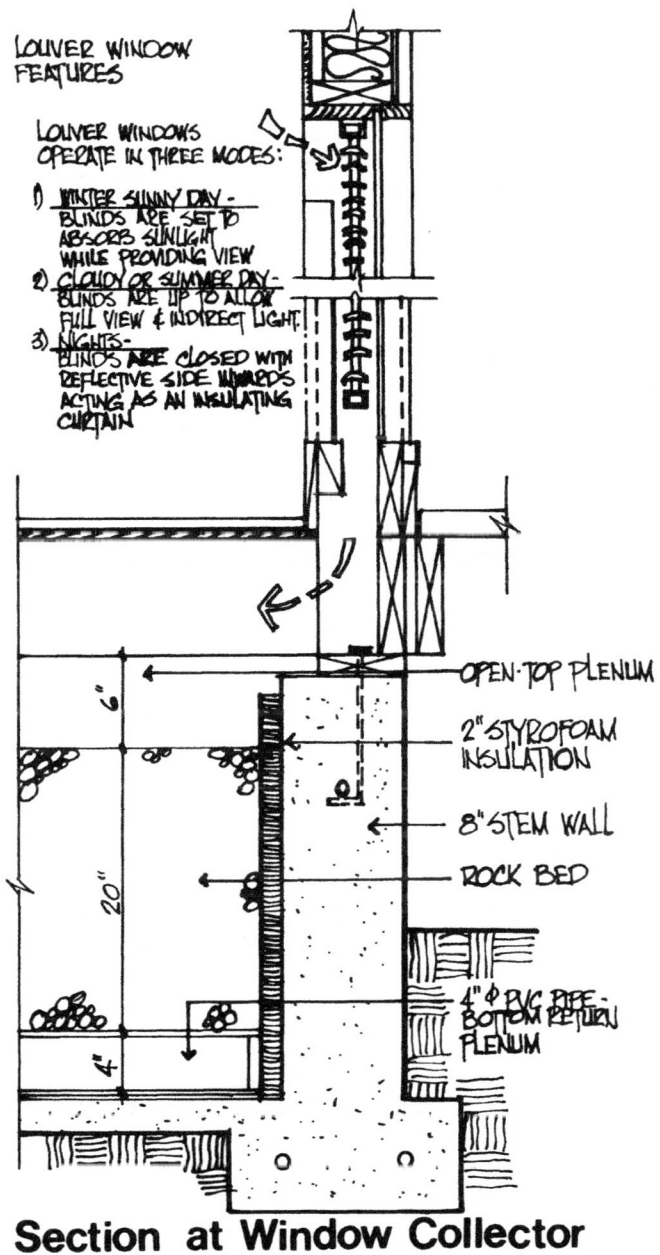

LOUVER WINDOW FEATURES

LOUVER WINDOWS OPERATE IN THREE MODES:

1) WINTER SUNNY DAY - BLINDS ARE SET TO ABSORB SUNLIGHT WHILE PROVIDING VIEW
2) CLOUDY OR SUMMER DAY - BLINDS ARE UP TO ALLOW FULL VIEW & INDIRECT LIGHT
3) NIGHTS - BLINDS ARE CLOSED WITH REFLECTIVE SIDE INWARDS ACTING AS AN INSULATING CURTAIN

OPEN-TOP PLENUM

2" STYROFOAM INSULATION

8" STEM WALL

ROCK BED

4" Ø PVC PIPE - BOTTOM RETURN PLENUM

Section at Window Collector

The passive heat distribution system functions much like the old radiant floor systems. The vertical charging of the rock bed is critical to insure the even heating of the total radiant floor area. With 100°F. air from the windows stored in the rocks, an 80°F. floor surface temperature can be expected. Unlike the older hot water pipe radiant floors, this floor cannot be turned on and off. Therefore, the area of the radiant floor must be designed to handle the average, rather than peak, heating load. This differs from climate to climate and for each house design. For example, this house has a heat loss of 532 BTUH and an average winter temperature of 45°F. ambient. This requires a 10,640 BTUH heat source. This house has 440 sq. ft. of 80°F. radiant floor producing approximately 24 BTUH/sq. ft. The total output of the floor therefore is 10,560 BTUH, enough to balance the average loads.

For unusual peak loads, floor registers can be opened to allow natural convection to deliver additional heat. Registers along the north exhaust manifold allow cool air to fall to the bottom of the bed while the warm air rises through the window plenum and into the living space. The bedroom above has an open balcony overlooking a two-story space, which allows the heat to rise naturally into this space. Unfortunately, the first floor bedroom has no radiant floor and will not receive heat from the solar system. Two Casablanca type fans high in the two-story spaces prevent any extreme temperature stratification. A simple wood burner will provide any backup heat necessary for the living/dining area and the bedroom above. The total south window area of 276 sq. ft. is 25% of the overall building area. For this climate such a collection area will provide a 80% solar participation in the heating loads. The storage capacity of 15,400 BTU/°F. or 55 BTU/°F. per sq. ft. of window is equivalent to about two days of storage.

With Venetian blinds and glass for collectors, rock bed under concrete for storage and a small fan, all the components of the system are standard market items with which both builders and home owners are familiar. The temperature output of this window is dependent on the airflow rate through it. By using the small fan, high temperatures—100°F.—reach the rock bed. These high temperatures increase the rock bed's storage capacity and the radiant output of the floor above. Finally, and perhaps most significantly, the system is applicable to standard light-frame structures and requires no expensive masonry construction or additional interior space for heat storage mass or solar collection.

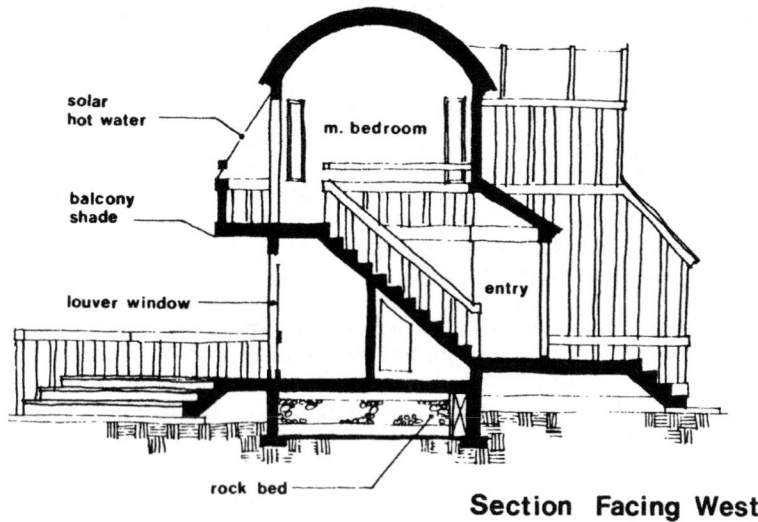

solar
hot water

m. bedroom

balcony
shade

louver window

entry

rock bed

Section Facing West

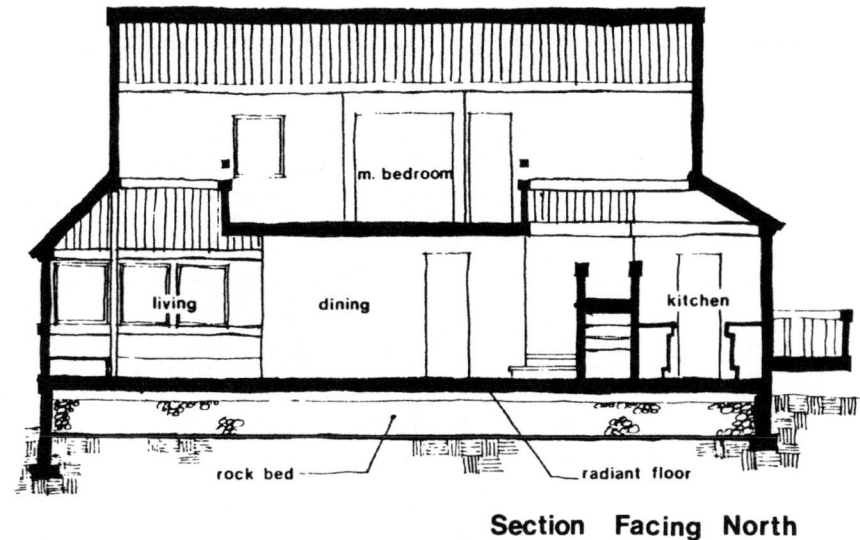

m. bedroom

living   dining   kitchen

rock bed   radiant floor

Section Facing North

COOLING

The Northern California climate does not experience large or consistent cooling requirements, but it does have occasional hot spells of temperatures in the 90's. Shading and ventilation are the primary cooling devices of this house. In addition, the rock bed is used occasionally to store night "coolth". In this mode, cool night air is blown through the bed when temperatures are below 60°F. During the next day, this coolth is blown back into the space while the interior heat is deposited in the rocks to be exhausted the following night.

The lower south windows are shaded in summer by the balcony and solar domestic hot water collectors above. The upper south windows are shaded by reflective Venetian blinds similar to those in the collector windows. The two-story spaces aid natural ventilation by exhausting the warm air high in the building and allowing cooler air from the north side to enter low. All the spaces have cross ventilation. Finally, the Casablanca fans can incease comfort during particularly still summer periods by creating air movement.

CONCLUSION

Although this building does not use purely passive features, its hybrid system offers some interesting alternatives. In cases where southern views are desired but the classic direct gain systems are inappropriate, this building demonstrates a technique in which the light and views are preserved while the glare is eliminated and solar heat is collected without overheating. For areas in which masonry construction is too expensive, the rock bed in crawl space may be a cost-effective alternative. The systems offers "controllability" through operation of the blinds and floor registers. All of the system's components such as the blinds, drainpipe, rocks, fan and thermostat, are common building elements. Its only disadvantages are that the system depends on electrical power to operate, it is not as well known as some other passive systems, and the windows and blinds require periodic cleaning.

This hybrid solar system could be applied to many homes of differing plans and styles. This particular building offers the economy of small size, tight plan, and minimized exterior. The two-story spaces not only allow natural ventilation and heat distribution but also provide for a more spacious interior atmosphere. The roof shape could be varied to meet market demands. The designer has presented an unusual solar system in an unusually shaped building. But the basic concepts have a broad appeal.

Site & Roof Plan

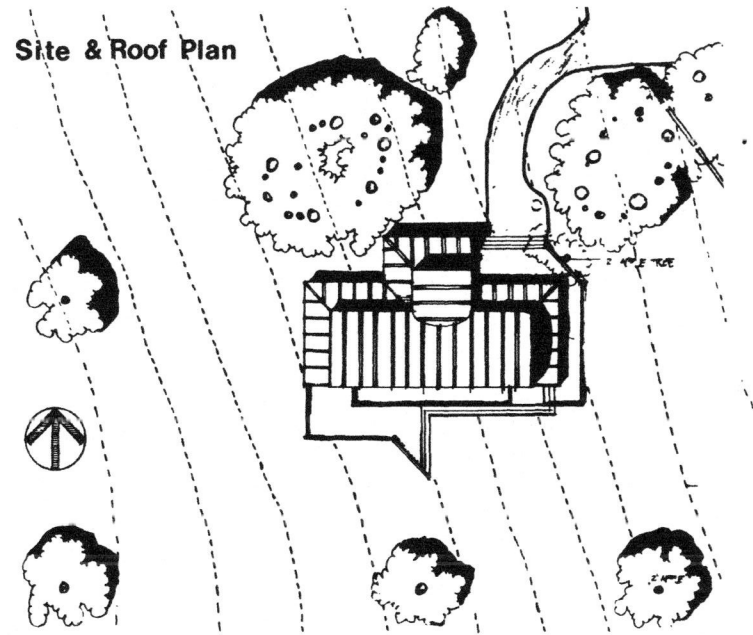

Princeton, NJ

Doug Kelbaugh, Environmint Partnership
Princeton, NJ

HEATED AREA: 2,310 FT²

NUMBER OF DEGREE DAYS: 4,911

NET THERMAL LOAD: 35.0 10⁶BTU/YR

AUXILIARY ENERGY: 3.22 BTU/DD/FT²

YEARLY SOLAR FRACTION: 77%

Retrofit.

CONTEXT

This solar addition is added to a 100-year old house in downtown Princeton, N.J. The existing house is on a narrow street in a residential neighborhood with a density of about 6 units per acre. The climate has about 5,000 heating degree days. Both the parent house and addition are two-story wood frame structures. The existing house has 3 small bedrooms. The addition includes a new master bedroom and bathroom on the second floor and a family room and small solarium on the first floor.

CONSERVATION

The south wall of the existing house faces roughly 15° east of true south, for which there is a 5%-10% penalty in solar collection. The neighboring house casts some shadow on the south wall of the existing building but the south wall of the addition is in sunlight most of the day. One 20-year old shade tree was removed from the south and new trees and shrubs have been planted on the north side for wind protection. Neighboring homes in close proximity also provide winter wind protection. The shape of the addition approaches a cube, and has a low exterior surface-to-floor ratio to keep heat loss down. The heat loss is about 10 BTU/DD/sq. ft., which is low considering the large amount of glass area and the size of the structure. The double-glazed solarium acts as a vestibule for the back door of the house. The addition's wall insulation, of 3-1/2" fiberglass insulation plus 1" urethane insulation, adds an R-factor of about 20. The north masonry wall with 4" of expanded polystyrene also has an R-factor of 20. There is an R-30 glass-fiber insulated roof. The perimeter of the concrete floor slab is insulated with 2" styrofoam to a depth of three feet. Windows are triple glazed on the east side but double glazed to allow higher solar transmission on the south side. There are no windows in the north wall of the addition. Multiple-layer shades that are drawn down at night to cover the inside of the windows on the second floor are calculated to increase the windows' R-factor to 4 or 5. The clerestory windows are insulated with a 1" foam board that swings down at night to both decrease the volume of the room as well as cut heat loss through the high glazing. It is operated manually with a window pole. Thermally-lined curtains close off the solarium and windows on the first floor.

The original house has already been retrofitted with weather-stripping and storm windows. Several new windows as well as a greenhouse window unit are added to the south wall of the existing house. The hydronic distribution pipes in the basement of the old house are to be retrofitted with foam insulation.

HEATING
The philosophy behind this addition is to expand the house in such a way that not only the new rooms will be solar heated but the excess solar heat in the addition can also be drawn off to heat the original house. Daytime overheating, normally a problem with direct gain systems, can be avoided by ducting the excess warm air that collects at the top of the second floor clerestory into the first floor rooms of the old house. Less thermal storage is necessary because not all the solar heat collected must be stored within the addition.

There are two primary solar heating systems, both direct gain. There is a solarium or lean-to greenhouse on the first floor and a clerestory or roof aperture system on the second floor. There is also direct solar heat gain through the many large south-facing windows in the south wall, including several new ones added to the south wall of the original house.

The 8-inch concrete floor slab in the solarium is covered with dark color quarry tile to absorb incoming sunlight. There are three 55-gallon water drums painted black that also absorb sunlight. Both the floor and the water drums act as batteries to store heat in the day and discharge heat at night. There is also a strip of dark colored tile along the south side of the family room to absorb the

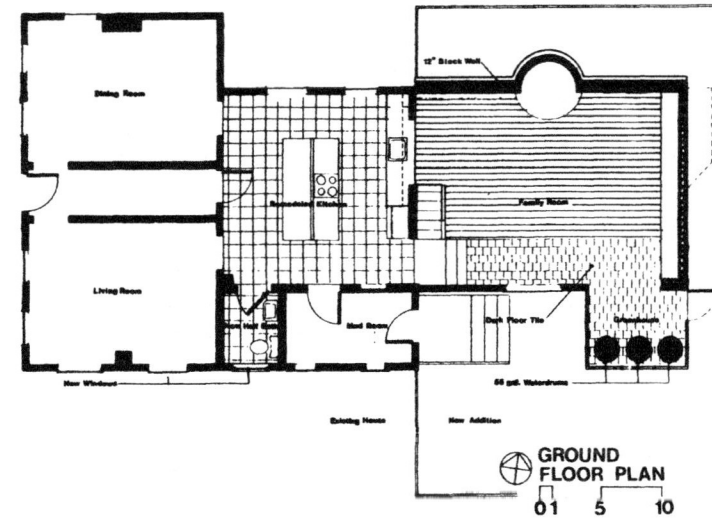

GROUND FLOOR PLAN
01 5 10

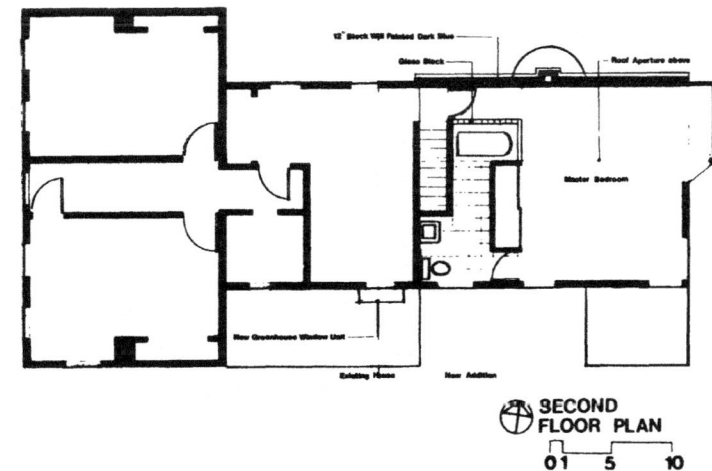

SECOND FLOOR PLAN
01 5 10

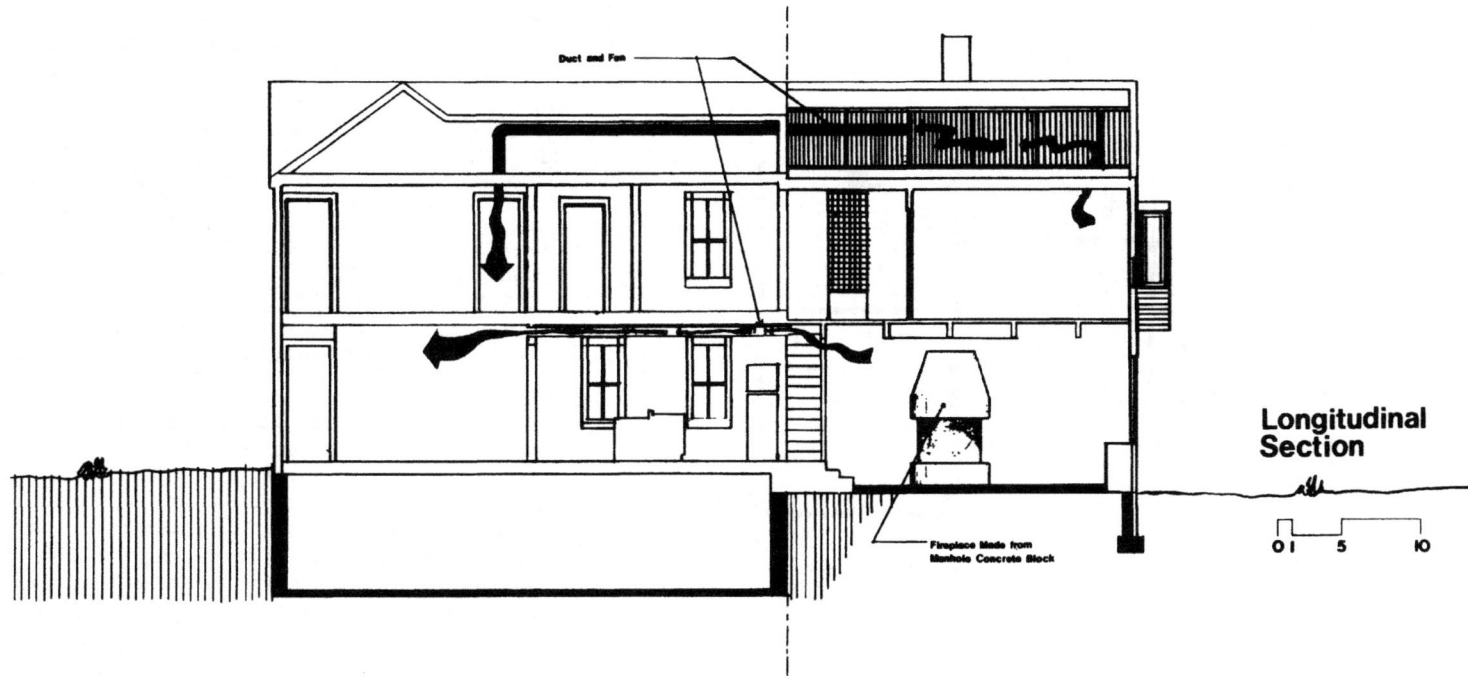

Duct and Fan

Longitudinal Section

Fireplace Made from Manhole Concrete Block

0 1   5   10

sunlight that enters through the low windows. There is a lined curtain which closes the solarium off from the family room at night. The heat stored in the walls and floor with the solarium is just calculated to keep the plants safe from freezing temperatures. On severely cold nights, the curtain can be left open a little to allow the solarium to rob more heat from the family room. Since the solarium is not occupied at night, it is kept as cool as the plants can tolerate so that its heat loss is as low as possible. In warmer climates, less thermal mass would be required in the solarium to keep the air from falling to temperatures harmful to the plants. The minimum temperature to which the solarium can be allowed to drop at night is also determined by the type of plants grown.

Thermal storage for the clerestory collector area on the second floor is the two-story, filled heavyweight concrete block wall, solid with mortar, which is painted a dark color on the second floor for solar absorption. The clerestory glazing is double-walled acrylic plastic held by aluminum glazing bars at the head and sill. The amount of solar radiation collected is enhanced by aluminized mylar reflectors stapled to the underside of the over-hang and

glued to the roof. The highly specular reflective material is also stapled to the sloped ceiling above the shutter in the bedroom to reflect the sunlight which is transmitted through the glazing on to the dark masonry wall. The moveable insulating shutter, made of 1″ insulation board, is also covered with a highly specular reflective material—aluminum foil.

With temperature swings of 10° to 12°F over 24 hours, the system is calculated to provide about 75% of the heating load of the addition from October through May. In addition, there is a duct with a small fan that draws off warm air from the ceiling of each floor and distributes it to the original house. The fans are automatically switched on by thermostats when the addition overheats.

A domestic hot water solar preheater is situated above the bathroom. It consists of an 80-gallon glass-lined water tank painted black that sits in a glazed insulated compartment on the roof. Water circulates through the tank and is heated before it reaches the conventional hot water heater in the basement. The conventional water heater thus receives preheated water and does not have to work as much.

COOLING

There are numerous windows throughout the addition which are strategically placed to maximize natural ventilation in summer. There are operable windows located both high and low in the solarium to induce ventilation. The thermal mass of the floor, the water drums and the back masonry wall will also absorb some heat to temper the high afternoon temperatures of summer.

CONCLUSION

In this retrofit, there were several compromises which had to be accepted. Shading of the solarium by the neighboring house and the clerestory by the original house could seriously limit afternoon solar collection. The window area facing east is excessive but is justified by the view to the back-yard. The fireplace, although it has a fresh-air intake, would consume less warm room air if it had glass doors that could be closed.

The design is an example of a passive solar retrofit that could be applied to any existing home with potential for expansion in an easterly or westerly direction. The application in a high-density neighborhood is encouraging. The architectural style of the addition attempts to integrate with the original house without simply imitating it.

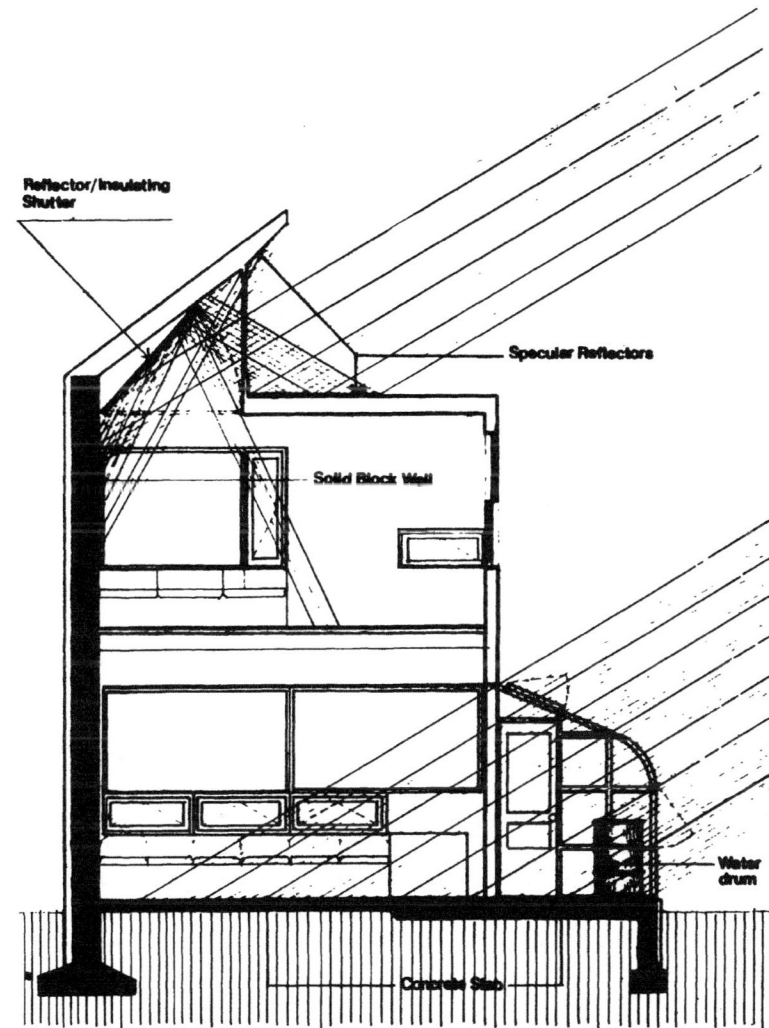

Reflector/Insulating Shutter

Specular Reflectors

Solid Block Wall

Water drum

Concrete Slab

**Cross Section
Noon, Jan 21**

0 1   5   10

CONTEXT

This 1,350 square foot house is located in Ft. Collins, Colorado at a latitude of 40°N. The house is situated on a gently sloping lot with striking views of the Rocky Mountains to the southwest. The general winter climate of the area is cold, dry, and sunny. The exterior design temperature for this locale is − 4°F with a yearly total of 6,600 degree days. Average January day-night temperature normally fluctuates between − 5°F and 30°F, and average humidity is 40%. Summer climatic conditions are very dry with an average relative humidity of 20%. Temperatures normally range from a low of 55°F at nighttime to 90°F high during the daytime, providing an average daily temperature swing of approximately 35°F.

The house is designed to be structurally simple and thermally efficient. The goal was to maximize usable space and utilize materials which are locally available and familiar to contractors. This floor is a 4″ concrete slab with a colored surface. Exterior walls are double wythe 4″ x 16″ x 4″ adobe-colored concrete slump block. Interior load bearing walls are 4″ x 16″ x 8″ colored concrete slump block; minor partitions and plumbing partitions are conventional stud framing. The overall architectural statement of the building is made by the conservative use of materials similar to the adobe and wood style of the southwestern United States.

Ft. Collins, CO

David G. Wagner & Associates
Ft. Collins, CO

HEATED AREA: 1,344 FT²

NUMBER OF DEGREE DAYS: 6,599

NET THERMAL LOAD: 41.1 10⁶BTU/YR

AUXILIARY ENERGY: 2.7 BTU/DD/FT²

YEARLY SOLAR FRACTION: 58%

CONSERVATION

This structure is oriented with its long side facing true south. It is set into the south-facing slope and the east, west, and north walls are heavily bermed, which is an efficient, cost-effective way to markedly reduce heat losses. The shape of the structure also plays an important role in the reduction of heat losses; the straightforward rectangular form of the dwelling minimizes the ratio of exterior walls to interior space, providing less exterior surface area through which heat may escape. An attached garage serves as a thermal buffer on the north and deflects prevailing northwest winds to further protect the house. The use of the garage as a winter entry also serves as a means of reducing losses due to infiltration.

The cavity of the double wythe slump blocks is filled with 2″ urethane foam to produce a wall with an effective R-value of 14; the 8″ ceiling joists filled with 5-1/2″ batt insulation create a roof R-value of 20. Effective perimeter insulation of 2″ polystyrene foam is used on all surfaces below grade. All windows have nighttime insulation with an effective R-value of 3.0.

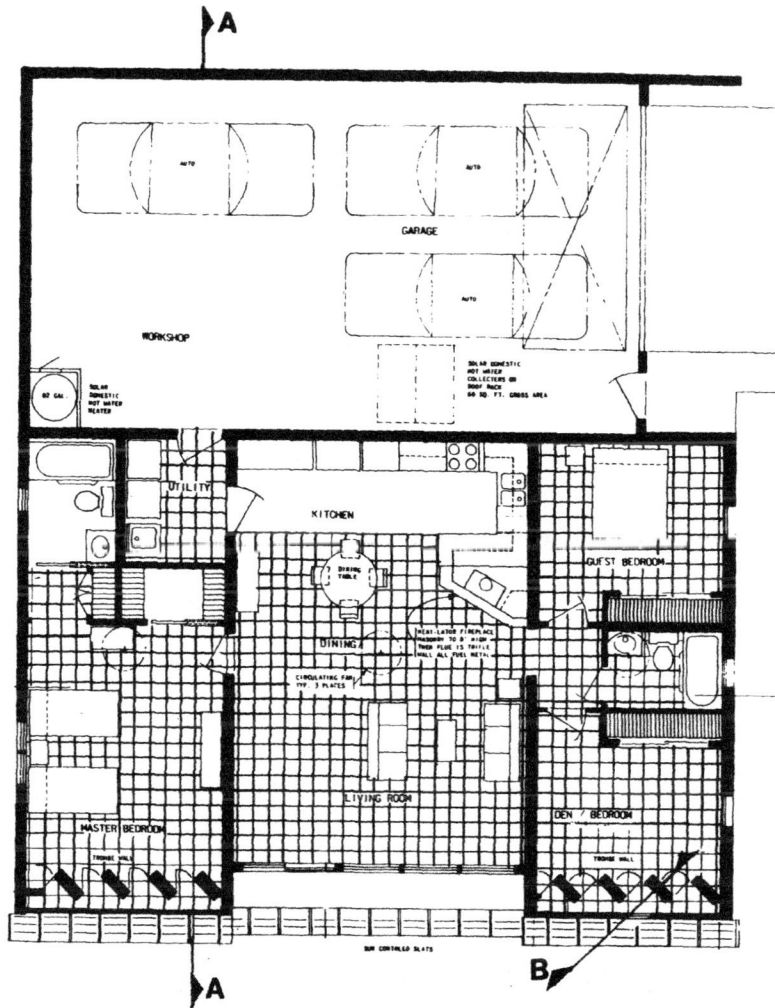

HEATING

The passive solar heating system for this dwelling is primarily direct gain with clerestory windows providing direct gain solar to rear portions of the house. The modified Trombe walls in the east and west wings provides morning direct gain into the interior of the structure, while in the afternoon Trombe wall storage is achieved. The masonry walls, the Trombe wall pillars, and the concrete floors all provide heat storage.

With 283 sq. ft. of vertical, south-facing double glazing, the solar system will provide approximately 83% of this home's heating needs. Out of a total yearly heating load of 41 million BTU (12,000 Kw), this means that 34 million BTU's (10,000 Kw) are provided by solar and the remaining 7 million BTU's (2,000 Kw) are provided by the auxiliary electric resistance baseboard units.

The sun penetrates the structure through the large expanse of direct gain and clerestory windows, strikes the light-colored ceilings and the buff-colored concrete floors, and is diffused over the floors and wall surfaces which store the excess heat for later use. As the room temperature falls below the surface temperature of the floors and walls, this stored heat is released into the room by radiation.

In each bedroom wing, a modified Trombe wall is formed by placing four 8" x 24" concrete slump block pillars behind the south window wall at a 50° angle to the glazing. The pillar orientation has been optimized to allow morning direct gain, afternoon Trombe wall storage, and continual light and view. In the morning, direct gain into each bedroom through the glazed areas provides immediate heat, with any excess heat being stored in the mass of the floors and walls. As the sun moves across the horizon during the day, afternoon sunlight is intercepted by the massive pillars creating, in effect, a Trombe wall. The sun directly hits the surface of the Trombe wall and slowly penetrates its mass; because of the amount of time it takes for the heat to travel through the wall, the inside surface of the wall reaches its maximum temperature in the early evening, providing a radiant heat source throughout the night.

The Trombe wall pillars are spaced to provide access for cleaning and manual operation of the venting windows. Wooden panel doors are hinged to the pillars for use as nighttime insulation over the glazed areas and temperature control within the living space. The doors can be opened for maximum daytime heat gain, partially closed to dampen daytime heat gain, and completely closed to maximize Trombe wall heat storage when direct heat gain is not needed within the home. Weatherstripping around the door

SECTION A-A

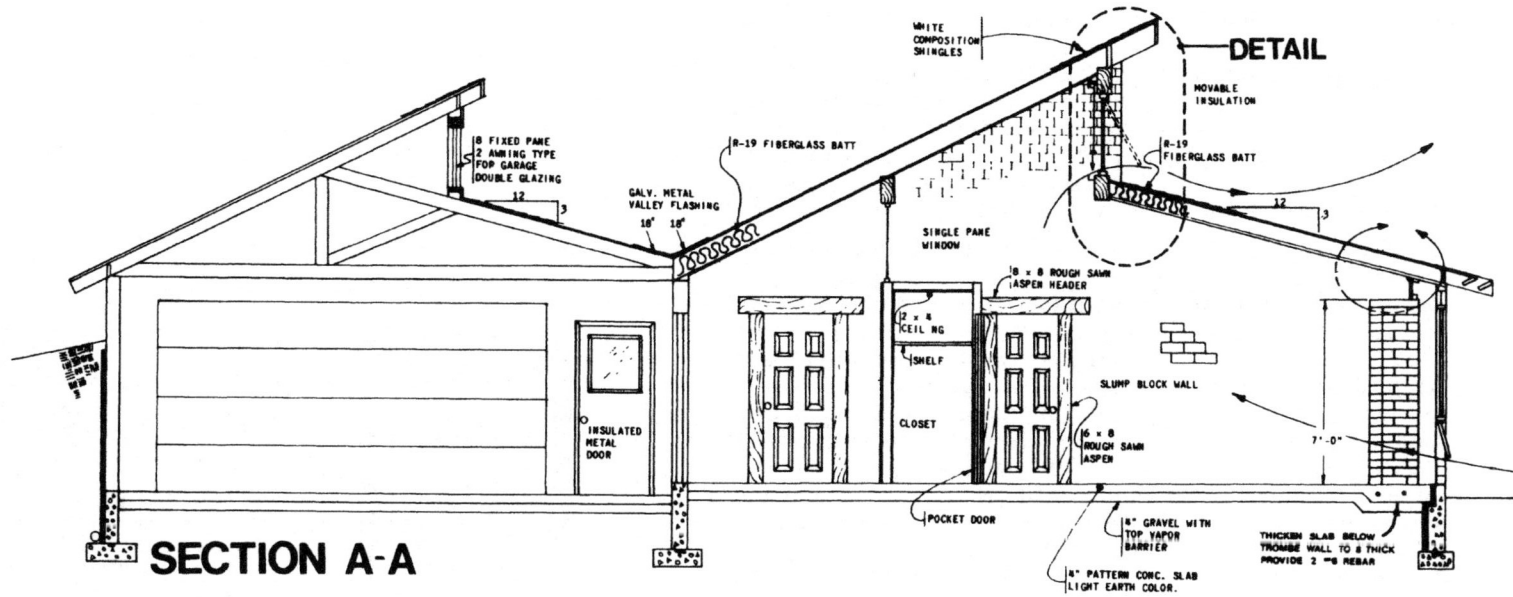

Labels in Section A-A:
- WHITE COMPOSITION SHINGLES
- DETAIL
- MOVABLE INSULATION
- R-19 FIBERGLASS BATT
- 8 FIXED PANE 2 AWNING TYPE FOR GARAGE DOUBLE GLAZING
- GALV. METAL VALLEY FLASHING
- R-19 FIBERGLASS BATT
- 12
- 3
- 18" 18"
- SINGLE PANE WINDOW
- 12
- 3
- 8 x 8 ROUGH SAWN ASPEN HEADER
- 2 x 8 CEILING
- SHELF
- SLUMP BLOCK WALL
- CLOSET
- 6 x 8 ROUGH SAWN ASPEN
- 7'-0"
- INSULATED METAL DOOR
- POCKET DOOR
- 4" GRAVEL WITH TOP VAPOR BARRIER
- THICKEN SLAB BELOW TROMBE WALL TO 8 THICK PROVIDE 2 =6 REBAR
- 4" PATTERN CONC. SLAB LIGHT EARTH COLOR.

SECTION B-B

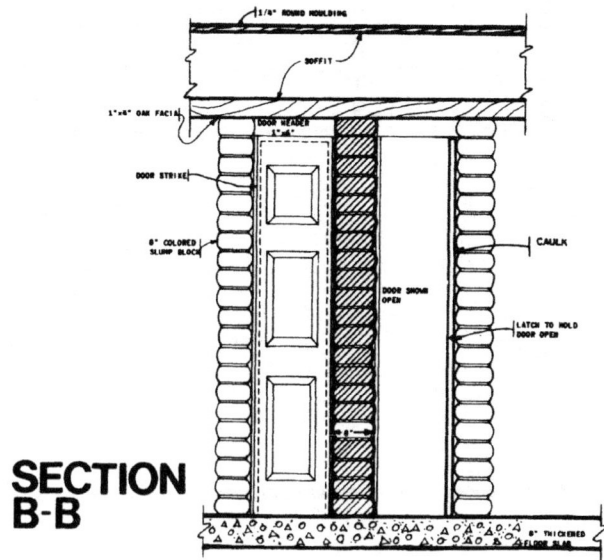

Labels in Section B-B:
- 1/4" ROUND MOULDING
- SOFFIT
- 1"x4" OAK FACIAL
- DOOR HEADER 1" x4"
- DOOR STRIKE
- 8" COLORED SLUMP BLOCK
- DOOR SHOWN OPEN
- CAULK
- LATCH TO HOLD DOOR OPEN
- 8" THICKENED FLOOR SLAB

frames and a soffit above the pillars help to isolate this space thermally.

Moveable insulation for the clerestory windows is a heat retention fabric commonly used in greenhouses. The fabric is wound on a tube located at the top of each clerestory bay and, when needed, is released by a gear motor drive to cover the entire bay of windows. Control is provided by a timeclock to allow the curtain to operate based on sunrise and sunset times throughout the year. Insulated drapes reduce nighttime losses through the remaining windows.

Three slow-speed ceiling propeller fans (100 watts each) protect against heat stratification and help to increase the convective flow of heat from the mass walls and floors. A standard differential thermostat with a sensor at floor level and at clerestory level activates the fans when the temperature at the top of the clerestory is higher than the floor temperature, and turns off the fans when the differential is decreased.

COOLING

In this location, with its dry climate and large day-night temperature fluctuations, careful design of the structure can maintain comfortable temperatures within a living space. By providing overhangs sized to fully intercept the sun in summer, no direct radiation is allowed to penetrate into the house. As the sun angle begins to decline after the summer solstice (June 21st), this shading factor gradually decreases, allowing an increasing amount of direct sunlight to fall on the glass. This simple method of passive control is quite effective until late summer and eary fall, when the heat loads for the building are still quite small and increasing heat gains may cause overheating.

The amount of building mass and the effect of berming in conjunction with adequate ventilation will do much to alleviate overheating. In climates with large day-night temperature swings, massive exterior walls can absorb much of the daytime heat, only radiating this heat at night when temperatures have dropped—a natural cooling and heating system. As excess heat accumulates, it is absorbed into the mass where it is lost quickly through bermed surface areas to the cooler temperature of the earth.

A continuous row of operable venting windows across the bottom of the south window wall and across the clerestory creates a system of induced ventilation, pulling air in from the outside to replace the hotter air escaping through the clerestory. This air movement in combination with the tempering effect of the mass helps to keep the interior temperature within a range acceptable in this dry climate.

CONCLUSION

The home uses its site to good advantage, making use of the existing vegetation and slope to provide thermal and wind buffers. The layout of the house relates well to the solar heating system; each room, including the garage-workshop, has immediate access to direct gain radiation. The amount of mass is adequate, and the use of materials for storage is appropriate. The variation of the Trombe wall is an interesting approach, one that deals well with problems inherent in either a direct gain or mass wall system; the problem of too much light or not enough light. The basic approach of the system is simple and straightforward, one that will be relatively easy to build.

A few slight changes might be appropriate. It would definitely be cost-effective in this climate to increase both the wall and roof insulation to achieve higher R-values and a smaller heat load, thus requiring a smaller percentage of south-facing glazing to achieve the same solar fraction. The airlock entry through the garage and

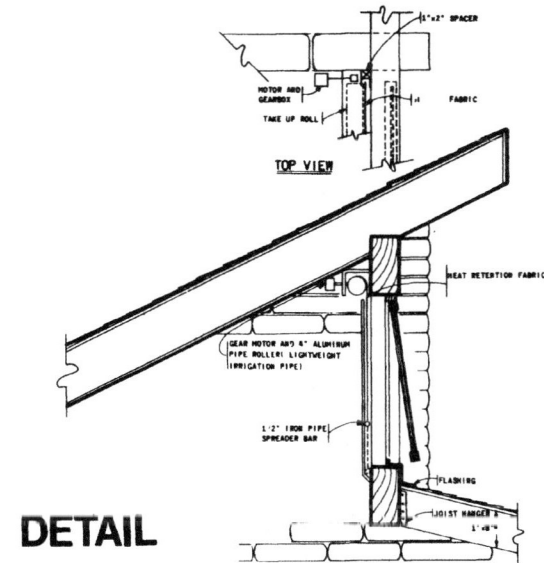

DETAIL

utility room is a poor compromise, and excessive use of the front entry through the sliding glass door will certainly increase heat losses. A definite concern with a large expanse of direct gain window is overheating and glare in the direct sun. A solution to these problems in this residence could be to exchange the direct gain system with the modified Trombe wall system, thus tempering overheating and glare in the living areas with the Trombe pillars and allowing overheating and glare to occur in the bedroom areas which are used less often during the day. Especially during August and September, overheating may be a problem, and a provision for exterior shading, possibly planting deciduous trees, would be an easy solution. Snowload in this climate must be considered, and provisions for raising the clerestory windows above snow level should be made.

This home is being built for sale and appears compatible with the local market. The house is to be priced in the $60,000-$70,000 range which is the median cost of new construction in the project's market area. The residence is designed for the middle income professional with a small family. Because the potential home buyer in Fort Collins is typically college-educated and directed toward energy consciousness, this home should meet with strong market acceptance.

The following project pages are devoted to a brief description of the other DIRECT SOLAR GAIN homes which were selected for awards. Each project is shown in perspective and accompanied by either a plan or a section. The project information extracted from the grant application is as follows:

Location of the home

Designer's name and firm
City and state of the designer*

HEATED AREA in square feet

NUMBER OF HEATING DEGREE DAYS

NET THERMAL LOAD in millions of British Thermal Units per year

AUXILIARY heating load in British Thermal Units per degree day per square foot

YEARLY SOLAR FRACTION: the percentage of heating energy provided by solar

COLLECTOR: Description and number of square feet

STORAGE: Description and Capacity in British Thermal Units per degree Fahrenheit

CONTROLS: Description

BACKUP: Type and capacity in British Thermal Units per hour

*Space limitations did not permit printing complete address. If you would like to contact the designer or builder concerning any of these projects, simply contact the National Solar Heating and Cooling Information Center (PROFESSIONALS FILE) by calling 800-523-2929 or 800-462-4983 (if calling from Pennsylvania), or by writing P.O. Box 1607, Rockville, MD 20850.

This project is designed to provide 60 units of attached housing for the staff and faculty of the Navajo Community College, located in the mountainous northeastern corner of Arizona. The area does not have a natural gas service but wood is plentiful; therefore, wood-burning stoves are used for backup heating. Living areas are oriented to the south with maximum windows, while bedrooms are placed to the north with minimum openings.

Private client.

Tsaile, AZ

William L. Burns, The Burns/Peters Group
Albuquerque, NM

HEATED AREA: 2-bdrm: 800 FT²; 3-bdrm: 936 FT²; 4-bdrm: 1107 FT²

NUMBER OF DEGREE DAYS: 7,467

NET THERMAL LOAD: 48 10⁶BTU/YR

AUXILIARY ENERGY: 4.6 BTU/DD/FT²

YEARLY SOLAR FRACTION: 59%

COLLECTOR: South-facing window wall, clerestories Area: 168 FT²

STORAGE: Concrete slab floor, concrete block walls Capacity: 2,400 BTU/°F

CONTROLS: Manually operable drapes and windows; automatic insulating louvers system

BACKUP: 30,000 BTUH wood fireplace; 25,000 BTUH electric furnace

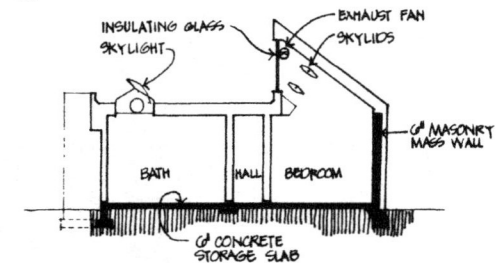

BUILDING SECTION

This is a 1-story, 3-bedroom home with stucco exterior and priced in the $45,000 range. Trees located on the west side of this level site provide summer shade.

Open market sale.

Cottonwood, AZ

Jim Raney, Sun System Engineering
West Sedona, AZ

HEATED AREA: 1,350 FT²

NUMBER OF DEGREE DAYS: 2,548

NET THERMAL LOAD: 11.8 10^6BTU/YR

AUXILIARY ENERGY: 2.29 BTU/DD/FT²

YEARLY SOLAR FRACTION: 94%

COLLECTOR: South-facing windows
Area: 156 FT²

STORAGE: Block and concrete floor, block and stone mass wall
Capacity: 6,574 BTU/°F

CONTROLS: Insulation panels, overhangs, manual vents

BACKUP: 10,000-100,000 BTUH wood stove

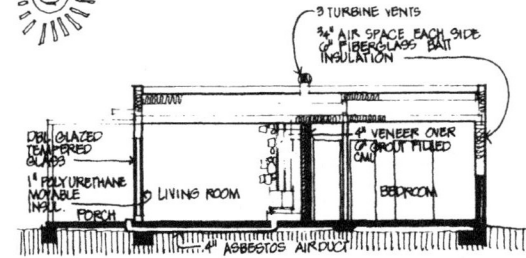

BUILDING SECTION

This new 2-story, 3-bedroom detached house is in the $93,000 price range. All entrances have overhangs and are recessed to provide winter wind protection. Sloped ceilings and lofts are intended to give a greater feeling of openness and a greatly reduced consumption of energy.

Open market sale.

Davis, CA

Paul Fellers
Davis, CA

HEATED AREA: 1,685 FT²

NUMBER OF DEGREE DAYS: 2,374

NET THERMAL LOAD: 20.5 10^6BTU/YR

AUXILIARY ENERGY: 2.64 BTU/DD/FT²

YEARLY SOLAR FRACTION: 82%

COLLECTOR: South-facing windows and skylights Area: 269 FT²

STORAGE: Concrete slab floor, water containers Capacity: 10,273 BTU/°F

CONTROLS: Operable drapes and shade screens

BACKUP: 40,000 BTUH gas furnace

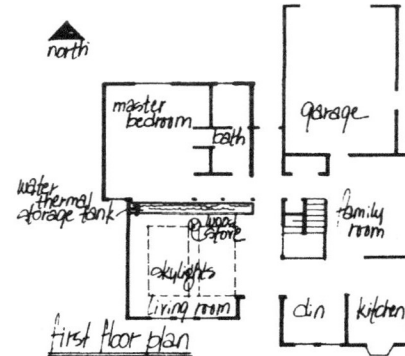

first floor plan

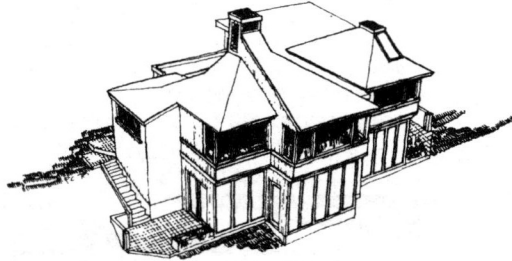

Dillon Beach, CA

Peter Calthorpe, Calthorpe/Fernau/Wilcox
San Francisco, CA

HEATED AREA: 1,800 FT²

NUMBER OF DEGREE DAYS: 3,413

NET THERMAL LOAD: 36.0 10⁶BTU/YR

AUXILIARY ENERGY: 3.28 BTU/DD/FT²

YEARLY SOLAR FRACTION: 89%

COLLECTOR: South-facing windows, tube wall
Area: 483 FT²

STORAGE: Concrete slab, water tube wall
Capacity: 15,232 BTU/°F

CONTROLS: Registers, belvedere, window
louvers and vents

BACKUP: 90,000 BTUH gas forced air furnace

This is a 2-level, 2-bedroom with studio, wood frame house. Located on a steep site, the home is benched into a hill. There are no trees on this site.

Private client.

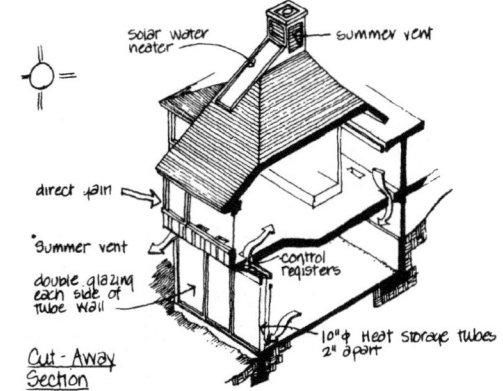

Georgetown, CA

Larry and Jacqueline Morgan
Georgetown, CA

HEATED AREA: 1,573 FT²

NUMBER OF DEGREE DAYS: 4,450

NET THERMAL LOAD: 56.9 10⁶BTU/YR

AUXILIARY ENERGY: 4.85 BTU/DD/FT²

YEARLY SOLAR FRACTION: 60%

COLLECTOR: Southeast-facing skylights,
greenhouse, and windows
Area: 369 FT²

STORAGE: Concrete and gravel floors, concrete
block wall, fireplace
Capacity: 7,273 BTU/°F

CONTROLS: Operable insulated shutters,
curtains, windows, and skylights
with shades

BACKUP: Heat circulating fireplaces, heat
pumps

This new 3-bedroom house is priced in the $75,000 range. Cedar siding and river rock and pebbles were used in the design to achieve a rustic appearance and to use native materials. The lower floor is set into the southerly sloping site for earth berm insulation.

Open market sale.

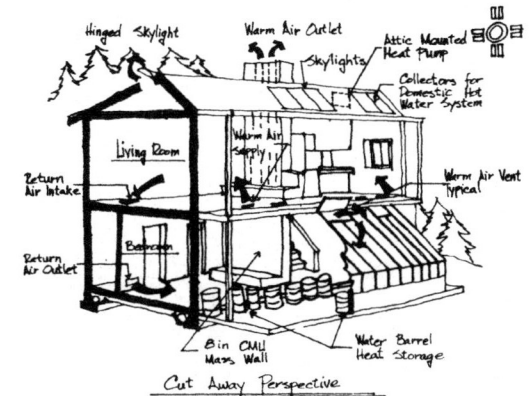

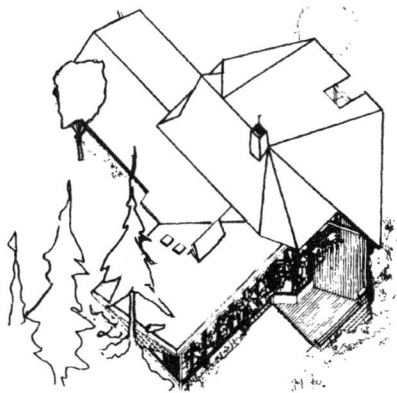

This 2-story, 3-bedroom building is priced in the $70,000 range. Shingled in redwood, the wood frame and slump stone structure is sunk 8 feet into a hillside meadow. A tall grove of redwoods to the west provides summer shading; forest to the north forms a natural windbreak.

Private client.

Miranda, CA

Jonathan Allan Stoumen,
Architect
Miranda, CA

HEATED AREA: 1,750 FT²

NUMBER OF DEGREE DAYS: 3,270

NET THERMAL LOAD: 15.0 10⁶BTU/YR

AUXILIARY ENERGY: 2.77 BTU/DD/FT²

YEARLY SOLAR FRACTION: 92%

COLLECTOR: South-facing windows
Area: 321 FT²

STORAGE: Concrete walls, piers, fireplace; tile, concrete and gravel floor; concrete, sand and water bench
Capacity: 23,576 BTU/°F

CONTROLS: Thermal curtains, vents, operable windows

BACKUP: 20,000 BTUH fireplace

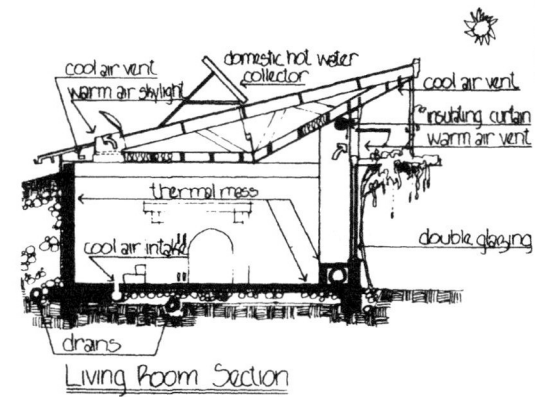

Living Room Section

This new, ranch-style, 3-bedroom detached house has an adobe plaster finish and is in the $70,000 price range. An earth roof, earth berming, and mature evergreens on the north side of the house provide winter wind protection; deciduous vines and overhangs to the south provide summer shading.

Private client.

Pacheco, CA

Lynn S. Nelson, The Habitat Center
San Francisco, CA

HEATED AREA: 1,739 FT²

NUMBER OF DEGREE DAYS: 2,697

NET THERMAL LOAD: 22.1 10⁶BTU/YR

AUXILIARY ENERGY: 8.04 BTU/DD/FT²

YEARLY SOLAR FRACTION: 92%

COLLECTOR: South-facing glass and clerestory
Area: 367 FT²

STORAGE: Brick and sand floor, adobe walls, pool Capacity: 77,506 BTU/°F

CONTROLS: Operable insulating curtains, air intake tubes, solarium shade, and clerestory hopper vents

BACKUP: 3,414 BTUH (per unit) electric baseboard; wood fireplace with thermograte

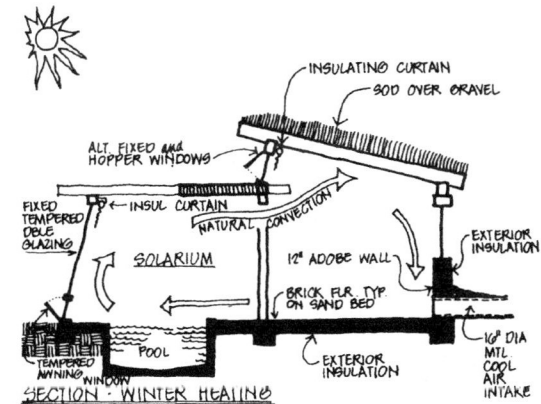

SECTION · WINTER HEATING

47

This new 4-bedroom is priced in the $62,000 range and has a traditional "bungalow" architectural style. Thirty percent of the lower level is below grade, and a sheltered entrance also reduces heat losses in the winter.

Open market sale.

Aurora, CO

David Elfring
Denver, CO

HEATED AREA: 2,000 FT²

NUMBER OF DEGREE DAYS: 5,936

NET THERMAL LOAD: 109 10⁶BTU/YR

AUXILIARY ENERGY: 5.5 BTU/DD/FT²

YEARLY SOLAR FRACTION: 40%

COLLECTOR: South-facing windows
              Area: 265 FT²

STORAGE: Water-filled steel tank, clay tile walls, brick floors, sand-filled concrete block outside walls
            Capacity: 7,414 BTU/°F

CONTROLS: Operable drapes, blinds, vents, and turbines

BACKUP: 75,000 BTUH electric forced-air heat pump

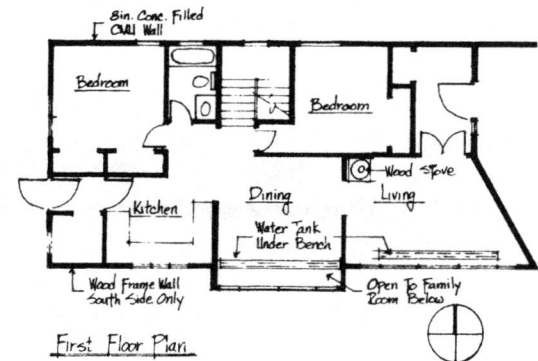

First Floor Plan

This contemporary 3-bedroom house is located on a flat site with good solar exposure. The house is priced in the $100,000 range. Atriums in the entryway and dining room have been utilized for moisture input in this dry climate. Trees will be planted on the south side of the house for summer shading.

Private client.

Boulder, CO

F. R. Rutz, Fenton A. Bain & Frederick Rutz
Longmont, CO

HEATED AREA: 3,260 FT²

NUMBER OF DEGREE DAYS: 6,202

NET THERMAL LOAD: 111 10⁶BTU/YR

AUXILIARY ENERGY: 4.63 BTU/DD/FT²

YEARLY SOLAR FRACTION: 45%

COLLECTOR: South-facing windows
              Area: 329 FT²

STORAGE: Concrete walls
            Capacity: 6,765 BTU/°F

CONTROLS: Operable ducts, insulation, windows, and skylights

BACKUP: 200,000 BTUH gas furnace; 40,000 BTUH wood fireplace

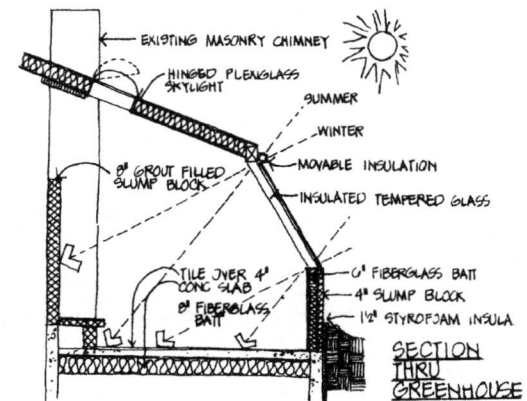

SECTION THRU GREENHOUSE

This is a retrofit on a 3-bedroom, 2-story house with drop siding exterior. Evergreens and the carport roof protect this home from winter winds. It is valued in the $80,000 range.

Retrofit.

Boulder, CO

Jeffrey Ellis, Superstructures
Boulder, CO

HEATED AREA: 1,734 FT²

NUMBER OF DEGREE DAYS: 5,446

NET THERMAL LOAD: 38.9 10⁶BTU/YR

AUXILIARY ENERGY: 3.70 BTU/DD/FT²

YEARLY SOLAR FRACTION: 52%

COLLECTOR: Two levels of south-facing windows; solarium windows and roof Area: 174 FT²

STORAGE: Concrete and rock floor, concrete walls exposed to air preheated by solarium Capacity: 4,123 BTU/°F

CONTROLS: Operable insulation panels

BACKUP: 64,000 BTUH gas furnace; fireplace

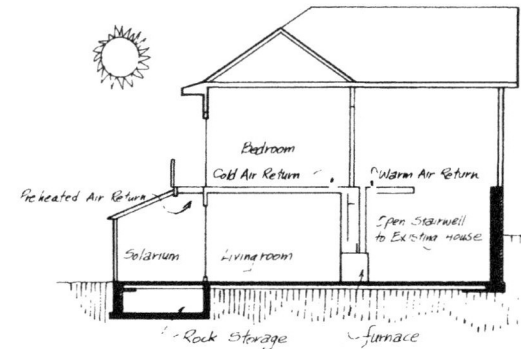

Section through living area

This new, 2-story, 4-bedroom house is priced in the $95,000 range. The building is excavated into the north side of the hill, and the entrance on the house's east side also prevents northwest winter wind infiltration. Coniferous planting on the north and northwest sides is planned.

Open market sale.

Boulder, CO

Stephen Sparn, Architect
Boulder, CO

HEATED AREA: 1,700 FT²

NUMBER OF DEGREE DAYS: 6,173

NET THERMAL LOAD: 88.3 10⁶BTU/YR

AUXILIARY ENERGY: 4.57 BTU/DD/FT²

YEARLY SOLAR FRACTION: 59%

COLLECTOR: South-facing windows, skylights, and solarium Area: 390 FT²

STORAGE: Water drums, concrete slab and walls Capacity: 7,910 BTU/°F

CONTROLS: Operable insulating louvers, vents, and windows

BACKUP: 56,000 BTUH electric baseboard

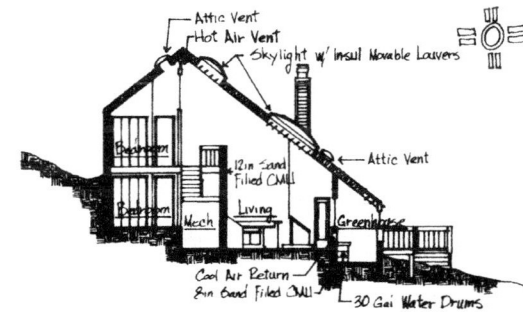

Building Section

This new, 2-story, 3-bedroom contemporary house is priced in the $105,000 range. Earth berming and a garage buffer zone on the building's north side prevent winter wind infiltration. Deciduous trees are to be planted to the south and west of the house. The design has a large, open first-floor area.

Open market sale.

Boulder, CO

Jim Logan and Paul Nicely,
Logan Construction, Boulder, CO

HEATED AREA: 2,304 FT²

NUMBER OF DEGREE DAYS: 5,803

NET THERMAL LOAD: 63.4 10⁶BTU/YR

AUXILIARY ENERGY: 2.37 BTU/DD/FT²

YEARLY SOLAR FRACTION: 77%

COLLECTOR: South-facing windows
　　　　　Area: 442 FT²

STORAGE: Concrete floor; concrete solar mass
　　　　　wall Capacity: 28,406 BTU/°F

CONTROLS: Automatic fans and curtains, with
　　　　　manual override; in summer,
　　　　　skylight over central hall to be left
　　　　　open

BACKUP: 40,000 BTUH gas hot water
　　　　　baseboard

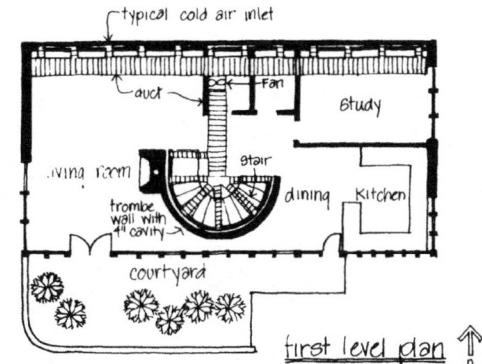

first level plan

This new 2-story, 3-bedroom detached frame and stucco house is in the $75,000 range. Trees will be planted to the north and west to form a natural windbreak, and to the south for summer shading.

Open market sale.

Colorado Springs, CO

Rick Cowlishaw, Natural Systems, Inc.
Colorado Springs, CO

HEATED AREA: 1,800 FT²

NUMBER OF DEGREE DAYS: 6,423

NET THERMAL LOAD: 84.6 10⁶BTU/YR

AUXILIARY ENERGY: 4.66 BTU/DD/FT²

YEARLY SOLAR FRACTION: 55%

COLLECTOR: South-facing windows and
　　　　　skylights Area: NA FT²

STORAGE: Brick fireplace and walls, concrete
　　　　　slab, rock bed
　　　　　Capacity: 39,000 BTU/°F

CONTROLS: Fan, dampers, operable
　　　　　windows/vents

BACKUP: 80,000 BTUH gas furnace

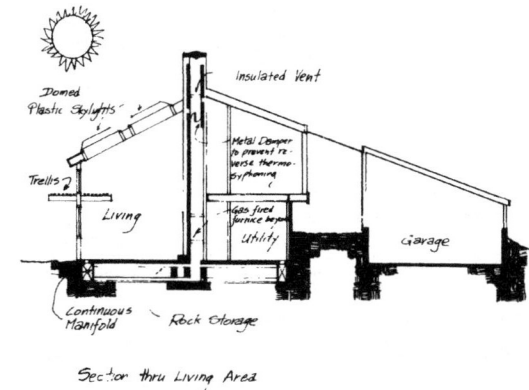

Section thru Living Area

This new, 3-bedroom detached house is in the $89,000 price range and is located in a planned community of three passive solar heated residences. A minimum of east, north, and west windows reduces heat loss in the winter and overheating in the summer.

Open market sale.

New Milford, CT

Stephen Lasar
New Milford, CT

HEATED AREA: 2,316 FT²

NUMBER OF DEGREE DAYS: 6,673

NET THERMAL LOAD: 92.4 10⁶BTU/YR

AUXILIARY ENERGY: 4.59 BTU/DD/FT²

YEARLY SOLAR FRACTION: 51%

COLLECTOR: South-facing skylights, windows, east, southeast, west facing windows Area: 407.3 FT²

STORAGE: Concrete floors, concrete block walls Capacity: 18,961 BTU/°F

CONTROLS: Natural convective air flow, auxiliary storage fan; manually opening doors and windows

BACKUP: Oil fired forced air furnace

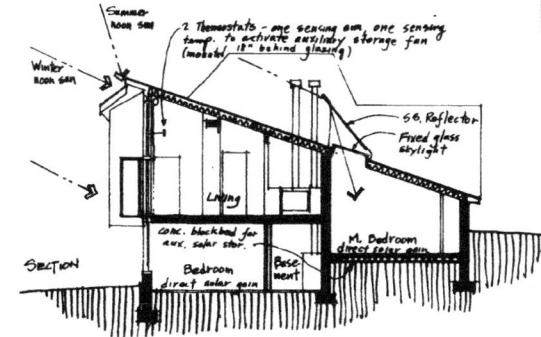

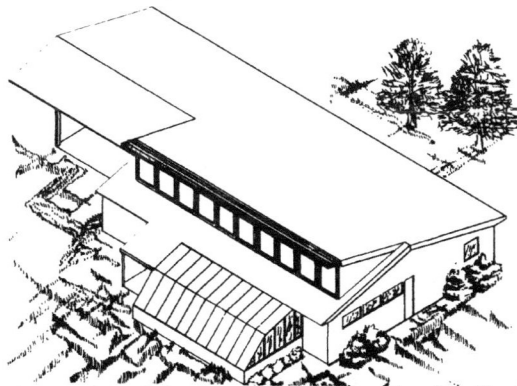

This 1-story, 3-bedroom house is priced in the $40,000 range. Winter wind infiltration is minimized by the use of earth berming and evergreen shrubs on the north side of the building, and by air-lock entries on the south side. Deciduous plantings and solar shades over roof monitors block out the summer sun but admit the winter sun.
Open market sale.

Columbus, GA

Thomas Lee Goodson, Goodson and Ahlquist Associates
Columbus, GA

HEATED AREA: 1,850 FT²

NUMBER OF DEGREE DAYS: 2,383

NET THERMAL LOAD: 33.6 10⁶BTU/YR

AUXILIARY ENERGY: 6.22 BTU/DD/FT²

YEARLY SOLAR FRACTION: 82%

COLLECTOR: Solarium and roof monitors Area: 480 FT²

STORAGE: Concrete floor, water-filled steel drums Capacity: 48,144 BTU/°F

CONTROLS: Operable shades and vents; ventilation; roof shading

BACKUP: 7,000 BTUH wood fireplace; 40,000 BTUH gas furnace

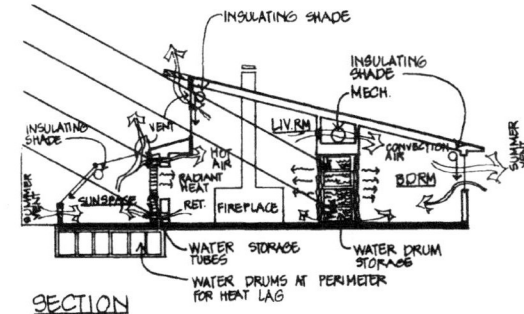

This new ranch-style house is protected from winter winds by the heavily wooded area surrounding the house, but the site slope orientation allows for maximum solar exposure. The lower level of the house is set into the slope. The price is in the $62,500 range.

Open market sale.

Norcross, GA

W. Jeff Floyd, Jr.
Atlanta, GA

HEATED AREA: 2,325 FT²

NUMBER OF DEGREE DAYS: 2,940

NET THERMAL LOAD: 195 10⁶BTU/YR

AUXILIARY ENERGY: 18.2 BTU/DD/FT²

YEARLY SOLAR FRACTION: 51%

COLLECTOR: South-facing glazing; south-facing solarium Area: 880 FT²

STORAGE: Concrete floor, water drums Capacity: 6,826 BTU/°F

CONTROLS: Automatic heat pump; operable insulating polyethylene over solarium

BACKUP: Wood fireplace; 30,000 BTUH electric heat pump; 26,000 BTUH wood Franklin stove

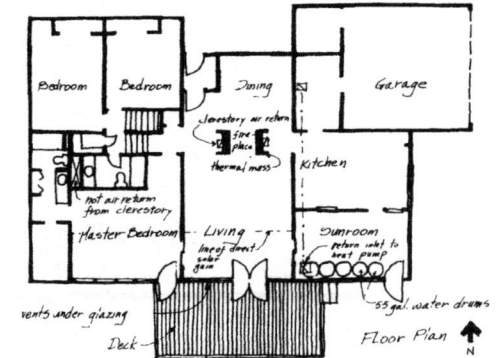

Floor Plan

This new 1-1/2-story, 3-bedroom detached house using a roof window to collect the sun's heat is in the $40,000 price range. Light vegetation on the site's flat terrain provides modest shading in the summer to prevent overheating.

Open market sale.

Indianapolis, IN

Robert L. Beecher II
Indianapolis, IN

HEATED AREA: 1,472 FT²

NUMBER OF DEGREE DAYS: 5,699

NET THERMAL LOAD: 65.7 10⁶BTU/YR

AUXILIARY ENERGY: 3.96 BTU/DD/FT²

YEARLY SOLAR FRACTION: 48%

COLLECTOR: South-facing windows, roof window Area: 220 FT²

STORAGE: Brick and concrete walls, concrete and gravel floor Capacity: 6,848 BTU/°F

CONTROLS: Vents, automatic roof insulator

BACKUP: 50,000 BTUH electric furnace; 70,000 BTUH fireplace

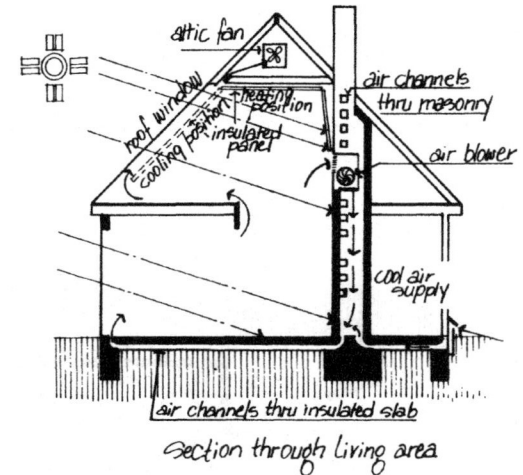

Section through Living area

This home, located on a gently sloping site, is one story with 2 bedrooms and a loft. The exterior is wood frame with cedar trim. There is an evergreen windbreak to the northwest. The house is priced in the $50,000 range.

Private client.

Salina, KS

C. F. Abercrombie
Salina, KS

HEATED AREA: 1,128 FT^2

NUMBER OF DEGREE DAYS: 5,628

NET THERMAL LOAD: 48.5 10^{6}BTU/YR

AUXILIARY ENERGY: 3.5 BTU/DD/FT^2

YEARLY SOLAR FRACTION: 76%

COLLECTOR: South-facing window, skylight
Area: 473 FT^2

STORAGE: Concrete floor and walls
Capacity: 26,097 BTU/°F

CONTROLS: Vents, louvers, window insulation, skylid, operable solarium door

BACKUP: 26,000 BTUH electric heat pump; 45,000 BTUH fireplace

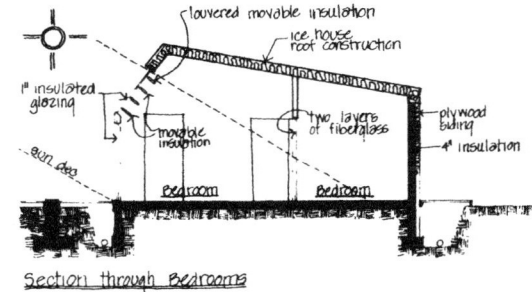

Section through Bedrooms

This 2-story, 3-bedroom contemporary home will be used by the builder as a prototype to be marketed through New England and the mid-Atlantic states as a pre-engineered package house. Trees will be planted to provide a windbreak, and the price of the house is in the $45,000 range.

Open market sale.

Amherst, MA

John J. Rossini, Architects, Inc.
Northampton, MA

HEATED AREA: 1,524 FT^2

NUMBER OF DEGREE DAYS: 6,575

NET THERMAL LOAD: 86.7 10^{6}BTU/YR

AUXILIARY ENERGY: 6.41 BTU/DD/FT^2

YEARLY SOLAR FRACTION: 26%

COLLECTOR: South-facing windows masonry walls Area: 189 FT^2

STORAGE: Concrete floor, masonry walls, rockbed Capacity: 17,345 BTU/°F

CONTROLS: Automatic fans, operable interior shutters

BACKUP: 15,000 BTUH electric baseboard

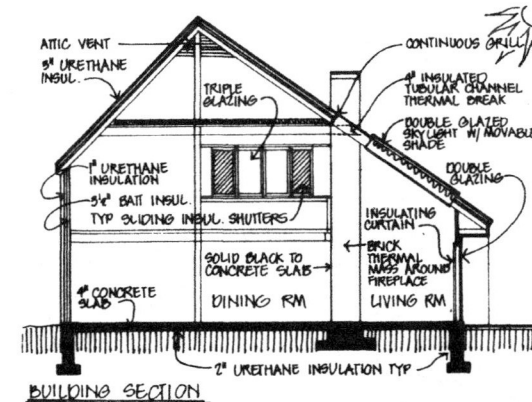

BUILDING SECTION

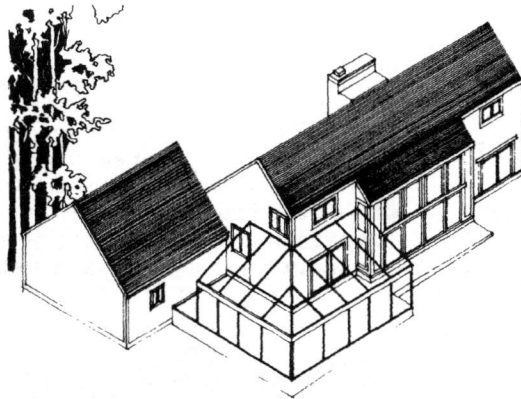

This new 2-story, 3-bedroom house is priced in the $94,800 range. This adaptation of the traditional Saltbox architectural design required the replanning of spaces on both floors; all habitable rooms are now arranged along the south face of the house.

Open market sale.

Andover, MA

Mollie Moran, Massdesign
Cambridge, MA

HEATED AREA: 1,887 FT²

NUMBER OF DEGREE DAYS: 5,750

NET THERMAL LOAD: 47.2 10⁶BTU/YR

AUXILIARY ENERGY: 3.60 BTU/DD/FT²

YEARLY SOLAR FRACTION: 64%

COLLECTOR: South-facing windows and walls
Area: 323 FT²

STORAGE: Concrete floor slab, gypsum board walls Capacity: 4,297 BTU/°F

CONTROLS: Operable shutters, windows, and dampers; automatic fans

BACKUP: Woodstove; electric heater

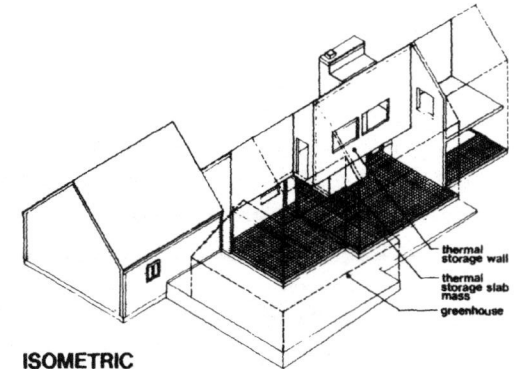

ISOMETRIC

This new, south-facing house is in the $80,000 price range. Earth berming and dense shrub pines on the north side of the house provide winter wind protection.

Open market sale.

Sharon, MA

Ole Hammarlund, Solsearch Architects
Cambridge, MA

HEATED AREA: 2,800 FT²

NUMBER OF DEGREE DAYS: 5,750

NET THERMAL LOAD: 99.2 10⁶BTU/YR

AUXILIARY ENERGY: 3.37 BTU/DD/FT²

YEARLY SOLAR FRACTION: 82%

COLLECTOR: South-facing windows, solar staircase Area: 1,029 FT²

STORAGE: Rockbed Capacity: 36,000 BTU/°F

CONTROLS: Automatic fans; operable windows

BACKUP: 40,000 BTUH woodstove; 25,000 BTUH woodstove; 50,000 BTUH oil water heater

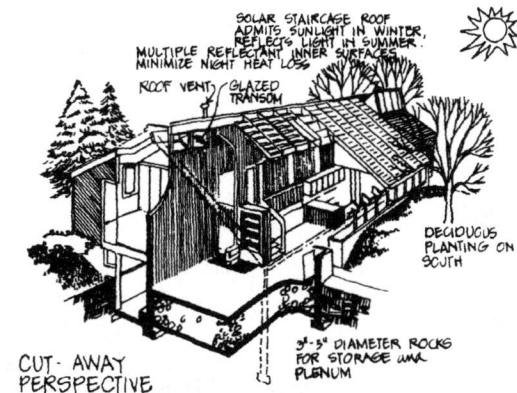

Located on a slight south slope, this contemporary 2-story, 4-bedroom wood frame house is priced in the $160,000 range. The exterior has cedar board siding.

Open market sale.

Gaithersburg, MD

John P. Ross, John Ross/Trellis and Watkins, Inc.
Bethesda, MD

HEATED AREA: 2,465 FT²

NUMBER OF DEGREE DAYS: 4,142

NET THERMAL LOAD: 51.1 10⁶BTU/YR

AUXILIARY ENERGY: 4.34 BTU/DD/FT²

YEARLY SOLAR FRACTION: 54%

COLLECTOR: South-facing windows Area: N/A

STORAGE: Masonry floor, brick walls, concrete block Capacity: 11,085 BTU/°F

CONTROLS: Insulating curtains, operable windows, dampers, fan

BACKUP: 27,150 BTUH and 11,500 BTUH electric heat pumps

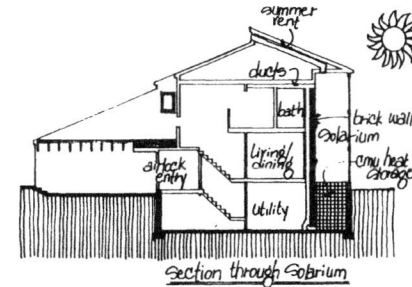

Section through Solarium

This 2-story, 2-bedroom cedar-shingled house is of the New England saltbox design. It is located in a heavily wooded area that gently slopes to the north and steeply slopes to the south toward a 30-foot dropoff to a stream. This home is valued in the $25,000 range.

Private client.

Wilton, ME

Conrad Heeschen, Sunsystems
Dryden, ME

HEATED AREA: 1,302 FT²

NUMBER OF DEGREE DAYS: 8,505

NET THERMAL LOAD: 39.1 10⁶BTU/YR

AUXILIARY ENERGY: 2.39 BTU/DD/FT²

YEARLY SOLAR FRACTION: 75%

COLLECTOR: Solarium, upper level windows Area: 130 FT²

STORAGE: Concrete walls and floors, water-filled cans, solarium bench Capacity: 19,243 BTU/°F

CONTROLS: Insulated shutters, fan, dampers, operable windows

BACKUP: 22,000 BTUH and 12,000 BTUH wood stoves

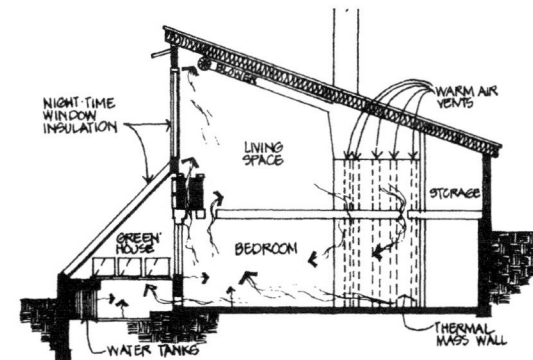

SECTION, HEATING MODE

DIRECT SOLAR GAIN

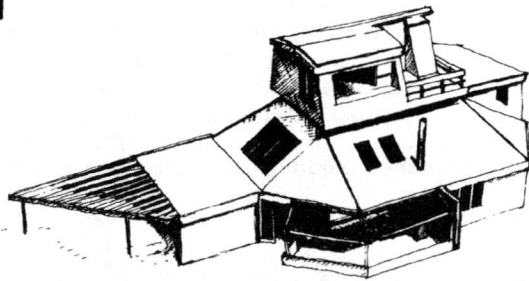

York, ME

Glen Langley, Philip S. Tambling/Glen Langley
Rye Beach, NH

HEATED AREA: 1,770 FT²

NUMBER OF DEGREE DAYS: 7,446

NET THERMAL LOAD: 86.6 10⁶BTU/YR

AUXILIARY ENERGY: 4.4 BTU/DD/FT²

YEARLY SOLAR FRACTION: 47%

COLLECTOR: South/southeast-facing solarium, southeast-facing skylights and windows Area: 255 FT²

STORAGE: Concrete floor, brick wall, rock storage Capacity: 15,084 BTU/°F

CONTROLS: Operable insulating curtains, awnings; automatic fans

BACKUP: 50,000 BTUH wood or coal stove; 30,000 BTUH electric radiant convector

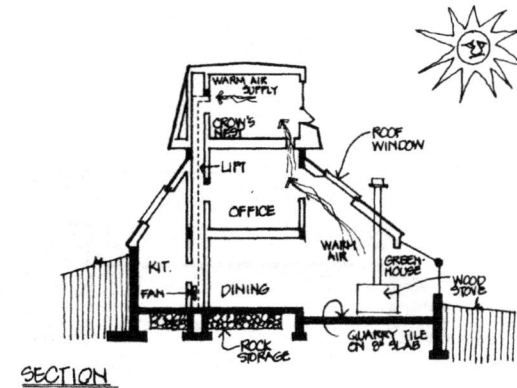

This 2-bedroom house is set into the face of a hill, with the northwest side bermed to the top of the first story. The house is oriented to have both a view of the Atlantic Ocean to the southeast and good solar collection from both the south and southwest. The price is in the $90,000 range.

Private client.

SECTION

Minneapolis, MN

Peter Pfister
Minneapolis, MN

HEATED AREA: 1,980 FT²

NUMBER OF DEGREE DAYS: 8,248

NET THERMAL LOAD: 126 10⁶BTU/YR

AUXILIARY ENERGY: 5.97 BTU/DD/FT²

YEARLY SOLAR FRACTION: 46%

COLLECTOR: South-facing windows, solarium, and solar mass wall Area: 258 FT²

STORAGE: Concrete floor, solar mass water wall Capacity: 6,275 BTU/°F

CONTROLS: Operable windows, window insulation, dampers, and valves

BACKUP: 128,000 BTUH gas boiler; wood fireplace

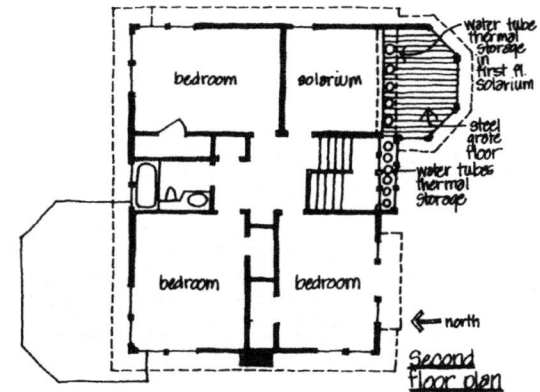

This 2-story, 2-bedroom detached retrofit has an evergreen tree on its north side for winter wind protection and a deciduous tree on its south side for summer sun protection.

Retrofit.

Second floor plan

Maple Grove, MN

L. W. Berg, Berg and Associates
Wayzata, NM

HEATED AREA: 1,180 FT²

NUMBER OF DEGREE DAYS: 8,062

NET THERMAL LOAD: 45.7 10⁶BTU/YR

AUXILIARY ENERGY: 2.23 BTU/DD/FT²

YEARLY SOLAR FRACTION: 87%

COLLECTOR: South-facing windows and
storage mass Area: 729.5 FT²

STORAGE: Concrete storage mass wall,
concrete floor
Capacity: 52,688 BTU/°F

CONTROLS: Automatic thermostat control of
fans; operable shades and louvers

BACKUP: 50,000 BTUH fireplace; 3,000 BTUH
radiant cooling panels

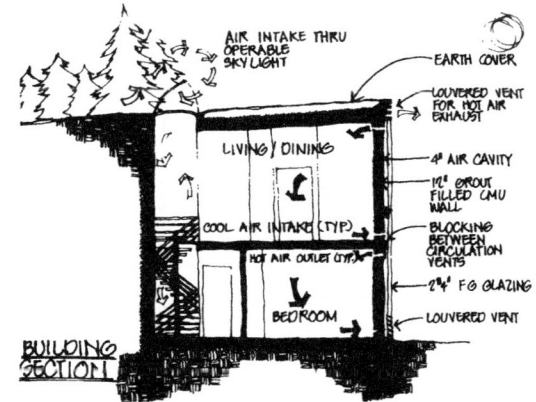

This new house is earth bermed on three sides
and has an earth roof; conifer trees to the
northwest and an air-lock/vestibule entrance
away from the winter wind also reduced winter heat losses. Southwest deciduous trees
provide summer shade. The house is in the
$68,000 price range.

Open market sale.

Minneapolis, MN

Michael Salvatore Cox
Richfield, MN

HEATED AREA: 1,320 FT²

NUMBER OF DEGREE DAYS: 8,382

NET THERMAL LOAD: 26.4 10⁶BTU/YR

AUXILIARY ENERGY: 2.25 BTU/DD/FT²

YEARLY SOLAR FRACTION: 80%

COLLECTOR: South-facing windows
Area: 300 FT²

STORAGE: Masonry walls and floors, rock bed
Capacity: 350.2 BTU/°F

CONTROLS: Operable shutters, vents, and
windows

BACKUP: 40,000 BTUH gas furnace; 40,000
BTUH fireplace

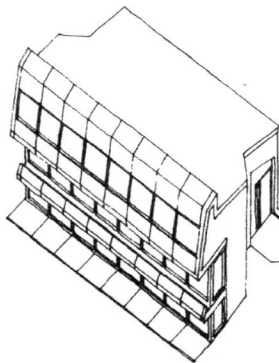

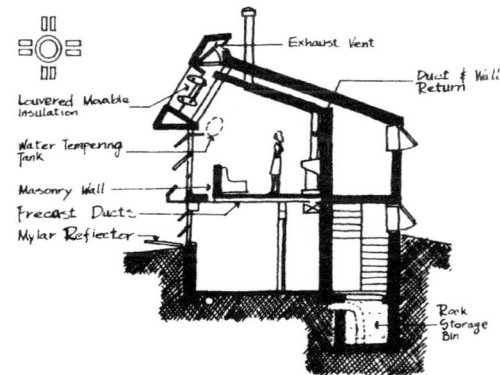

This 3-bedroom house with lap board siding is
located in a residential neighborhood. Trees
on this level lot will provide summer shading
and a winter windbreak. Its price is in the
$60,000 range.

Private client.

57

This 2-story, 2-bedroom house with concrete walls is built off the crest of a hill for north wind protection. The home is located in a wooded rural area and is priced in the $65,000 range.

Open market sale.

Lawson, MO

John Hueser
Kansas City, MO

HEATED AREA: 2,264 FT²

NUMBER OF DEGREE DAYS: 4,711

NET THERMAL LOAD: 64.5 10⁶BTU/YR

AUXILIARY ENERGY: 3.34 BTU/DD/FT²

YEARLY SOLAR FRACTION: 71%

COLLECTOR: South-facing windows
　　　　　　Area: 567 FT²

STORAGE: Concrete floors, walls and ceiling
　　　　　Capacity: 20,483 BTU/°F

CONTROLS: Operable insulated shutters, vents
　　　　　and skylights

BACKUP: 51,000 BTUH electric heat pump

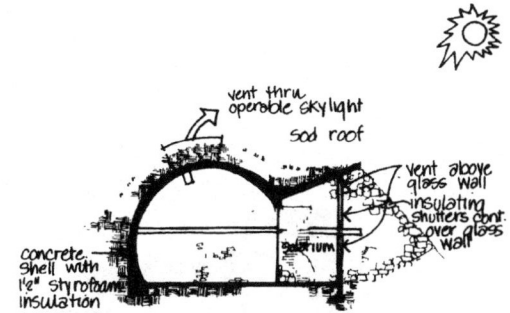

vent thru operable skylight
sod roof
vent above glass wall
insulating shutters cont. over glass wall
concrete shell with 1½" styrofoam insulation

Section thru Living Area

This contemporary 1-story, 2-bedroom wood frame house on its north sloping lot will be priced in the $40,000 range. Located on a 2-acre, heavily wooded site, the home will be well protected from winter winds.

Private client.

Wake Forest, NC

William R. Watkins, Jr., Sunshelter Design
Raleigh, NC

HEATED AREA: 1,604 FT²

NUMBER OF DEGREE DAYS: 3,352

NET THERMAL LOAD: 32.5 10⁶BTU/YR

AUXILIARY ENERGY: 4.64 BTU/DD/FT²

YEARLY SOLAR FRACTION: 55%

COLLECTOR: South-facing windows, roof glass
　　　　　　Area: 309.1 FT²

STORAGE: Concrete and brick floors and walls
　　　　　Capacity: 5,325.6 BTU/°F

CONTROLS: Moveable insulation, ceiling fan,
　　　　　ridge vents, operable windows

BACKUP: 39,500 BTUH wood stove

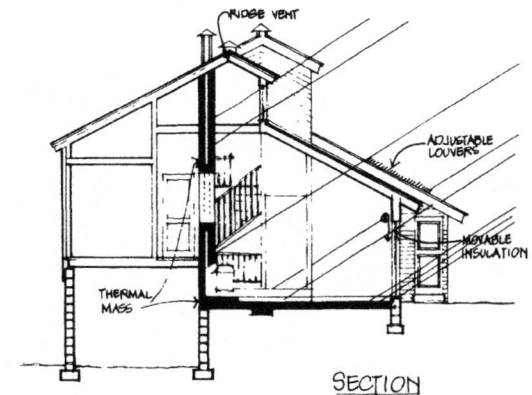

RIDGE VENT
ADJUSTABLE LOUVERS
MOVABLE INSULATION
THERMAL MASS

SECTION

This contemporary 3-bedroom house has a Spanish/stucco style of construction and is priced in the $120,000 range. Irrigated trees will be planted on the east and west sides of the house. Overlooking lofts and the interior living space are spacious.

Open market sale.

Las Vegas, NV

Bob Chase, Maxwell Starkman and Associates
Beverly Hills, CA

HEATED AREA: 2,312 FT²

NUMBER OF DEGREE DAYS: 2,625

NET THERMAL LOAD: 41.4 10⁶BTU/YR

AUXILIARY ENERGY: 5.39 BTU/DD/FT²

YEARLY SOLAR FRACTION: 60%

COLLECTOR: South-facing greenhouse, window, and Beadwall™ Area: 291 FT²

STORAGE: Water tank, concrete floor Capacity: 7,513 BTU/°F

CONTROLS: Automatic moveable Beadwall™

BACKUP: Wood or coal fireplace with exchanger

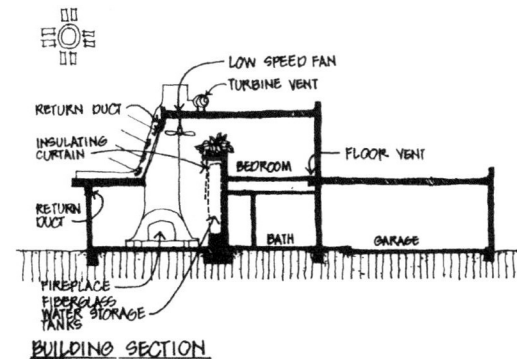

BUILDING SECTION

Set into a south-sloping hillside, this 2-story, 3-bedroom masonry and wood frame house is in the $75,000 range. It is sheltered by dense evergreen growth to the west and north.

Private client.

North Conway, NH

Steven J. Strong, Solar Design Associates
Canton, MA

HEATED AREA: 1,980 FT²

NUMBER OF DEGREE DAYS: 7,613

NET THERMAL LOAD: 28.8 10⁶BTU/YR

AUXILIARY ENERGY: 2.03 BTU/DD/FT²

YEARLY SOLAR FRACTION: 72%

COLLECTOR: South-facing clerestory windows and solarium Area: 235 FT²

STORAGE: Stone and concrete wall and floors, stone central masonry core, greenhouse earth beds Capacity: 15,338 BTU/°F

CONTROLS: Insulating curtains, operable windows

BACKUP: 20,000 BTUH fireplace; 25,000 BTUH active solar; 40,000 BTUH electric

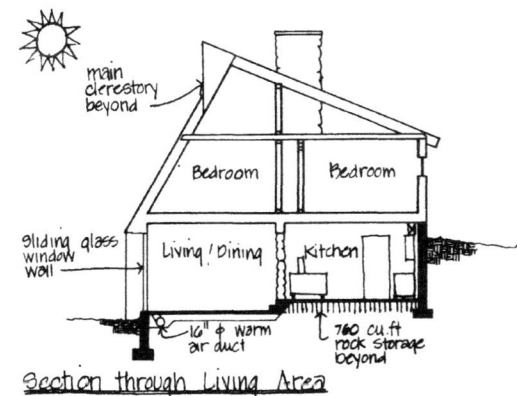

Section through Living Area

This new 2-story, 2-bedroom detached house is on a gently sloping site and is priced in the $60,000 range. The heavily wooded area provides winter wind protection and summer shading for the house.

Private client.

Somersworth, NH

Carl L. Fike
Somersworth, NH

HEATED AREA: 2,388 FT²

NUMBER OF DEGREE DAYS: 7,252

NET THERMAL LOAD: 55.8 10⁶BTU/YR

AUXILIARY ENERGY: 1.46 BTU/DD/FT²

YEARLY SOLAR FRACTION: 80%

COLLECTOR: South-facing windows
Area: 295 FT²

STORAGE: Concrete and tile floor
Capacity: 10,840 BTU/°F

CONTROLS: Thermal shutters

BACKUP: 90,000 BTUH wood stove; 105,000 BTUH oil furnace

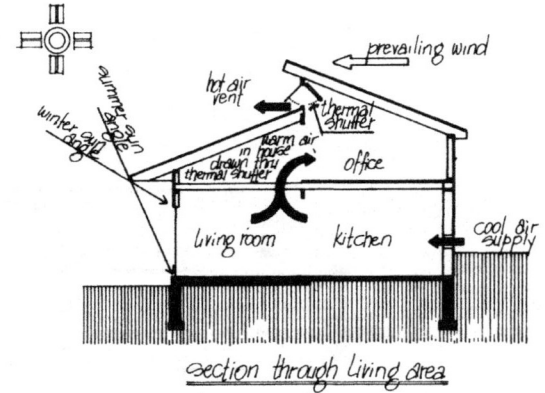

section through living area

El Rito, NM

Robert Nicolais/Quentin C. Wilson
La Madera, NM

HEATED AREA: 1,231 FT²

NUMBER OF DEGREE DAYS: 5,605

NET THERMAL LOAD: 76.6 10⁶BTU/YR

AUXILIARY ENERGY: 5.9 BTU/DD/FT²

YEARLY SOLAR FRACTION: 71%

COLLECTOR: Clerestory, south-facing windows, solarium Area: 440.5 FT²

STORAGE: Concrete floors, adobe walls
Capacity: 9,099 BTU/°F

CONTROLS: Operable clerestory and solarium roof insulation

BACKUP: 20,000 BTUH electric baseboard; 30,000 BTUH fireplace

This 1-story, 2-bedroom detached home on its gently sloping site is priced in the $45,000 range. It is a high mass adobe structure with attached solarium. Vegetation on the site consists of sagebrush and grass; there are no trees.

Open market sale.

section

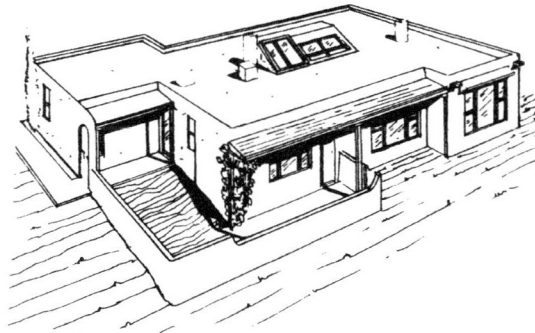

This new 1-story, 3-bedroom detached house is on a slightly sloping site and is in the $70,000 price range. Trees will be planted to provide shading during the summer as well as winter wind protection.

Open market sale.

Las Cruces, NM

Monika Lumsdaine, E & M Lumsdaine, Solar Consultants
Las Cruces, NM

HEATED AREA: 1,835 FT²

NUMBER OF DEGREE DAYS: 3,260

NET THERMAL LOAD: 20.0 10⁶BTU/YR

AUXILIARY ENERGY: 1.5 BTU/DD/FT²

YEARLY SOLAR FRACTION: 92%

COLLECTOR: South-facing windows,
              Area: 154 FT²

STORAGE: Concrete slab and interior walls,
             rock bed Capacity: 46,589.76 BTU/°F

CONTROLS: Shutters, fan, operable windows

BACKUP: Wood stove; wood fireplace

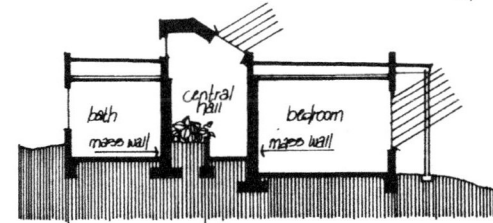

section through bedroom

This new 1-story, 3-bedroom home has a stucco exterior and will be priced in the $65,000 range. The house is located on a relatively level, lightly vegetated site.

Private client.

Moriarty, NM

William L. Burns, The Burns/Peters Group
Albuquerque, NM

HEATED AREA: 2,024 FT²

NUMBER OF DEGREE DAYS: 5,559

NET THERMAL LOAD: 81 10⁶BTU/YR

AUXILIARY ENERGY: 3.9 BTU/DD/FT²

YEARLY SOLAR FRACTION: 71%

COLLECTOR: Clerestory window and
              south-facing window
              Area: 400 FT²

STORAGE: Brick and concrete floors, adobe
             walls Capacity: 11,300 BTU/°F

CONTROLS: Removable insulation, operable
              windows

BACKUP: 30,000 BTUH fireplace; 55,000 BTUH
            electric furnace

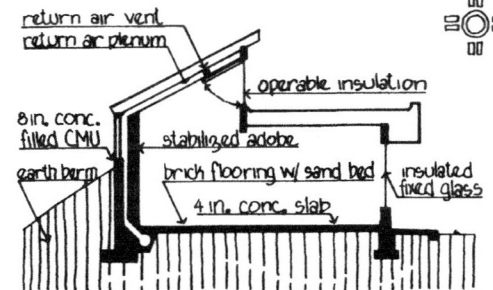

Building Section

DIRECT SOLAR GAIN

This new house conforms to the traditional "Pueblo" architectural style of Santa Fe, NM, and is priced in the $106,000 range. Earth berming along the north wall along with northern vegetation windbreaks decrease winter heat losses.

Private client.

Santa Fe, NM

Ken Brooks, Rocky Mountain Sun Power
Santa Fe, NM

HEATED AREA: 1,738 FT²

NUMBER OF DEGREE DAYS: 5,913

NET THERMAL LOAD: 85.1 10⁶BTU/YR

AUXILIARY ENERGY: 3.43 BTU/DD/FT²

YEARLY SOLAR FRACTION: 54%

COLLECTOR: South-facing windows
Area: 318 FT²

STORAGE: 4" slab on sand; concrete insulating block and interior masonry
Capacity: 8,743 BTU/°F

CONTROLS: Operable windows

BACKUP: 36,000 BTUH electric baseboard

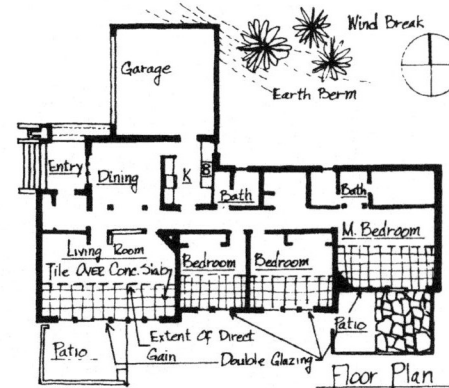

This is a retrofit and renovation that will join two existing 1-story adobe houses. The two houses, which are located in the Santa Fe Historical District, were originally purchased for $30,000.
Retrofit.

Santa Fe, NM

Jim Hays, Jim and Laura Hays/Bob Slattery
Santa Fe, NM

HEATED AREA: 1,600 FT²

NUMBER OF DEGREE DAYS: 5,720

NET THERMAL LOAD: N/A

AUXILIARY ENERGY: N/A

YEARLY SOLAR FRACTION: N/A

COLLECTOR: South-facing windows, Trombe wall, grow hole Area: N/A

STORAGE: Adobe walls, brick floors, wood ceiling Capacity: N/A

CONTROLS: Drapes, operable doors and windows, skylight covers

BACKUP: Fireplace; wood stove; gas heater

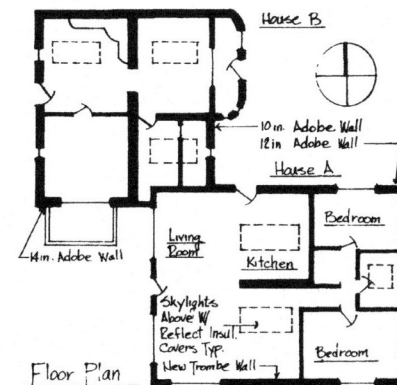

Bronx, NY

Thomas C. Regino, Energy Resources Group
Albin Associates
New York, NY

HEATED AREA: 3,000 FT²

NUMBER OF DEGREE DAYS: 4,778

NET THERMAL LOAD: 60.9 10⁶BTU/YR

AUXILIARY ENERGY: 4.42 BTU/DD/FT²

YEARLY SOLAR FRACTION: 27%

COLLECTOR: Solarium Area: 161 FT²

STORAGE: Brick walls, gypsum wall, concrete
floors Capacity: 1,946 BTU/°F

CONTROLS: Insulation barriers

BACKUP: 31,500 BTUH electric heat pump

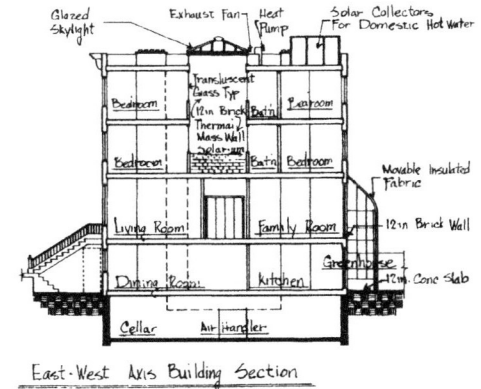

East-West Axis Building Section

This older, inner city building is located in the
Mott Haven Historic District. It is a 3-story,
4-bedroom brick row house, and after reno-
vations it will be priced in the $45,000 range.

Retrofit.

Woodhull, NY

Anne Hersh, Connell and Hersh Architects
Corning, NY

HEATED AREA: 1,231 FT²

NUMBER OF DEGREE DAYS: 6,532

NET THERMAL LOAD: 68.0 10⁶BTU/YR

AUXILIARY ENERGY: 3.36 BTU/DD/FT²

YEARLY SOLAR FRACTION: 72.6%

COLLECTOR: South-facing windows
Area: 570 FT²

STORAGE: Concrete floor and wall
Capacity: 25,447 BTU/°F

CONTROLS: Automatic motorized insulating
doors; operable doors and
windows

BACKUP: 70,000 BTUH electric baseboard
heater

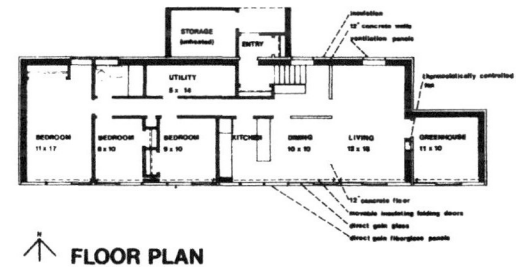

↑ **FLOOR PLAN**

This new 1-story, 3-bedroom house is priced in
the $40,000 range. Earth berms regulate the
house's inside temperatures in both the winter and summer. The exterior siding is local
lumber, which helps to blend into its natural
setting.

Open market sale.

This contemporary resort structure has a compact interior, with special attention given to public areas. The price is in the $40,000 range, and vegetation on the north side of the home will provide winter wind protection.

Open market sale.

East Hampton, NY

Alfred De Vido, Alfred De Vido Associates, Architects
New York, NY

HEATED AREA: 900 FT²

NUMBER OF DEGREE DAYS: 5,400

NET THERMAL LOAD: 56.8 10⁶BTU/YR

AUXILIARY ENERGY: 4.63 BTU/DD/FT²

YEARLY SOLAR FRACTION: 74%

COLLECTOR: South-facing windows, solarium
Area: 540 FT²

STORAGE: Concrete wall, brick on sand floor
Capacity: 20,372 BTU/°F

CONTROLS: Skylid, shutters, vents, operable windows

BACKUP: 20,000 BTUH electric baseboard; fireplace

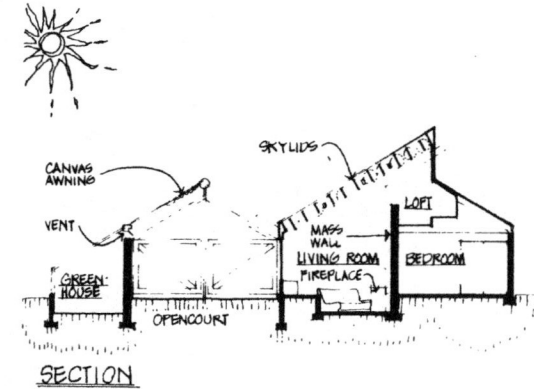

SECTION

This 2-bedroom house is designed to allow the homeowner to add future additions which would expand the house to three bedrooms. Earth berming, an earth roof, and Scotch pine trees on the north side protect the house from winter winds. The price is in the $69,600 range.

Private client.

Fredericktown, OH

Richard D. Strayer
Dublin, OH

HEATED AREA: 1,620 FT²

NUMBER OF DEGREE DAYS: 5,543

NET THERMAL LOAD: 22.6 10⁶BTU/YR

AUXILIARY ENERGY: 0.53 BTU/DD/FT²

YEARLY SOLAR FRACTION: 95%

COLLECTOR: South-facing windows and skylights Area: 430 FT²

STORAGE: Concrete floor, gravel floor, concrete wall, stone rockbed
Capacity: 47,786 BTU/°F

CONTROLS: Operable insulating shutters, windows, automatic clerestory shutters, earth/air supply and hot air exhaust

BACKUP: 48,000 BTUH wood fireplace

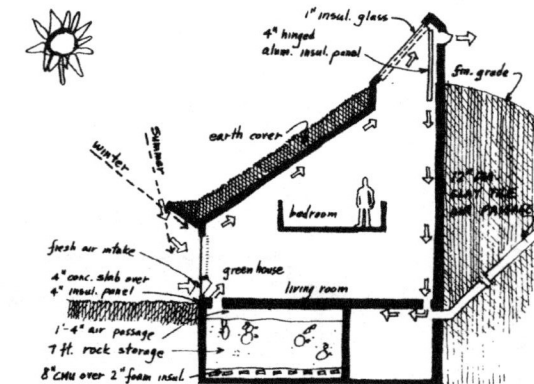

This 1-story, 3-bedroom wood frame house is priced in the $75,000 range. Trees on the southwest corner of this level lot will form a natural windbreak. Trellis vines will shade the summer sun.

Open market sale.

Eugene, OR

Lee Kersh
Eugene, OR

HEATED AREA: 1,218 FT²

NUMBER OF DEGREE DAYS: 4,599

NET THERMAL LOAD: 4.63 10⁶BTU/YR

AUXILIARY ENERGY: 2.98 BTU/DD/FT²

YEARLY SOLAR FRACTION: 87%

COLLECTOR: South-facing windows and solarium Area: 210 FT²

STORAGE: Concrete-filled blocks and clerestory wall Capacity: 5,701 BTU/°F

CONTROLS: Operable doors and windows, vents

BACKUP: 20,000 BTUH electric radiant ceiling; 20,000 BTUH wood stove

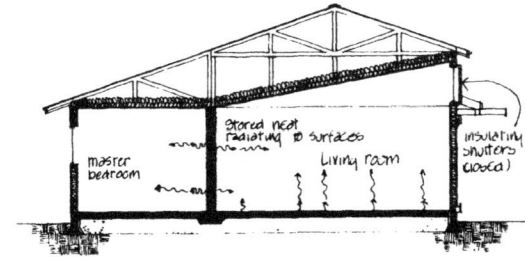

Section · winter night

This contemporary 3-bedroom house uses standard building methods and is priced in the $97,000 range. Earth berms and an east entrance protect the house from winter winds. A locust tree shades the house in the summer.

Open market sale.

West Salem, OR

Thomas A. Meados
Eugene, OR

HEATED AREA: 1,779 FT²

NUMBER OF DEGREE DAYS: 4,599

NET THERMAL LOAD: 46.3 10⁶BTU/YR

AUXILIARY ENERGY: 3.69 BTU/DD/FT²

YEARLY SOLAR FRACTION: 73%

COLLECTOR: East-, south-, and southwest-facing windows; south-, and west-facing greenhouse; south-facing storage mass wall Area: 523 FT²

STORAGE: Brick floors and wall, rock storage, bedroom storage mass wall Capacity: 19,864 BTU/°F

CONTROLS: Operable insulating panels, louvers, and vents

BACKUP: 68,000 BTUH electric furnace

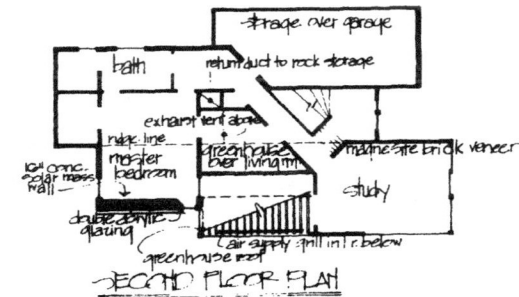

SECOND FLOOR PLAN

This new 2-story, 2-bedroom detached house with loft is a contemporary frame design in the $70,000 price range. Trees on the slightly sloping site form a natural windbreak. A grape arbor shades the south windows during the summer.

Private client.

Carlisle, PA

Michael V. Arnold and TEA
Harrisville, NH

HEATED AREA: 2,145 FT²

NUMBER OF DEGREE DAYS: 5,176

NET THERMAL LOAD: 73.6 10⁶BTU/YR

AUXILIARY ENERGY: 5.90 BTU/DD/FT²

YEARLY SOLAR FRACTION: 35%

COLLECTOR: South-facing window, skylights and solarium Area: 343 FT²

STORAGE: Masonry fireplace, brick floor, solar mass wall Capacity: 8,605 BTU/°F

CONTROLS: Draperies and window opening

BACKUP: 90,000 BTUH oil furnace

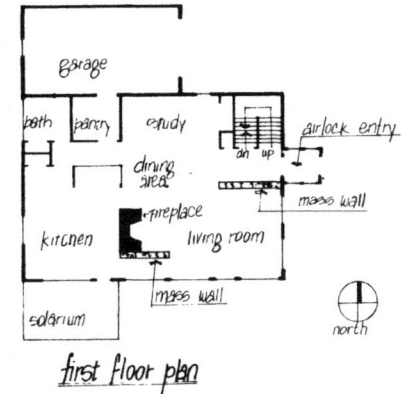

first floor plan

This new 1-story, 3-bedroom house is priced in the $75,000 range. The adobe blocks were made on the site by the owner. North and west earth berms and protected south and east entries reduce winter wind infiltration.

Private client.

Bovina, TX

David Smith, Design Services
Austin, TX

HEATED AREA: 3,000 FT²

NUMBER OF DEGREE DAYS: 3,911

NET THERMAL LOAD: 59.8 10⁶BTU/YR

AUXILIARY ENERGY: 42.0 BTU/DD/FT²

YEARLY SOLAR FRACTION: 47.1%

COLLECTOR: South-facing windows Area: 320 FT²

STORAGE: Adobe walls, concrete and brick floor Capacity: 17,580 BTU/°F

CONTROLS: Operable shutters, windows, exhaust ports and fans

BACKUP: 30,000 BTUH fireplace; 35,000 BTUH electric heat pump

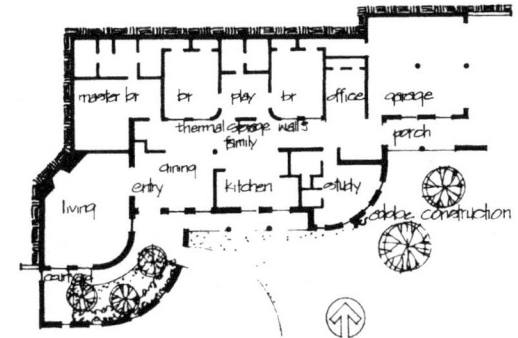

This is a 2-story, 3-bedroom wood frame house with brick veneer and a clay tile roof. The site is relatively flat with trees located to the north and south. It is priced in the $75,000 range.

Open market sale.

Spring, TX

Richard A. West, Spartan Technologies, Inc.
Houston, TX

HEATED AREA: 2,390 FT²

NUMBER OF DEGREE DAYS: 1,434

NET THERMAL LOAD: 16.2 10⁶BTU/YR

AUXILIARY ENERGY: 4.64 BTU/DD/FT²

YEARLY SOLAR FRACTION: 63%

COLLECTOR: Solarium Area: 318 FT²

STORAGE: Concrete floor, water bench, concrete solarium Capacity: 15,351 BTU/°F

CONTROLS: Operable windows

BACKUP: 24,000 BTUH gas furnace; fireplace

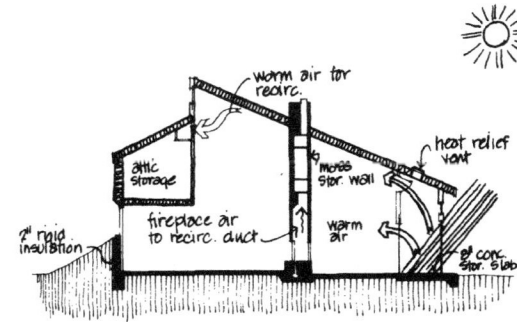

This new 2-story, 3-bedroom house is designed to conform to the local traditional architectural style with brick walls, cedar siding, and a cedar shake roof. The first floor is extensively earth bermed, and evergreen trees around the entrances also prevent winter wind infiltration. The price is in the $79,000 range.

Open market sale.

Winchester, VA

Sam Cravotta, Star Tannery Design Studio
Star Tannery, VA

HEATED AREA: 1,950 FT²

NUMBER OF DEGREE DAYS: 5,662

NET THERMAL LOAD: 36.8 10⁶BTU/YR

AUXILIARY ENERGY: 1.28 BTU/DD/FT²

YEARLY SOLAR FRACTION: 82%

COLLECTOR: South-facing skylights and windows Area: 310 FT²

STORAGE: Water storage mass floor, concrete floor, masonry walls, solar heated pool Capacity: 29,428 BTU/°F

CONTROLS: Vents, fans, skylight, shades, and Insulation

BACKUP: 10,000 BTUH wood stove; 4,266 BTUH electric forced-air heater; 3,200-5,000 BTUH electric baseboard heating

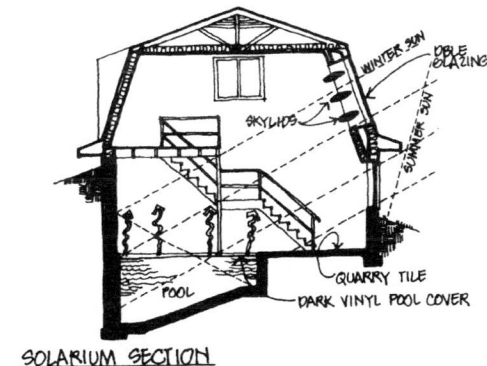

SOLARIUM SECTION

This house is a 2-level structure containing 3 bedrooms, with foundation and exterior walls constructed of native sandstone. Located on a small slope, the home will be shielded from winter winds by a dense forest to the north. It is priced in the $80,000 range.

Private client.

Wytheville, VA

James Guy, Barbara Guy and Christopher Umberger
Wytheville, VA

HEATED AREA: 2,072 FT²

NUMBER OF DEGREE DAYS: 4,675

NET THERMAL LOAD: 101 10⁶BTU/YR

AUXILIARY ENERGY: 6.4 BTU/DD/FT²

YEARLY SOLAR FRACTION: 49%

COLLECTOR: Clerestory and south-facing windows Area: 430 FT²

STORAGE: Solar mass wall and floor Capacity: 12,296 BTU/°F

CONTROLS: Insulated shutters, ridge vent, awnings

BACKUP: 30,000 to 50,000 BTUH wood stove

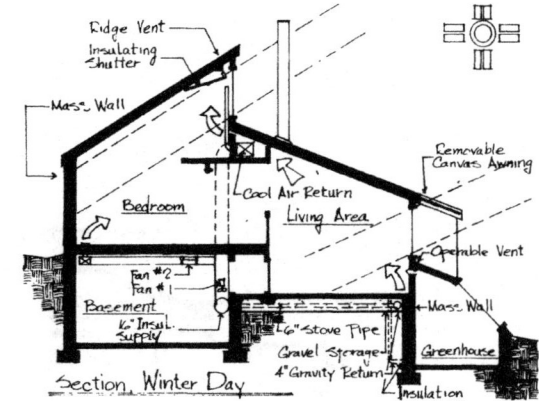

Section, Winter Day

This new 1-story, 4-bedroom detached house on its modestly sloping site is priced in the $70,000 range. The area is heavily wooded, which provides summer shading and winter wind protection.

Open market sale.

Shelburne, VT

Douglas C. Taff, Parallax, Inc.
Hinesburg, VT

HEATED AREA: 2,056 FT²

NUMBER OF DEGREE DAYS: 7,727

NET THERMAL LOAD: 75.5 10⁶BTU/YR

AUXILIARY ENERGY: 3.68 BTU/DD.FT²

YEARLY SOLAR FRACTION: 63%

COLLECTOR: South-facing windows Area: 518.1 FT²

STORAGE: Solar mass floor and walls Capacity: 20,476 BTU/°F

CONTROLS: Thermal drapes, fans, operable windows and doors

BACKUP: 50,000 BTUH wood stove; 48,000 BTUH electric baseboard

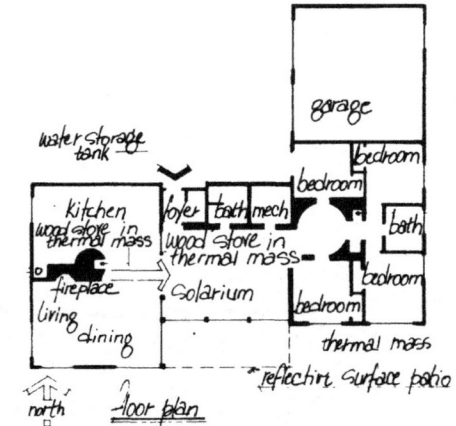

floor plan

This new, 2-story, 3-bedroom house has cedar siding and is priced in the $65,000 range. Earth berming and a future evergreen windbreak on the north side of the house will redirect winter winds.

Private client.

Amery, WI

Peter Pfister
Minneapolis, MN

HEATED AREA: 1,770 FT²

NUMBER OF DEGREE DAYS: 8,248

NET THERMAL LOAD: 61.5 10⁶BTU/YR

AUXILIARY ENERGY: 3.26 BTU/DD/FT²

YEARLY SOLAR FRACTION: 54%

COLLECTOR: South-facing window and
solarium Area: 254 FT²

STORAGE: Concrete/sand floor and wall,
Capacity: 7,490 BTU/°F

CONTROLS: Manual and automatic moveable
window insulation units, furnace
fan, windows, and turbine
ventilator

BACKUP: 40,000 BTUH oil/wood combination
forced air furnace; 25,000-40,000
BTUH wood fireplace

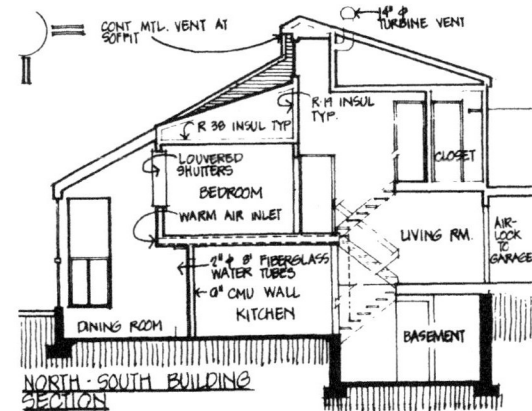

NORTH-SOUTH BUILDING SECTION

These new ranch-style houses are priced in the $66,000 range. All bedrooms are located on the building's north side to fully utilize winter sun for other areas of the house that are used more often. A south-facing airlock entrance reduces winter wind infiltration.

Open market sale.

Beloit, WI

James Cardwell
Portsmouth, NH

HEATED AREA: 3,136 FT²

NUMBER OF DEGREE DAYS: 7,863

NET THERMAL LOAD: 64.4 10⁶BTU/YR

AUXILIARY ENERGY: 2.93 BTU/DD/FT²

YEARLY SOLAR FRACTION: 38%

COLLECTOR: South-facing window
Area: 210 FT²

STORAGE: Concrete floors and walls
Capacity: 18,292 BTU/°F

CONTROLS: Operable shutters, blower, and
windows

BACKUP: 110,000 BTUH gas furnace

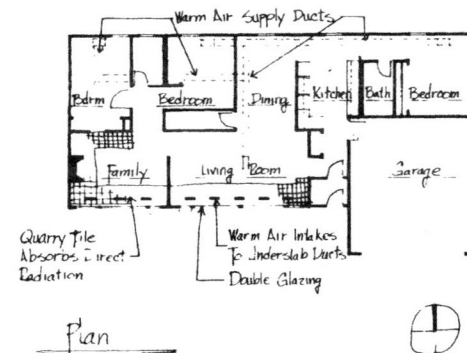

Plan

This traditional ranch-style house is to be built in a subdivision where all of the houses are pre-cut; broad market exposure and a high potential for repeatability are the builder/designer's goals, along with energy and material conservation. Earth berming and protected entrance locations minimize winter wind infiltration. The price is in the $52,000 range.

Open market sale.

Chilton, WI

Douglas Steege
Madison, WI

HEATED AREA: 1,318 FT²

NUMBER OF DEGREE DAYS: 7,852

NET THERMAL LOAD: 62.4 10⁶BTU/YR

AUXILIARY ENERGY: 4.92 BTU/DD/FT²

YEARLY SOLAR FRACTION: 46%

COLLECTOR: South-, east-, and west-facing windows Area: varies seasonally

STORAGE: Concrete floor and stairs, concrete block filled with concrete Capacity: 10,354 BTU/°F

CONTROLS: Motorized insulating curtains and shutters

BACKUP: 50,000 BTUH wood stove; 60,000 BTUH oil furnace

First Floor Plan

CHAPTER 4
INDIRECT SOLAR GAIN

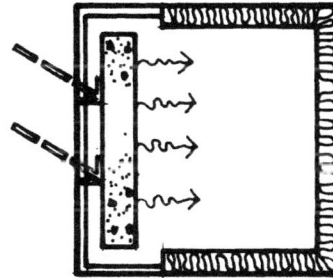

DEFINITION
In the INDIRECT SOLAR GAIN buildings that follow, the fabric of the house continues to collect and store solar energy, but the sun's rays do not travel through the living space to reach the storage mass. This minimizes the Direct Gain limitation in which solar collection temperatures are set by occupant comfort needs. Thus in the Indirect Solar Gain concept, a storage mass collects and stores heat directly from the sun, and then transfers heat to the living space.

There are two types of Indirect Solar Gain systems awarded in this competition: The Mass Storage Wall (known as a Trombe Wall) and the Water Storage Wall. In both types the sun's rays are intercepted directly beyond the collector glazing by a massive wall which serves as heat storage. However, the difference between storage materials and methods of containing those materials suggest that both concepts be discussed separately.

MASS STORAGE WALL (TROMBE WALL)

REQUIREMENTS AND VARIATIONS
The required elements of the Mass Trombe system involve only a large glazed collector area and a solid storage mass directly behind it. However the award winning examples display a variety of sophistications. The range of storage materials identified in Mass Trombe passive solar buildings include concrete, adobe, stone, and composites of brick and block. The property to consider in deciding on storage construction is the method of distribution inherent in massing materials with different heat storage capacities and emission properties. Radiant distribution from a storage mass to a living space can be almost immediate or it can be delayed up to twelve hours, depending on the thickness and time lag property of the storage material chosen. Distribution of air by natural convection is also viable with the Mass Trombe system, since the volume of air in the intervening space between glazing and storage mass

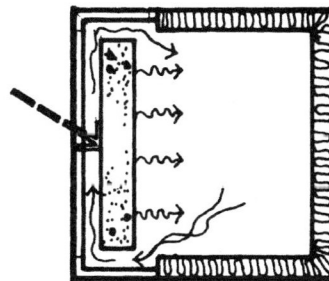

is being heated to high temperatures and seeks constant means of escape. Adding opening or vents at the top of the storage mass allows the hot air to force itself into the living space for a quick transfer of heat, drawing cooler room air through lower vents back into the collector air space. If the vents are controllable dampers, convective heat distribution can be shut off or started at will. Storage placement and construction, as well as storage openings and dampers, are variables which determine the Mass Trombe's efficiency in storing and distributing solar heat to the house.

CONTROLS

As in the Direct Gain building type, controls for the operation of the Mass Trombe building type are important, though less crucial since the living space is not directly influenced by solar collection. For optimum efficiency in the winter, external moveable insulation, or other insulation alternatives, should be included to protect the storage mass from wasteful heat loss to the overcast or night sky. In the summer, unwanted heating of the storage mass should be prevented by shading the glazed area with overhangs, by closing the external insulation, or by opening external dampers and vents. A Mass Trombe Wall has the potential to provide induced ventilation for summer cooling of the living space by including exhaust vents at the top of the glazed area. Solar-heated air in the collector air space will force its way outside, drawing air from the living space to replace it. Another opening must be provided within the living space for replacement air—preferably from a shaded or cooler area. This continual air movement exhausts hot air from the house, drawing in cooler air for ventilation.

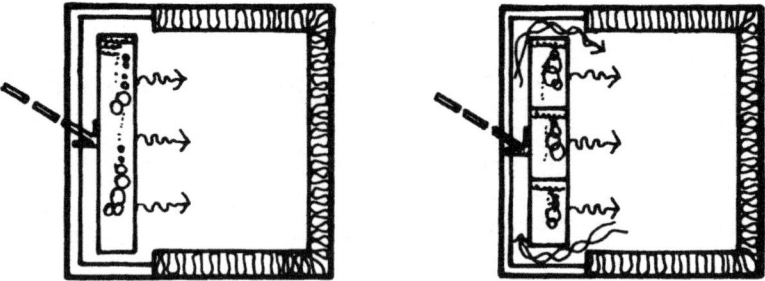

WATER STORAGE WALL

REQUIREMENTS AND VARIATIONS

The requirements for the Water Storage Wall are again a large glazed area and an adjacent massive heat storage. However, the storage is water, or another liquid, contained in a variety of containers, each representing different heat exchange surfaces to storage mass ratios. Larger storage volumes provide greater and longer term heat storage capacity, while smaller contained volumes provide greater heat exchange surfaces and thus faster distribution. This tradeoff between heat exchange surface versus storage mass has not as yet been explored in depth. However, many container variations have been built, including components such as tin cans, bottles, tubes, bins, barrels, drums, bags, and complete water walls. The selection and interrelationship of storage materials, then, is necessary to the effective operation of the Water Storage Wall.

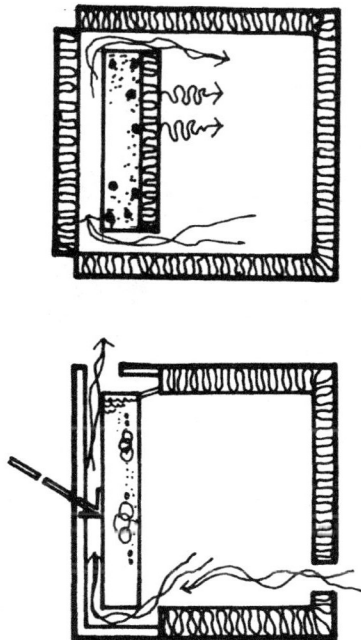

CONTROLS

In considering the control of heat distribution in a Water Wall, one must be aware that thermal transfer is rapid within a convective body of water, and radiant distribution for a solar-heated water storage wall to a living space is almost immediate. This is in contrast to the longer time-lag property of the Mass Trombe Wall. In climates where heat is undesirable until the cooler evening hours, the Water Wall system requires some storage-distribution control. The addition of daytime insulation to the outside and possible high and low vents for immediate air circulation to the living spaces provides one control solution. Aside from the choices between radiant and convective distribution, the controls for the Water Wall are similar to those of the Mass Trombe Wall. Overheating of the storage mass should be prevented by shading, and extravagant heat loss from the storage mass should be prevented by exterior insulation. In addition, the use of operable vents which open to the outside will induce summer ventilation as described for the Mass Trombe passive solar building.

The important issue to understand with INDIRECT GAIN passive solar homes is that the occupant will be in direct contact with the storage and distribution components of solar heating but is no longer faced with the collection component.

HOUSE II

HOUSE I

Duffield, VA

Adolphus Chester, Architectural Design Branch, TVA
Knoxville, TN

HEATED AREA: 1,374 FT^2

NUMBER OF DEGREE DAYS: 4,121

NET THERMAL LOAD: 24.2 10^{6}BTU/YR

AUXILIARY ENERGY: 1.54 BTU/DD/FT^2

YEARLY SOLAR FRACTION: 74%

Open market sale.

Duffield, VA

Danny Brewer, Architectural Design Branch, TVA
Knoxville, TN

HEATED AREA: 1,441 FT^2

NUMBER OF DEGREE DAYS: 4,121

NET THERMAL LOAD: 24.3 10^{6}BTU/YR

AUXILIARY ENERGY: 1.2 BTU/DD/FT^2

YEARLY SOLAR FRACTION: 81%

Open market sale.

CONTEXT

The following two schemes were concurrently designed and conceived by the Architectural Design Branch of the Tennessee Valley Authority to serve as models of energy-efficient and passively heated residential construction in the Tennessee Valley. Both homes are to be part of Thomas Village, a new development in western Virginia to relocate the occupants of nearby flood-prone areas. These designs are intended to be repeated in other subdivisions and to serve as prototypes for a solar building program throughout the seven-state area served by the Tennessee Valley Authority. The character and density of the surrounding development is suburban, although future plans include higher density developments.

Both designs are for modest, compact, three-bedroom homes nestled on south-sloping sites. These plans can be modified and adapted for flat sites or other site conditions as well. The design of both units is based on conventional building materials and construction skills available regionally. By adopting standard local finish materials and incorporating amenities that are expected in the new home market such as carports or garages, attached greenhouses, expansion space and low upkeep, the designers have assured a high degree of consumer acceptance.

Western Virginia is characterized by a temperate climate of 4,100 heating degree days in the winter and about 1,100 cooling degree days in the summer. These two schemes, which incorporate Trombe walls, water walls, greenhouses, and direct gain spaces, show the wide variety of design solutions for natural heating and cooling which this climate allows. In general, ease of construction has been stressed, occasionally at the expense of thermal efficiency. Cost effectiveness has been the principal design decision-making tool, and the result has been an efficient home for the lowest home-buying dollar.

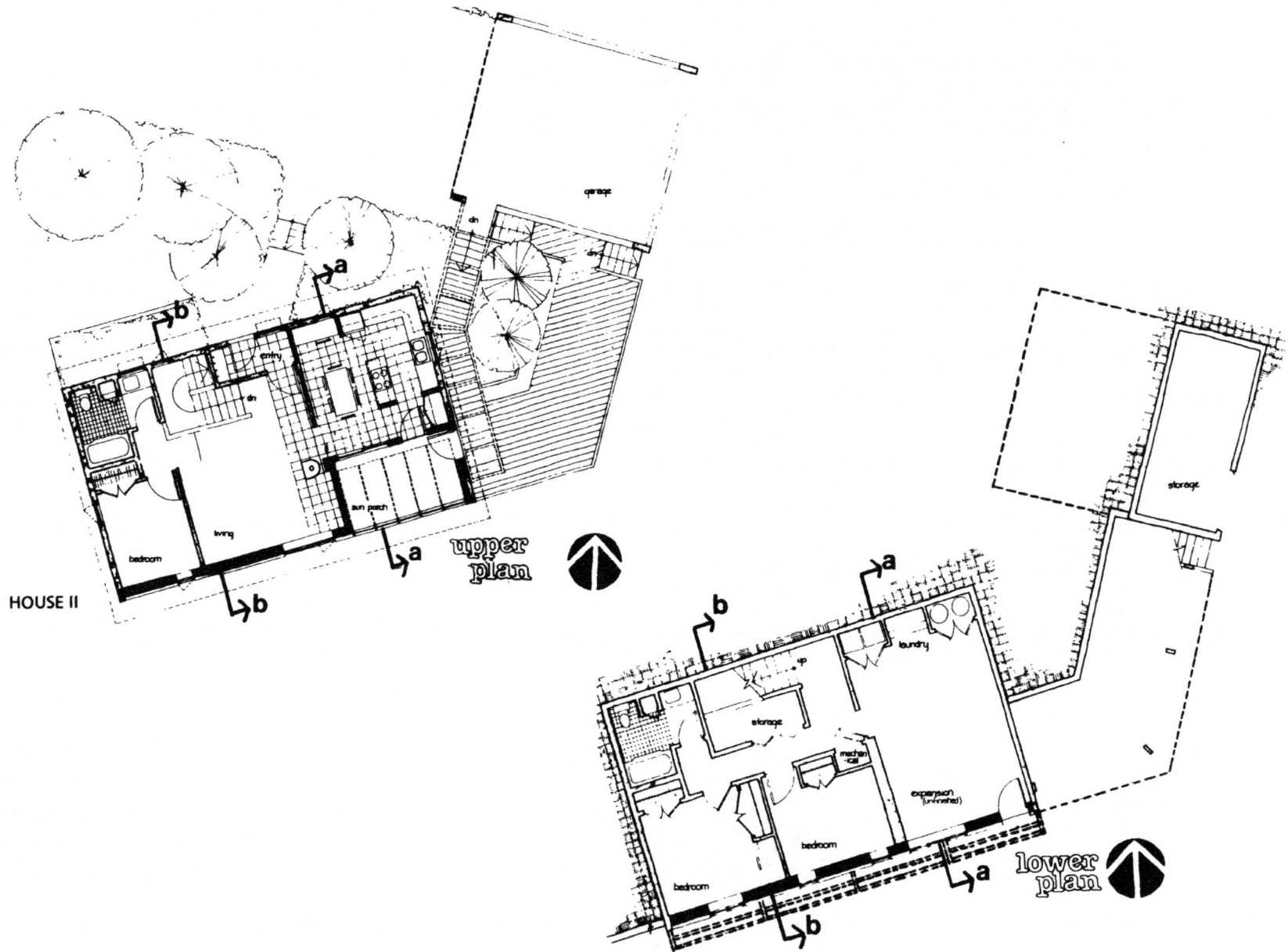

HOUSE II

upper plan

lower plan

HOUSE II
CONSERVATION

Much attention has been focused on building features which conserve energy within the home. This design demonstrates some excellent conservation features achieved outside the home through landscaping, sitework, and building placement. Not only are heating and cooling loads in the home reduced by this outside design, but outdoor spaces usable throughout most of the year are created as well. This home is dug into a south-sloping hillside, creating a summer court on the north side and a winter court on the south side. The north side court takes advantage of the northeast summer breezes for cooling and is shaded by the house and evergreen trees. These evergreens also protect the building from cold winds in winter. On the south side, the entry walk expands into a winter court, protected on the north by the house, east by an earth berm, and west by the garage. As the deciduous trees to the south lose their leaves in the fall, this court becomes a comfortable, sheltered, and sunny space.

In addition to these tempered outdoor spaces, the building enjoys several other conservation features. For summer, the placement of the garage on the west side shades the building from the severe summer afternoon sun. The light-colored roof reduces summer heat gain. The earth berming on the north and east cuts heat loss and gain through a large percentage of the exterior wall. For winter, the light-colored south court reflects additional solar energy into the Trombe wall and atrium. Infiltration is reduced by the sun-tempered airlock vestibule between the garage and kitchen, and by routing the main entrance through the atrium. An attic space provides an R-30 insulation value while high vents exhaust any heat buildup in the atrium and house during summer. The atrium itself is a primary energy-conserving feature, providing light, views and solar heat to much of the interior without the normal heat loss of exterior windows. In fact, the building uses only 30 sq. ft. of double pane exterior glass, in addition to the 170 sq. ft. which faces into the atrium. These conservation features, the landscaping, outdoor courts, and the atrium buffer zone create a more desirable home as well as saving energy. This residence has a modest 1,374 sq. ft. for its three-bedroom, two-and-a-half-bathroom interior, and a simple low-cost building form; but the sensitive addition of these features expands the useful space and value of this home without significantly increasing the cost.

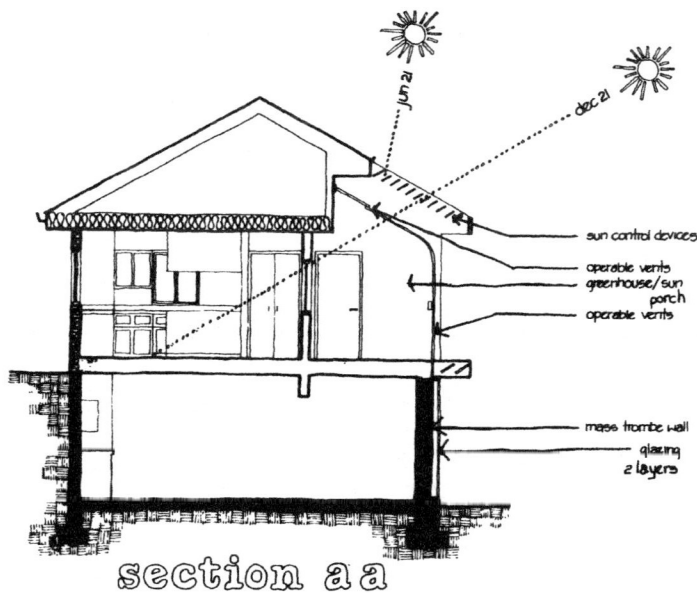

section aa

HOUSE II

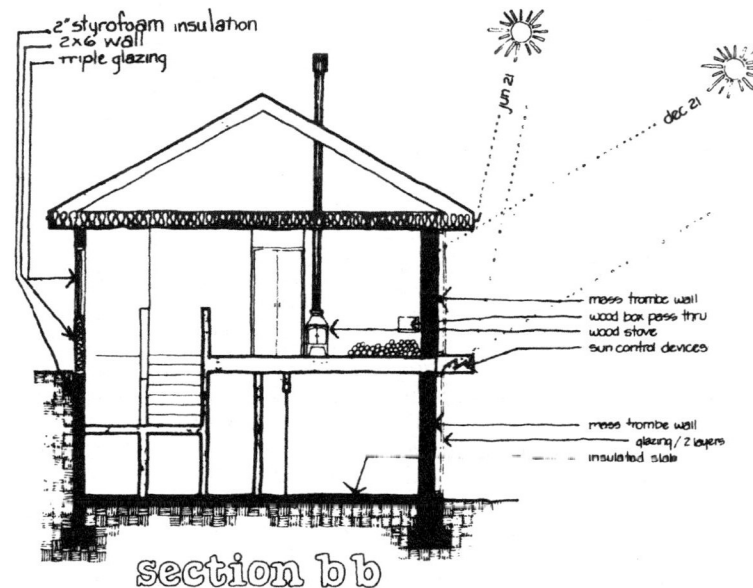

section bb

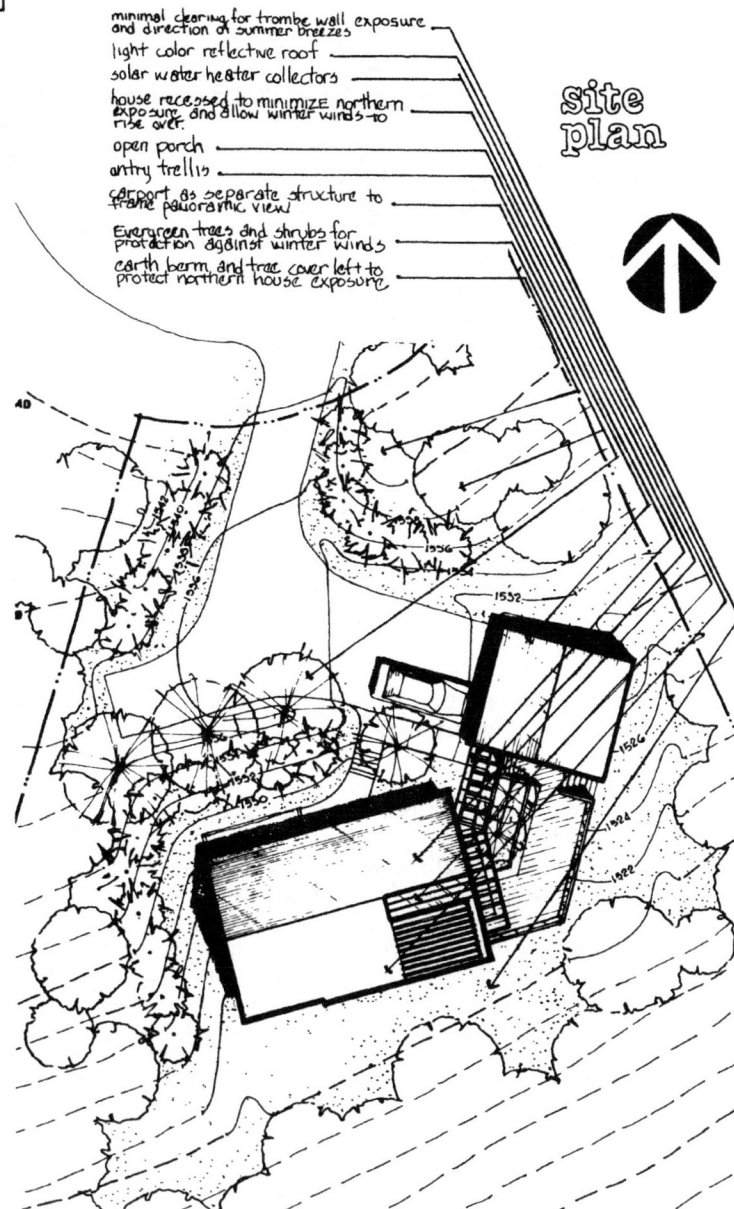

minimal clearing for trombe wall exposure and direction of summer breezes

light color reflective roof

solar water heater collectors

house recessed to minimize northern exposure and allow winter winds to rise over

open porch

entry trellis

carport as separate structure to frame panoramic view

Evergreen trees and shrubs for protection against winter winds

earth berm and tree cover left to protect northern house exposure

site plan

HOUSE II

HEATING

Two distinct passive solar systems are at work in this home: a Trombe wall and a solar atrium. The Trombe wall, located adjacent to the kitchen, dining room and two bedrooms, is 18' long and two stories high, providing about 240 sq. ft. of collector area. The wall consists of a 12" filled concrete block mass covered by two layers of Kalwall™ "Sun-Lite" fiberglass glazing. Absent are the common vent openings used to convect heat directly into the space during the sunny hours of the day. For a mild climate such as this, the need for heat during the day is minimal and can be easily satisfied by direct gain through the south windows and atrium. Eliminating these vents reduces construction expense and makes the wall a more effective heat source at night. The lack of vents creates higher daytime temperatures on the exterior side of the wall, causing a warmer inside temperature at night. A desirable modification of this Trombe wall would be the addition of vents exhausting to the outside any heat buildup during hot summer periods. Even though the wall is fully shaded by a roof eave through the summer, the diffuse light and hot air temperatures tend to heat the wall. With no exterior vents, the wall will conduct some unwanted heat into the interior of the house.

The Trombe wall collector is accompanied by a multi-purpose atrium space. Located by the living and dining room on the lower level, this solar-heated space provides a buffer zone for the interior spaces. Part of the sunlight entering the atrium is absorbed by the massive concrete floor and east wall during the day and released at night. The resultant temperature swing in this space is greater than that desired for living spaces, but considerably less than the outdoor ambient temperature swing. This buffer zone serves well as an entry vestibule, plant room, and all-purpose flexible space, and gives the otherwise tightly planned house a spacious quality.

Some of the light entering the atrium passes through a second layer of glass to directly heat the master bedroom and living room For nighttime heating, three 18" diameter, 10' tall water storage tubes are placed on the south side of the atrium. This mass is not only charged by the sunlight, but is integrated with the fireplace in such a way as to store the fire's excess heat output. The fireplace also has a convection chamber which can directly exhaust heat either to the livingroom or master bedroom above. The room layout is cleverly arranged to provide each "lived-in" room with either a Trombe wall or atrium exposure. The bathrooms, utility, and circulation spaces are located on the northeast to service and buffer the other living spaces.

COOLING

Beyond the cooling effects of the landscaping features previously discussed, this building employs several low-cost but effective cooling strategies. The simplest is shading. The roof eaves on the south side, where 90% of the glass is located, fully shade this facade through the summer months. The fixed shading slats over the atrium are so angled as to allow low winter sunlight to penetrate and to intercept the high summer light. Like all fixed south shades, these devices have one disadvantage compared with operable shades. Designed to shade 100% in June, they will only partly shade in September when there still is a considerable cooling load. Or inversely, designed to expose 100% of the window in December, they will partially shade the collectors in February and March, months which can require maximum heating. This is caused by the fact that the winter heating and summer cooling season is offset several months behind the sun's cycles.

Natural ventilation provides a second cooling strategy. Most significantly, the two-story space of the atrium coupled with the attic vents create a stack effect to exhaust hot air. With open windows low in the atrium and house, the air is drawn through the space up into the warmer attic and out the vents on the east and west. A disadvantage of the earth berming of the lower floor is the lack of cross ventilation in the livingroom. However, the concrete walls and water tubes provide a tempering effect on the temperature swing in the house by absorbing some excess heat during the day which is discharged during the cooler nights. The backup system is an efficient 1½-ton heat pump which can be used for summer cooling as well as winter heating.

HOUSE I
CONSERVATION

This two-story, mass Trombe wall home, oriented east of due south, is entered from the north through an airlock vestibule which prevents the infiltration of cold outdoor air. This vestibule opens into a compact living area that is heated by a Trombe wall and woodstove. This 1,800 sq. ft. house, of which 1,441 sq. ft. is to be heated, has 900 sq. ft. on the main floor level, which is the minimum allowable in the subdivision. The plan should have wide appeal since the upper-level bedroom can be utilized as a den/study space by families who require only two bedrooms. While the house is small, there is substantial room for expansion into the lower basement level, about one-third of which is unfinished.

As always, conservation is the first consideration. This home is constructed with walls of 2 x 6 framing and full thickness insulation, while the roof trusses provide space for R-30 insulation levels.

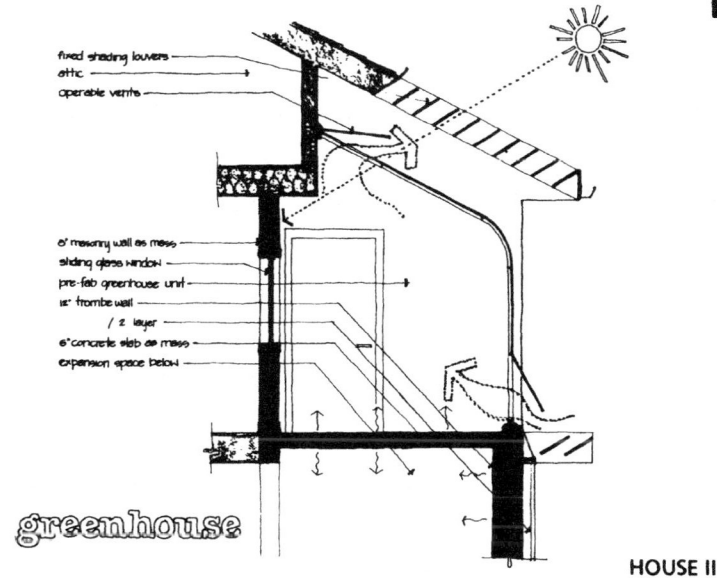

greenhouse

HOUSE II

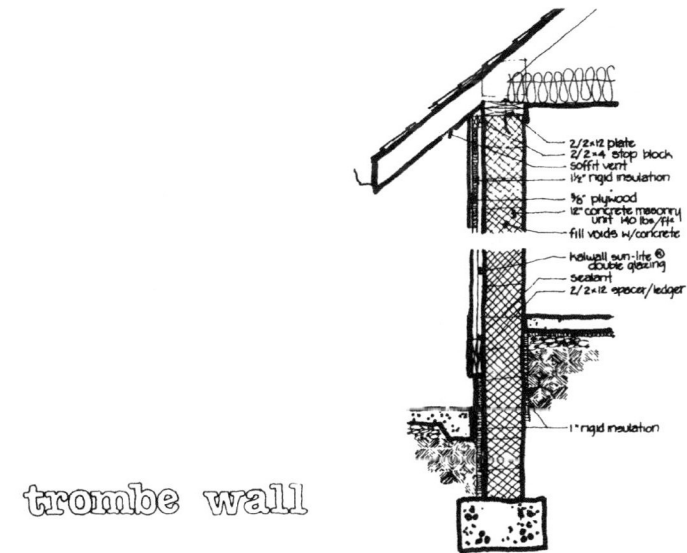

trombe wall

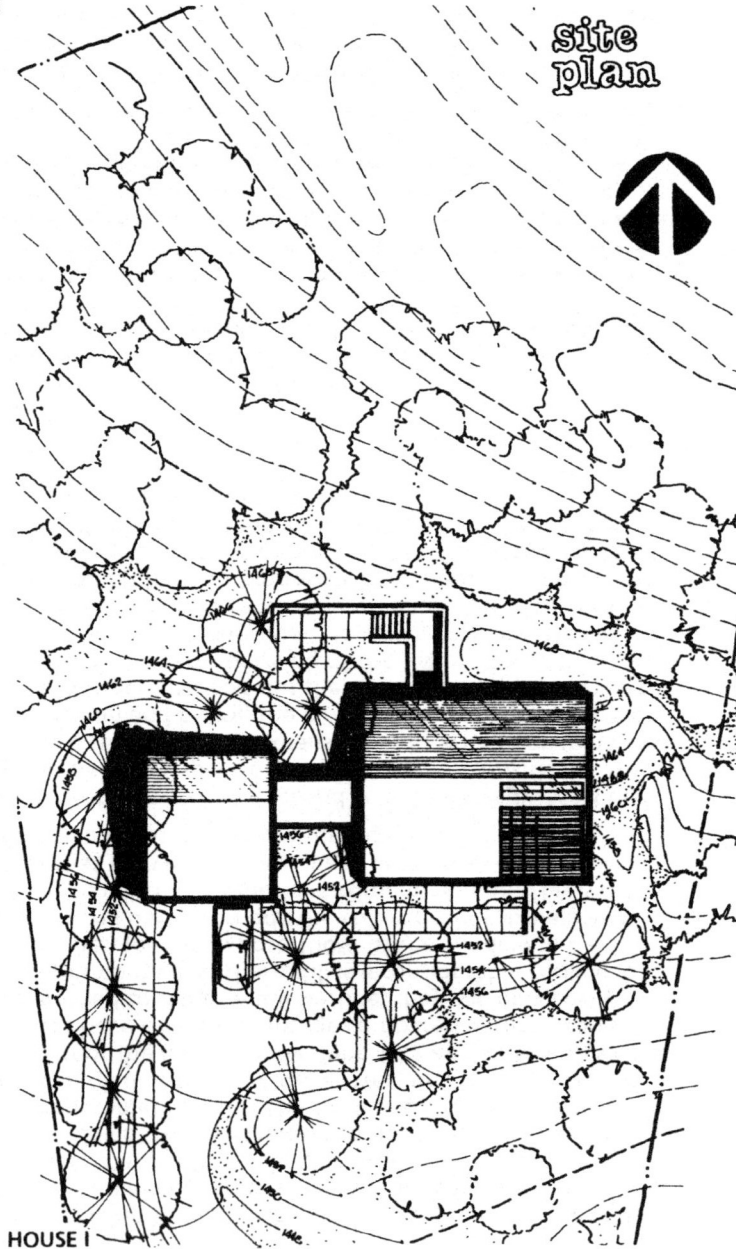

site plan

HOUSE I

There is a continuous external blanket of foundation insulation, composed of 2" of styrofoam which is carried to a distance of at least 2' below grade. Beneath that depth, 2' of insulation is used.

Double glazing is provided for windows on the east, west, and south, and triple glazing is specified for the north windows. On north elevations, the additional cost of the third glazing layer must be carefully balanced against the additional insulating benefits, while on the south, east, and west elevations the use of extra glazing must also be balanced against the poorer transmission of sunlight. Glazed areas on the east, west, and north which receive little or no winter sun are kept to a minimum, although provisions are made to allow openings for cross ventilation. Unlike a number of other schemes, no moveable insulation is indicated for the windows or for the unvented Trombe wall.

The attached carport is turned off axis from the house and creates a trellised deck and porch area which provides a second entry. The greenhouse/sun porch room also acts as an unheated entrance vestibule space and should again inhibit cold air infiltration. Perhaps more importantly, it adds interest to a simple home design at a minimum of additional cost.

HEATING

Departing from more conventional Trombe wall designs, this scheme uses a non-vented Trombe wall to heat the primary spaces in the house. Without a daytime convective loop into the rooms, all heating of the spaces is by radiation and convection off the inside surface of the Trombe wall. This wall is constructed of 12" reinforced and filled concrete block and covered by two layers of fiberglass-reinforced plastic framed in treated wood mullions. A potential problem is the absence of exterior venting for the heated Trombe wall air cavity. While shading is provided to alleviate most of the overheating problem, it would still be useful to allow for venting this cavity in summer.

Experience has shown that two-story Trombe wall systems develop some temperature stratification. The second floor is invariably 3°-5°F warmer than the first floor. In most designs, the bedroom spaces are on the second floor, receiving the unnecessary benefit of higher temperatures. In this design, however, the bedroom spaces are primarily on the lower level, allowing the major living spaces to remain warmer. This stratification phenomenon does not generally occur in conventional, air-heated homes because the mechanical systems, which are equipped with registers, return air ducts, and warm air blowers, mix the air in the home. Provisions can be made to allow this house to also redistribute air. For

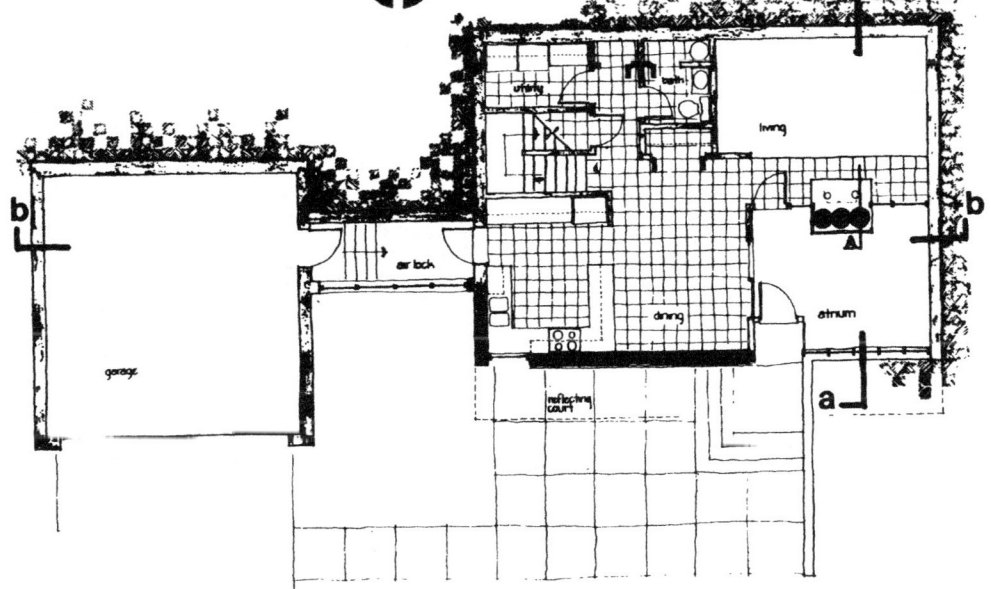

lower level

a

b

b

pantry

bath

living

air lock

dining

atrium

garage

reflecting court

a

HOUSE I

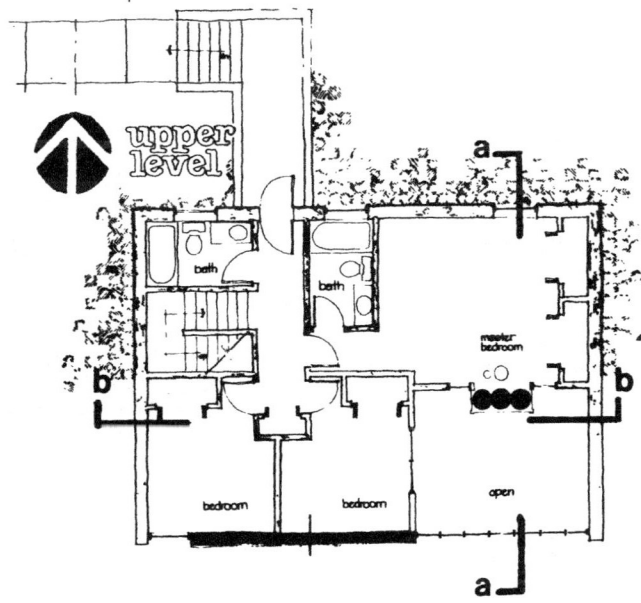

upper level

a

b

b

bath

bath

master bedroom

bedroom

bedroom

open

a

example, a small fan and duct arrangement can be designed to extract the warmest air from the second floor and redistribute it to the first floor.

The greenhouse/sun porch provides some additional heat to the 8" masonry wall which borders the kitchen. The slab of the greenhouse is 6" of concrete, which acts as thermal storage for sunlight entering the greenhouse, although requiring some additional structure and costs. The greenhouse can be allowed greater swings in temperature than the house and can act as a buffer zone to inhibit loss at night.

The passive solar components of this house are calculated to provide about 85% of the heating load. Both a wood stove and an electric heat pump provide auxiliary heat as needed.

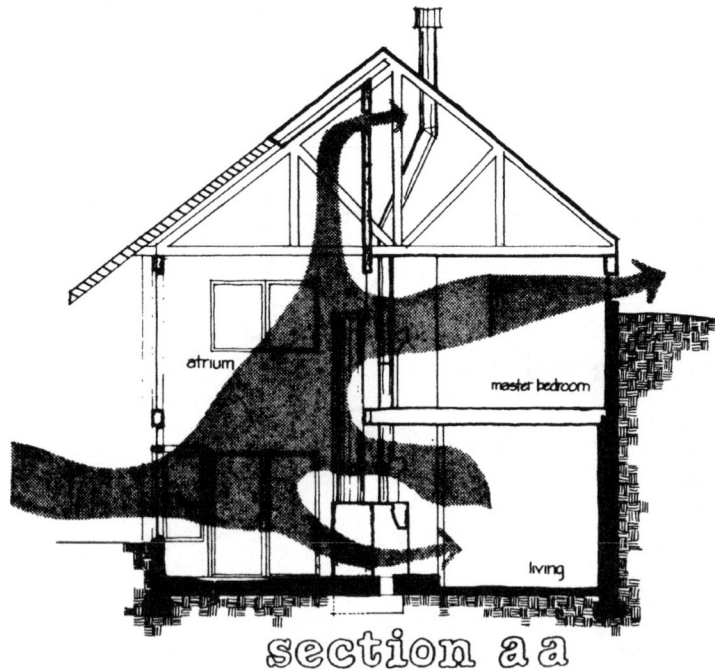

COOLING

Since the summertime brings substantial periods of overheating to the western Virginia countryside, special concern is taken to provide sufficient natural ventilation. Unlike many Trombe wall schemes, this design allows operable windows to penetrate through the Trombe wall directly to the outside. By using this detail, each room is provided with at least two windows, usually on different walls. This is the optimum situation for cross ventilation.

Care is taken to provide fixed horizontal louvers on the south elevation which keep direct sunlight off the fiberglass-reinforced plastic of the Trombe wall system. The pre-fabricated greenhouse unit on the upper floor level is also nicely shaded in the overheated season by the fixed louvers which are designed to match the roof line. Additional protection against overheating is provided by operable vents high and low in the greenhouse.

A light-colored roof is specified to reflect radiation from the high summer sun and to keep surface temperatures at a minimum. An entry trellis and nearby deciduous trees also help shade the home in the summertime.

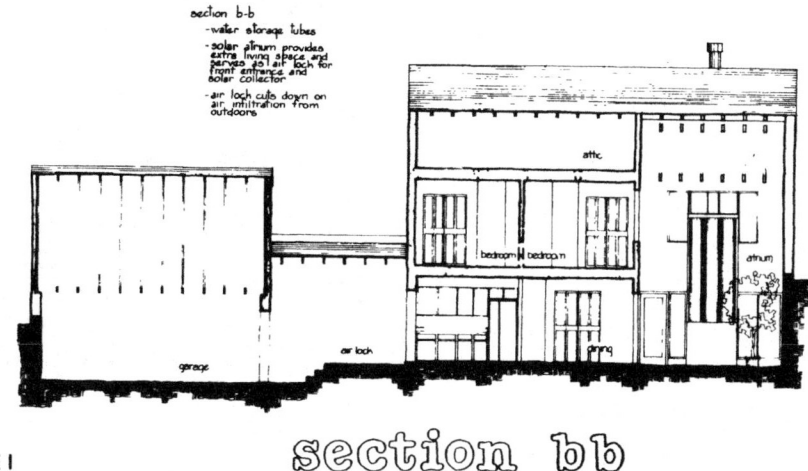

atrium

master bedroom

living

section aa

HOUSE I

section b-b
- water storage tubes
- solar atrium provides
  extra living space and
  serves as air lock for
  front entrance and
  solar collector
- air lock cuts down on
  air infiltration from
  outdoors

attic

bedroom bedroom

atrium

garage

air lock

section bb

passive components:

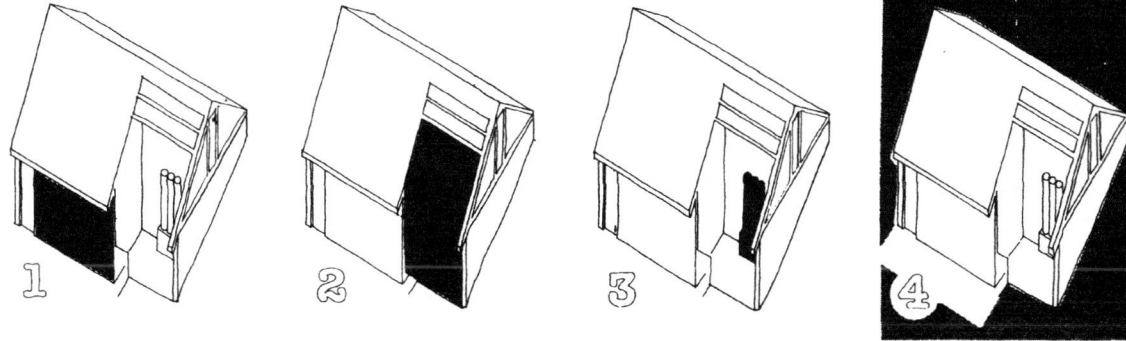

1 mass trombe wall
2 solar atrium
3 water storage tubes
4 earth integration

CONCLUSION

These two houses developed for the Tennessee Valley Authority are extremely fine examples of integrated passive design. The compact floor plans combined with standard, low-cost building techniques and materials make these buildings broadly applicable and desirable models. The combination of atrium and Trombe wall shown in both these designs demonstrates an approach which has many advantages over purer systems. While the Trombe wall provides a simple, effective passive heat source, the atrium supplements this with light and ventilation as well as solar heat. In addition, the atrium, in this supplementary role, is more cost effective and useful as an entry and buffer zone. Buildings which depend totally on atriums often have too much space and expense invested in these thermally intermediate zones, while those buildings which depend totally on Trombe walls sometimes suffer from lack of light, views, and a sense of spaciousness. These designs demonstrate different ways to extract the best of both systems in an inventive but controlled manner.

Chapel Hill, NC

Ted Hoskins and Daniel Koenigshofer, Integrated Energy Systems, Inc.
Chapel Hill, NC

HEATED AREA: 1,135 FT²

NUMBER OF DEGREE DAYS: 3,454

NET THERMAL LOAD: 45.2 10⁶BTU/YR

AUXILIARY ENERGY: 5.77 BTU/DD/FT²

YEARLY SOLAR FRACTION: 66% .

Private client.

CONTEXT
This passive solar home, located in the progressive university community of Chapel Hill, North Carolina, has been designed by the engineering firm, Integrated Energy Systems, as a response·to the requirements of a very energy conscious family.

The home has been constructed on an east-sloping site eight miles outside Chapel Hill. By careful placement of the house among existing evergreen and deciduous trees, the designers have minimized its impact upon the site while making full use of the windbreak effect provided by the trees and topography. The evergreen trees located on the northwest side of the home block the winter wind, while the deciduous trees on the south side provide summer shading. Earth berming on the north side of the home provides additional protection against cold winter wind.

This area of North Carolina experiences equal heating and cooling seasons, so it was necessary for the passive solar system in the house to be able to provide some cooling as well as most of the heating. Throughout the evening hours and into the early morning the temperature is usually the lowest, making it possible to cool with natural ventilation. The nights are also the hours of highest humidity, however, so it was also necessary to provide for the movement of limited amounts of air to counteract the humidity.

The house itself is constructed of readily available materials: concrete block foundation walls, wood framing, exterior stained wood siding, and interior drywall. Windows are standard windows found on most typical speculative houses.

CONSERVATION

Basic energy conservation techniques have been incorporated into the home to minimize heat loss in winter and heat gain in summer. For example, the home is oriented with the majority of its double-glazed windows to the south, and the entry with its vestibule is located on the side away from prevailing winds. The basic shape of the home with its low wall exposure on the north and higher wall exposure to the south provides for greater surface area for solar collection to the south and also serves to deflect the winter wind on the north. Shading devices which double as vent covers open to provide some protection against potential overheating from the sun. Extra insulation in the walls and adequate insulation in the ceiling reduced the overall heat loss. The internal arrangement and open plan of the living spaces provides for free circulation of heated and cooled air throughout the year, and also provides visual continuity between the rooms which increases the apparent size of the home.

HEATING

By incorporating all of the energy conserving techniques into the house design, the designers have reduced the yearly heating and cooling load, with the result that the passive solar system can provide a larger proportion of these loads. The passive solar system designed and built into this residence is a Solar Mass Wall concept, often referred to as a Trombe wall. This Trombe wall consists of an 8" concrete block wall painted black with double glazing attached to the painted exterior surface. Regular windows mounted in the Trombe wall behind its glazing allow sunlight to penetrate into the home's interior. In this way the designers have provided visual contact with the exterior and direct solar heat gain to the living spaces. The direct gain windows and Trombe wall have been sized so as to provide a major portion of the heating and cooling requirements.

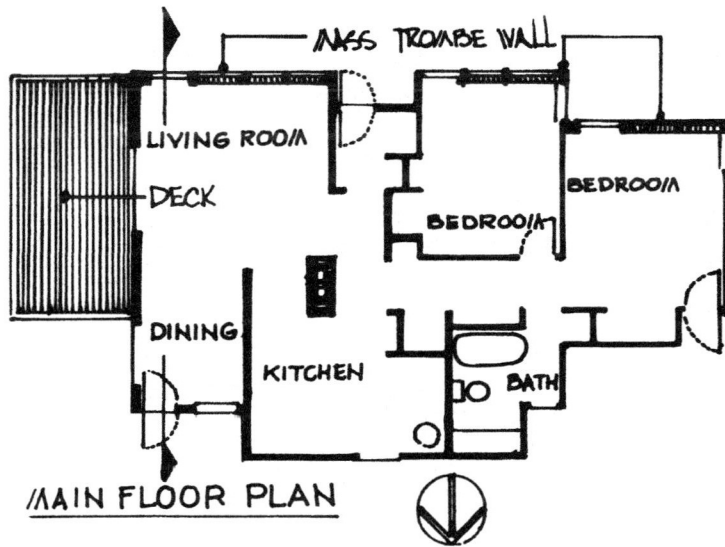

MASS TROMBE WALL

LIVING ROOM

DECK

BEDROOM

BEDROOM

DINING

KITCHEN

BATH

MAIN FLOOR PLAN

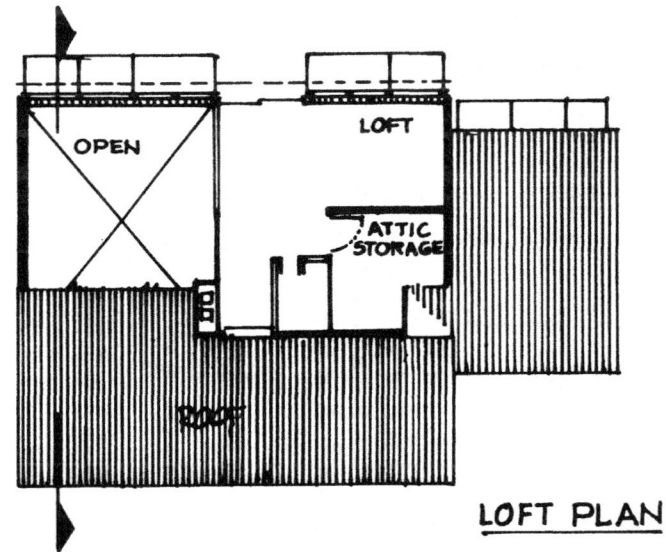

OPEN

LOFT

ATTIC STORAGE

ROOF

LOFT PLAN

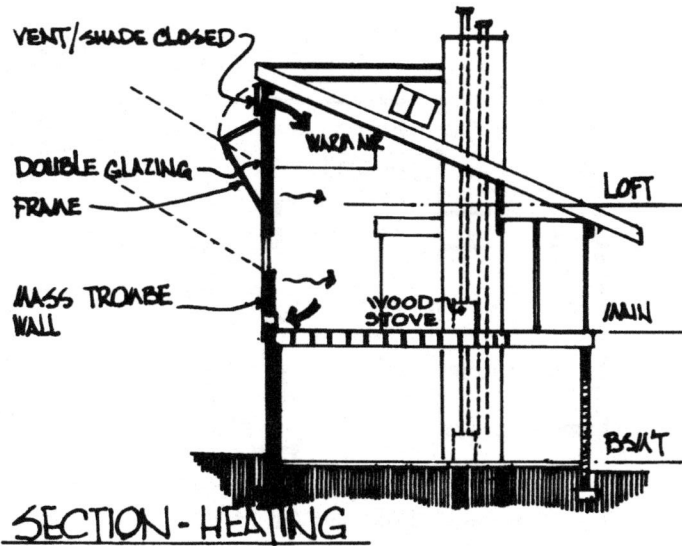

SECTION - HEATING

Labels: VENT/SHADE CLOSED, DOUBLE GLAZING, FRAME, MASS TROMBE WALL, WARM AIR, WOOD STOVE, LOFT, MAIN, BSMT

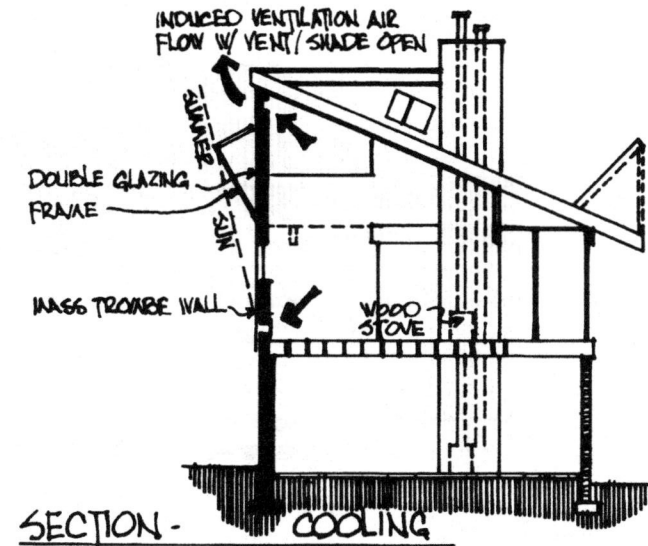

SECTION - COOLING

Labels: INDUCED VENTILATION AIR FLOW W/ VENT/SHADE OPEN, DOUBLE GLAZING, FRAME, MASS TROMBE WALL, WOOD STOVE, SUMMER SUN

The Trombe wall heats and cools by using the process of natural convection of heated air. On a sunny day, the sunlight penetrates the exterior glazing, striking the black concrete block wall surface, which absorbs the sunlight and becomes hot. As the wall heats up, some heat is given off and begins to rise in the 1-1/2" space between the wall and outer glazing. On sunny days during the heating season, vents in the top and bottom of the wall are opened. The warm air rises up, goes out the top vents and in so doing pulls cooler air in the bottom vents. As the cooler air is heated, it rises again and the cycle repeats itself. At night, the vents are closed; this keeps warm air from being pulled from the house into the space in front of the Trombe wall where the heat would be conducted through the glazing and lost to the cold night air. Depending upon the temperature of the outer surface of the Trombe wall, the thickness of the wall, and the outside air temperature, a certain amount of heat will be conducted through the wall to the interior surface. Here heat will radiate to the cooler interior rooms. By morning, the wall will have cooled down and if the sun is shining the process will again repeat itself.

COOLING

In the late spring, summer, and early fall, high humidity levels in combination with warmer air temperatures present a cooling problem. To provide as much cooling potential as possible through the use of natural and induced ventilation, the designers have carefully placed operable windows and exterior Trombe wall vents. In late spring when heating is no longer required, the induction of outside air for cooling is provided by closing the high interior vents, leaving the lower vents open and opening the outside Trombe wall vent. Sunlight that is not blocked by the shading devices (open vent covers) strikes the masonry wall, and the air in the air space is heated and rises. But instead of the air being directed into the living spaces, it flows out the exterior vent, pulling in cooler air from the north side of the home through the house. However, this air movement will provide cooling only up to the point when the combination of high temperatures and humidity cannot be counteracted by the use of induced air. When this happens, the occupants can either try and adjust to the climate or provide the necessary cooling by mechanical means. Certainly in this climatic area, this situation will occur at times, and then it is

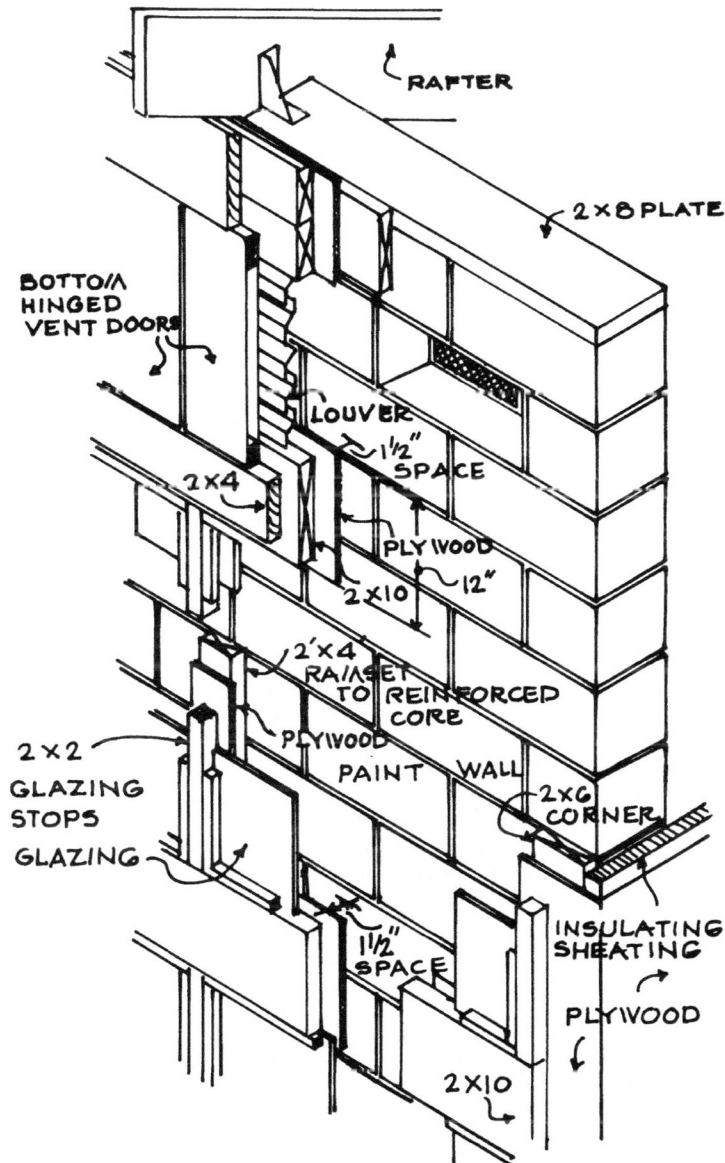

RAFTER

2×8 PLATE

BOTTOM HINGED VENT DOORS

LOUVER

1½" SPACE

2×4

PLYWOOD

12"

2×10

2'×4 RAMSET TO REINFORCED CORE

PLYWOOD

PAINT WALL

2×2 GLAZING STOPS

GLAZING

2×6 CORNER

1½" SPACE

INSULATING SHEATING

PLYWOOD

2×10

ISOMETRIC OF TROMBE CONSTRUCTION

up to the individual occupants to determine their own thermal comfort requirements and decide how they will provide additional cooling.

CONCLUSION

The details of the Trombe wall are made up of standard construction materials. The Trombe wall glazing consists of standard size double-glazed sliding door replacement glass, and glazing supports are standard wood frame sections installed by using a standard ramset gun. The vents and louvers are also standard items that have been purchased "off-the-shelf". These details should help dispel any myth that passive solar systems are highly sophisticated and complicated to install, since, with an understanding of the basic concept of the Trombe wall and familiarity with these standardized details, any builder or owner-builder could construct this passive solar system.

By their total design approach, the designers have responded not only to the needs of their clients, but have provided a very honest and straightforward solution. The proper siting, orientation, response to climate, use of standard construction materlals, open plan, and system design make this a successful passive solar home. To further enhance the home's successful operation, some additional exterior shading devices (roller, shades, etc.) could be used on the Trombe wall during the cooling season and the Trombe wall could be thicker. This would protect against overheating. The use of a 1-1/2" air space between the Trombe wall glazing and storage wall is narrower than most so far, since most Trombe walls have been designed with a 3"-6" space. By using a 1-1/2" air space, the designers have chosen to increase the air velocity over the wall and in so doing have reduced the air temperature supplied to the living spaces. It will be interesting to see how this space performs. There has been little information published on the optional glazing space in Trombe walls.

The open plan and flexible siting potential of this home make it an attractive unit for builder speculation. At first, because of the uniqueness of the design, some market hesitation may be evident. But with the successful functioning of the home after a full year's occupancy and operation and with the added social, psychological, and economic benefits of the design, the home should have a wide appeal to the progressive university community in which it is located.

Strasburg, VA

Victor Habib, One Design, Inc.
Winchester, VA

HEATED AREA: 1,168 FT²

NUMBER OF DEGREE DAYS: 5,978

NET THERMAL LOAD: 19.9 10⁶BTU/YR

AUXILIARY ENERGY: 2.16 BTU/DD/FT²

YEARLY SOLAR FRACTION: 92%

Open market sale.

CONTEXT

As the cost of materials and labor continues to climb, there is an expanding market for compact, resource-conscious and readily buildable single-family design. This scheme which is designed for the cold, 6,000 degree day mountain region of Virginia is a modest three-bedroom, 1-1/2 bath, single-family detached dwelling. With a total heated square footage of only 1,168 square feet and basically conventional wood frame construction, this water wall solar house design can be reproduced at a cost which places it at the low price end of the new housing market. In addition, by incorporating standard colonial features such as shutters, weather vanes, replica gas-light entry lights, and picket fences, the homeowner is not asked to forego any of the symbolic elements they have come to expect in their homes.

Departing from traditional designs, however, this home takes on different appearances as the time of the day and the thermal needs change. When the exterior moveable insulation panels are in the retracted position, the southern facade of the home will appear to be predominately glazed with both windows and water walls. When solar radiation is no longer available, the house changes its appearance. The insulating panels come down, and the homeowner is protected in his home secure in the knowledge that he and his solar storage mass are surrounded by a colonial insulating blanket.

While the home design is simple, it is not without sophistication. A modestly peaked roof, for example, covers the dwelling in a slightly "L"-shaped plan which articulates the sleeping wing and provides both interest and an exterior appearance of spaciousness to an otherwise modestly sized building.

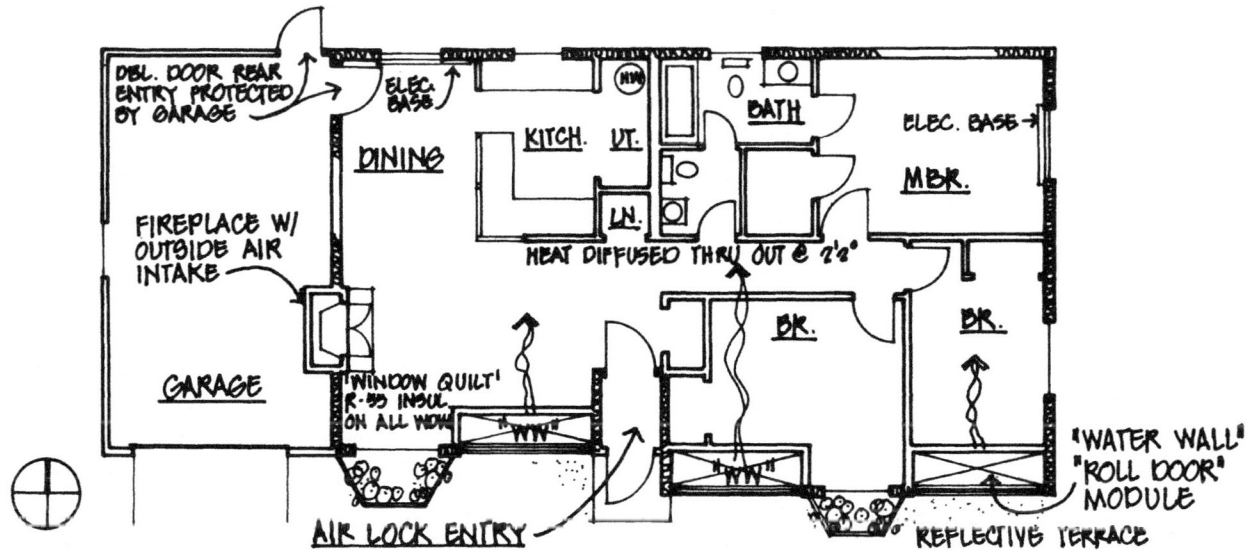

FLOOR PLAN FRONT FACING SOLAR

(Labels on plan:)
DBL. DOOR REAR ENTRY PROTECTED BY GARAGE
ELEC. BASE
BATH
ELEC. BASE →
DINING
KITCH. UT.
MBR.
LN.
FIREPLACE W/ OUTSIDE AIR INTAKE
HEAT DIFFUSED THRU OUT @ 7'2"
BR.
BR.
GARAGE
"WINDOW QUILT" R-55 INSUL. ON ALL WDW
WW
WW
"WATER WALL" "ROLL DOOR" MODULE
AIR LOCK ENTRY
REFLECTIVE TERRACE

CONSERVATION

Just as people must walk before they run, so must homes conserve energy before they can collect energy.

1. In the house site plan there is an appreciation of the value of vegetation. For example, evergreen windbreaks are planted to the northwest of the building to slow the prevailing cold winter winds. This should help decrease the infiltration rate of cold air into the house.

2. An often overlooked source of energy conservation in this design is the basic decision to keep square footage to a minimum and to keep the plan compact. By building few square feet and by keeping wasted circulation space to a minimum, the heat loss and the demand for space heat will be minimized.

3. The "New American" uses a number of details which are becoming standard energy conscious construction techniques. These include 2 x 6 wood frame wall construction to allow larger thickness of fiberglass insulation, 11" insulation between the attic rafters, double glazing on all window areas with nighttime insulating quilts, and insulated doors. In order to reduce infiltration, a vestibule is added to the front entrance while the unheated garage acts as an effective vestibule for the rear entrance to the house.

4. Care is taken to provide a continuous perimeter insulating detail, a significant component of total building heat loss in slab-on-grade construction. Conservation-minded controls are also specified for the home, including the use of an automatic nighttime setback on the electric hot water heater which is housed in a corner of the kitchen space.

By keeping the total auxiliary home heat requirement low through conservation and solar collection, it is economically feasible to turn to electric resistance baseboard as the auxiliary heating source. This eliminates the need for a mechanical equipment space which eliminates square footage and thus decreases project cost. If a similar system was conceived for a locale with advantages of off-peak electric rates, the thermal storage might be designed to allow thermal charging at off-peak hours to provide for upcoming overcast weather.

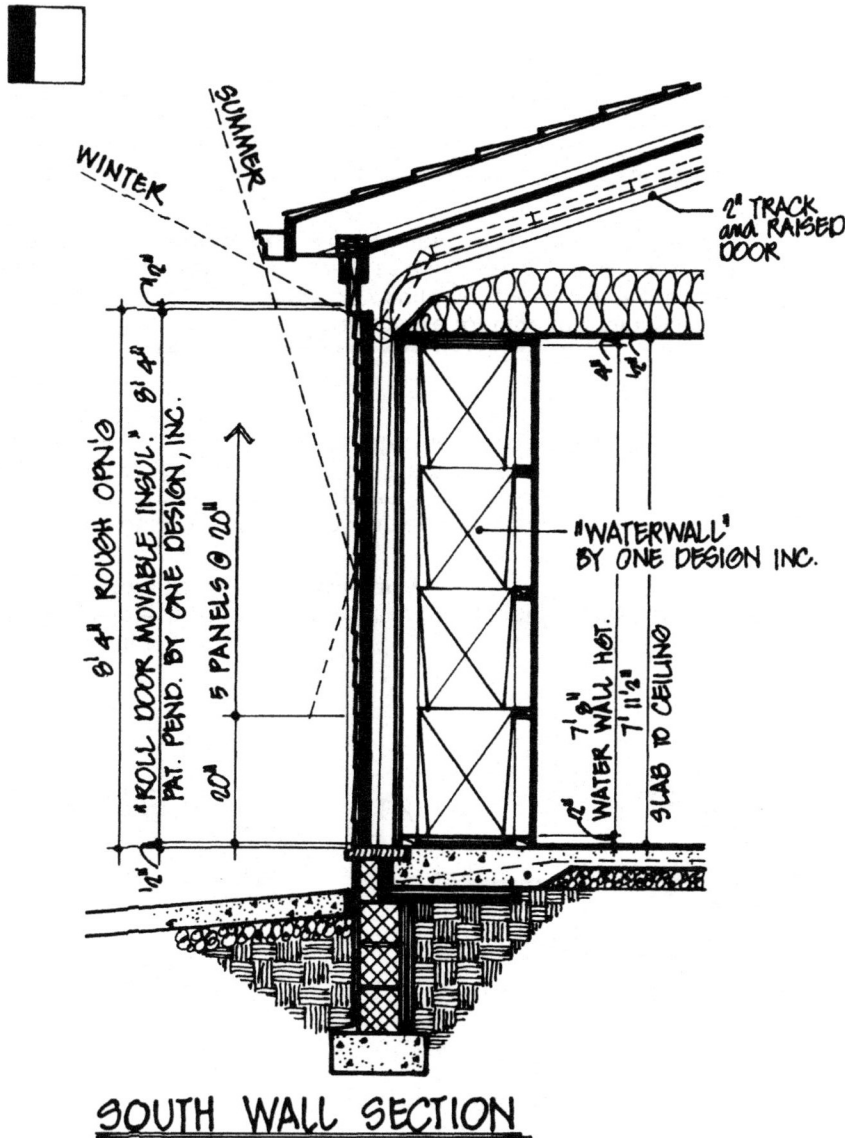

SUMMER

WINTER

2" TRACK
and RAISED
DOOR

8' 4" ROUGH OPN'G

"ROLL DOOR MOVABLE INSUL." 8' 4"
PAT. PEND. BY ONE DESIGN, INC.

5 PANELS @ 20"

20"

"WATERWALL"
BY ONE DESIGN INC.

WATER WALL HGT. 7' 11½"

SLAB TO CEILING

SOUTH WALL SECTION

HEATING

The solar heating system is characterized by two basic systems: windows and water walls. Forty-five square feet of large windows allow "direct gain" solar energy into the house to provide some immediate heating during the daytime. In order to provide heating at times when sunlight is not available such as in the evening, it was necessary to make provisions for additional solar collection area *and* coupled solar storage. The solar storage mass in this scheme is provided by large quantities of water. The water wall containers used in this application are composed of premolded fiberglass troughs which nest for shipment, but stack tightly one upon the other in application to preclude evaporation loss. Solar radiation enters through the 192 square feet of single pane vertical glass and is converted to heat at the black surface of the water wall. The heated water circulates throughout, raising the average temperature of the storage. Even when it is necessary to introduce new elements into the construction process, such as the nighttime moveable insulation for the water wall, these elements are derived from standard construction details. In this case, conventional garage overhead-sliding door assemblies with insulated door panels are modified to provide exterior insulation. The sliding doors retract into modified ceiling pre-fab roof trusses to expose the glass areas and water storage.

For each square foot of glazing there is approximately 1-1/3 cubic feet of water with a storage capacity of about 35 BTU per square feet of glazing.

After you have collected the solar energy and stored it in sufficient solar storage mass, it is necessary to properly distribute the energy to the space. In this case, instead of allowing the water storage to radiate directly to the room as is the general practice, a wood frame wall which is finished in the interior with a thin sheet of gypsum board is constructed between the water storage and the space. Although this slows the flow of heat into the room somewhat, the conventionally finished inside wall should appeal to the mainstream of the housing market. In this case, superior performance might have been achieved by providing top and bottom vents into the water wall cavity, thus providing a daytime convective loop, possibly assisted by a fan to speed air flow. An operable louvered shutter wall or a fabric panel wall would also allow for easier convective distribution to assist the radiant distribution from the water storage mass. An essential part of the controls of the heating system in this case is the use of nighttime moveable insulation to retain the heat within the water wall module. In this case the insulated sliding door is rolled down from the attic space, providing a tight and elegant front facade. One weakness of this scheme is that remote rooms must rely on convection of air from rooms adjacent to the water walls and the very modest re-radiation of heat from party walls between the relatively warmer south rooms and the relatively cooler north rooms. Concern is expressed for this problem by the designers in their placement of the spaces which need less heat, such as the bathroom and the kitchen, on the north wall while reserving access to the south wall to rooms which need more heat.

In most solar designs and in most climates, it is not cost-effective to seek very high solar fractions. This design accepts a reasonable solar contribution of perhaps one-half to two-thirds of the seasonal heating demand of this very well-insulated small home.

COOLING

In a 6,000 degree day climate, reasonable cooling can be achieved a large percentage of the time by allowing provisions for adequate ventilation. This design makes use of the well-proven attic fan which will increase the air flow velocity in the house and create comfortable conditions during most of the overheated season. In fact, the ventilation grill over the bedroom wing is prominently displayed and used as a positive design feature. Natural cross ventilation is also available in some rooms, but this aspect of the design could be improved.

To avoid summer overheating, the moveable insulation outside the water storage mass is left in the closed position during the daytime. The interior water storage mass can thus act as a heat sink to alleviate some of the peak daytime heating conditions. In climates with substantial day-night temperature swings, the heat absorbed in the mass during the hot summer day can be re-radiated at night by opening the insulation at night and exposing the water storage mass to the cooler night sky, thus cooling the water mass for the next hot day.

CONCLUSION

One of the most commonly heard concerns about passive solar design is its inability to produce designs which are acceptable to the large numbers of the American people at prices which are within their reach. This design overcomes both problems and presents a home which is replicable and adaptable.

Of course, there are still some technical and construction concerns. For instance, it is unusual to find it this cold in a climate with a latitude so far south. Consequently, the sun is somewhat higher in the sky in the winter than one might wish and the use of the reflective terrace is important.

Also, a modification to the design which would allow all primary rooms to be adjacent to a water wall would be beneficial. The

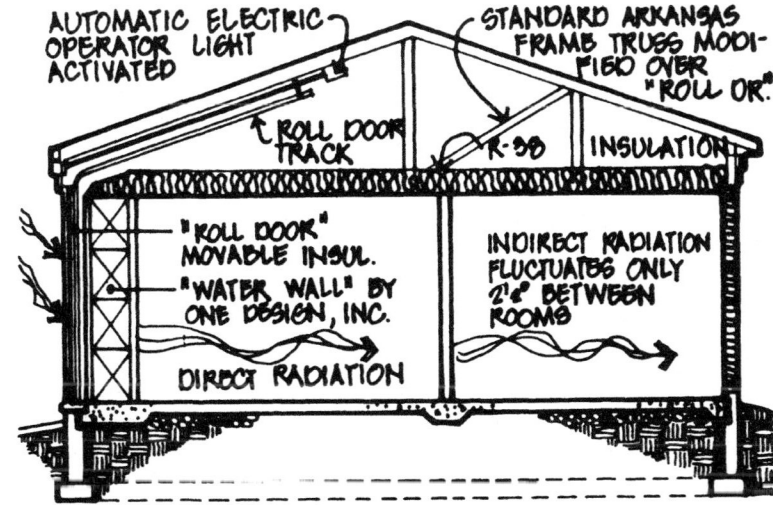

BUILDING SECTION

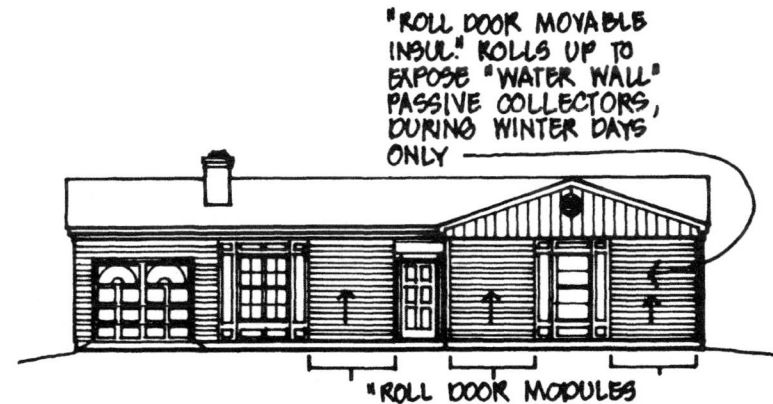

SOUTH ELEVATION
FRONT FACING SOLAR FACADE

designer here has simply made a trade-off between ease and economy of construction and improved thermal performance.

Basically, however, this design is truly for the "New American". It is a trimmed-down version of a slice of the suburbs—a home for tomorrow at a price for today.

Santa Fe, NM

Susan Nichols, Communico
Sante Fe, NM

HEATED AREA: 968 FT²

NUMBER OF DEGREE DAYS: 5,586

NET THERMAL LOAD: 44.3 10⁶BTU/YR

AUXILIARY ENERGY: 3.49 BTU/DD/FT²

YEARLY SOLAR FRACTION: 91%

Open market sale.

CONTEXT

Four of these 1,000 sq. ft., two bedroom homes are to be built in La Vereda, a nineteen unit planned environmental community located in Santa Fe, New Mexico. The ten acre site, half of which is to be left in open space, is all south sloping with views of the surrounding mountains to the south and southwest.

Santa Fe, at a latitude of 36°N and an elevation of 7,000 feet, experiences a sunny, cold winter with 70% available sunshine and a heating season of 5,586 degree days. Snowfall can be heavy, but rarely remains on the ground for more than three to four days. The summer climate is hot and dry with the average daytime temperature in the 80°-90°F range; because of the high elevation, nighttime temperatures drop to 55°-60°F.

The Spanish Pueblo style architecture is one indigenous to the Santa Fe area - flat roofs, thick stuccoed walls, tile floors, wood decks and beamed ceilings. This house integrates the elements of a passive solar system with these design features. Trombe walls, interior mass walls and floors are incorporated into the conventional 2x6 structural frame of the house to give an "adobe" feeling. The cost advantages of frame construction and the thermal advantages of an R-25 wall are realized by allowing the mass required for the direct gain clerestory windows to be inside the 2x6 insulated exterior along the north walls. Construction will require no special materials or skills; it is simply a matter of putting the same components together in a slightly different manner.

CONSERVATION

Energy conservation plays a major role in the design and building of a solar home. Proper orientation, careful siting and sufficient insulation are all necessary to maximize solar gain and retain it within a structure. The nineteen homes in La Vereda are grouped into three clusters to allow each house the proper south orientation without the look of soldiers in formation. Each dwelling is situated below the northern ridge line and dropped below the tree line of dense juniper and piñon coverage as protection against the northwest winter winds. The airlock entry of this model is located on the east side of the structure to allow the garage to serve as an additional wind buffer. The location of the garage, entry, and utility room along the north wall of the dwelling serve to create an effective thermal barrier by creating a transition from heated spaces to unheated spaces and then the outside.

The house is designed to be very energy efficient. The stem walls, 3'-4' above the finished floor to allow for berming, are 8" concrete block with all cells filled with concrete, waterproofed with plastic roofing tar, and enclosed in a vapor membrane of 6 mil polyvinyl and 2" of rigid polystyrene. All perimeters are insulated with 2" of rigid polystyrene which will not deteriorate when in contact with moisture. When insulation is in direct contact with the earth the choice of materials is important. The 2x6 frame walls are insulated with 5-1/2" fiberglass batt and sealed with a 6 mil polyvinyl vapor barrier before application of the dry wall, resulting in an R-value of 25. Ceilings, depending on the buyer's choice, will be either beams and decking with 5" of rigid dense styrofoam with a total R-value of 34 or 12" joists with blown-in cellulose fiber with an R-value of 44. All windows are double glazed with the exception of the direct gain windows in the Trombe wall which are triple glazed. The clerestory windows and skylights are protected against nighttime heat losses by hinged insulating shutters. By building a "heatilator"-type fireplace and introducing outside air for combustion, the efficiency of the fireplace is increased and becomes a net thermal gain. Venting for the stove and dryer through simple heat exchangers is internal to the structure, so that the heat generated by these appliances is not lost to the outside.

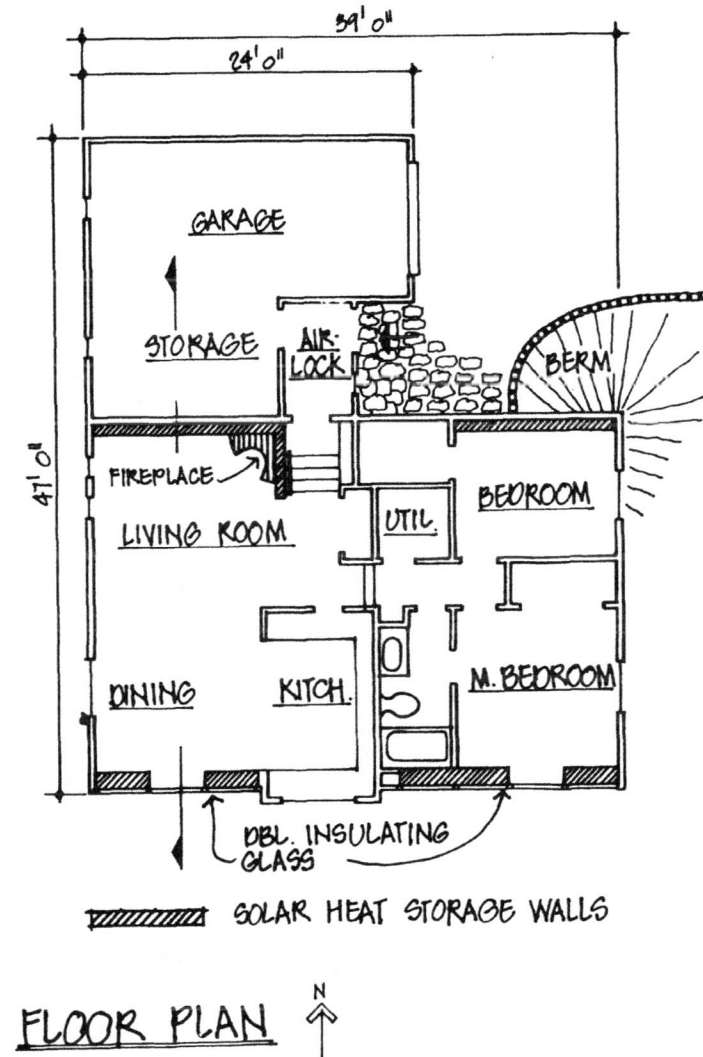

FLOOR PLAN

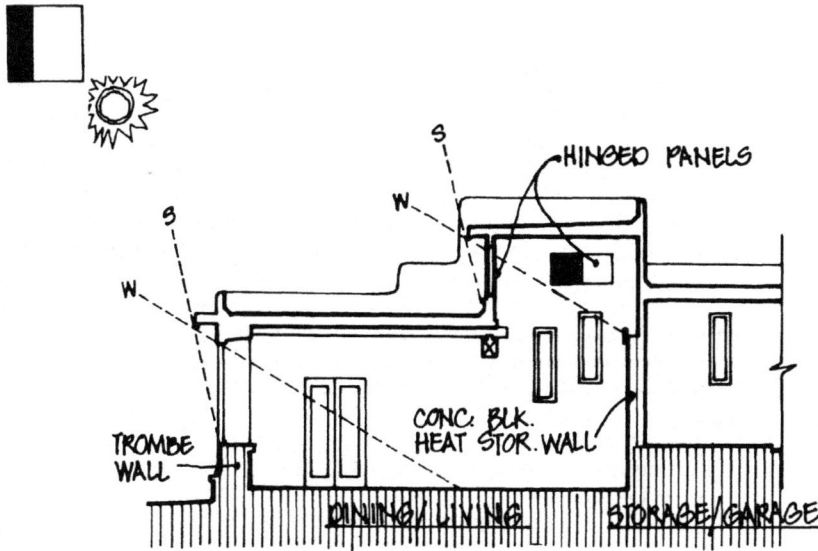

CROSS SECTION

Diagram labels: HINGED PANELS, CONC. BLK. HEAT STOR. WALL, TROMBE WALL, DINING/LIVING, STORAGE/GARAGE, S, W

HEATING SYSTEM

The heat for this home is provided by a combination of unvented Trombe walls and direct gain windows which work in conjunction with one another to provide a stable thermal environment. As the sun strikes the south wall of the structure, heat gain is admitted immediately through the direct gain windows in the Trombe walls and clerestory, providing the source of daytime heat. Because of the time lag effect of the Trombe wall or the amount of time it takes for the heat to travel through the 16" of solid concrete, the inside surface of the wall will reach its maximum temperature in the late evening, providing the source of nighttime heat.

During clear days, excess heat generated by the direct gain windows is stored in the mass of the floors, the stemwalls, the fireplace, and the interior skin of 8" concrete block along the north wall. By painting the interior walls a light color, the light striking them is diffused for a more even distribution of heat throughout the mass. All distribution of heat from the storage mass into the structure is by natural means; as the temperature in the building drops below that of the mass surfaces, stored heat is radiated back into the rooms to maintain thermal comfort.

During cloudy days, the passive solar system will use the diffuse radiation penetrating the cloud cover to provide some heat gain within the house. On a day when an active solar system would not turn on, this passive system with its natural means of collection, storage, and distribution continues to generate heat within the structure.

Given average winter conditions, the solar system will provide 87% of the home's heating needs over the heating season, with the remaining 13% provided by the back-up electric baseboard units. Of a yearly total heating load of 8,700 Kw, the passive system will generate 7,570 Kw and electricity will generate 1,130 Kw. At a cost of $.045/Kw in Santa Fe, the annual heating bill will be $50, and an annual savings of $340 will be realized. This performance is based on the assumption that the temperature of the house will be maintained at or above 65°F, and if, in fact, the temperature is allowed to drop below 65°F at night, the amount of back-up heat required will be reduced, and a larger annual savings can be realized.

There is 155 sq. ft. of double glazed Trombe wall pierced with 60 sq. ft. of triple glazed direct gain window and 175 sq. ft. of night insulated double glazing on the south facade of the structure.

The Trombe wall is 16" of solid concrete providing 200 pounds of storage mass per square foot of glazing and a heat capacity of 6,500 BTU/°F. The rest of the mass consists of quarry tile on a 4" slab, the 8" concrete stem walls, the north storage walls of 6" concrete filled block, and the fireplace. This serves as the heat reservoir for the 235 sq. ft. of direct gain with a mass of 530 lbs. per sq. ft. of glazing and a heat capacity of 26,250 BTU/°F. A larger amount of mass is required for the direct gain system than for the Trombe wall. An adequate amount of mass in relation to the glazing is important as a temperature control; the more mass, the more even the temperature fluctuation within the space.

COOLING SYSTEM

In Sante Fe, cooling is not a real concern. With the dry climate and large day-night temperature fluctuations, a carefully designed and well insulated structure is able to maintain a comfortable temperature throughout the summer. Overhangs above the Trombe walls and clerestories are sized to completely cut off the sun on June 21 and provide full shading to the south collector walls. This shading factor decreases as the sun angle declines after the summer solstice, allowing an increasing amount of direct sunlight to fall on the glass. During the months of September and October with their shallow solar angles and lower heating loads, some overheating may occur. However, the amount of building mass and the effect of berming in conjunction with adequate ventilation does much to temper any overheating. As excess heat accumulates, it is absorbed into the mass where it is lost more quickly through bermed surface areas to the cooler temperature of the earth. As the night sky cools, conductive losses of excess heat occur through exterior surfaces and convective losses occur through open vents which are located high up in the clerestory. Air is pulled in from the outside to replace the warmer air escaping from these vents, and this air movement in combination with the tempering effect of the mass maintains an acceptable temperature.

CONCLUSION

This combination of passive solar systems is a cost effective approach that offers advantages of both systems. The direct gain windows give good natural light in controlled amounts through the windows in the Trombe wall and clerestory. The Trombe wall is important in producing an even source of heat and is effective as a nighttime heat source.

The percentage of solar participation is high in a climate that is cold and sunny. An option in a colder, cloudy climate is to add nighttime insulation and/or a reflective surface in front of the south-facing glass. Overheating is not a real concern in Santa Fe, but in a climate where the day-night temperature swing is small during the late summer and early fall, some sort of shading is necessary, possibly planting deciduous trees.

The home is designed for the specific market segment that has shown a definite interest in passive solar, the young professional single or couple with few or no children and the retired couple who face living on a fixed income. While they cannot afford a large, expensive solar home, they can afford a smaller, solar home with the possibility for future expansion. The house is priced in the $65,000-$70,000 range, the lower end of the high quality, hand-crafted home market in this location in Santa Fe. The price position of the home in the area, the location of the project, the design of the home, and the reliability of the solar system should assure good market acceptance.

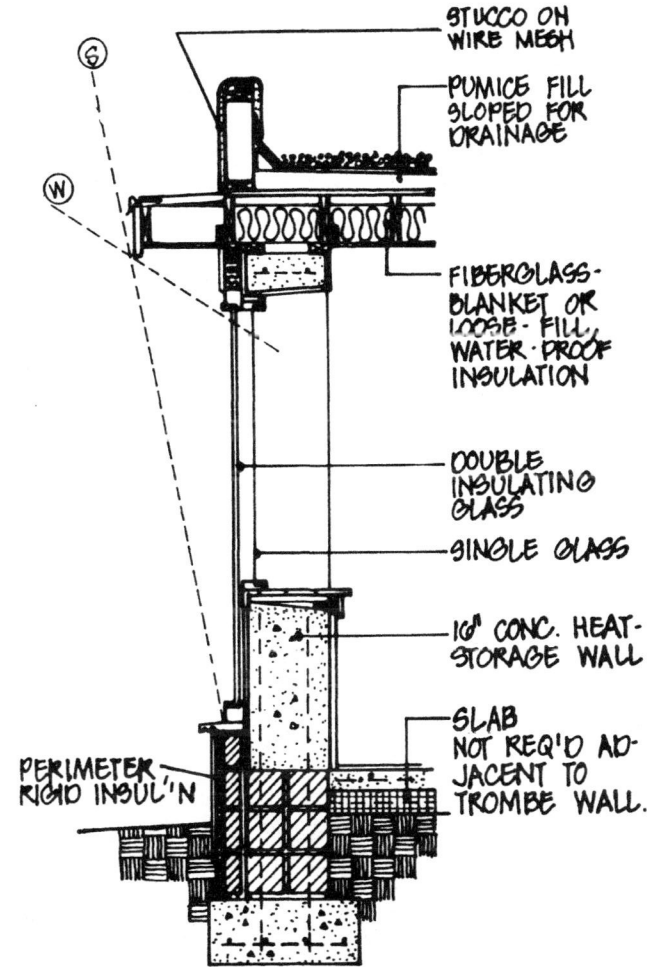

STUCCO ON WIRE MESH

PUMICE FILL SLOPED FOR DRAINAGE

FIBERGLASS-BLANKET OR LOOSE-FILL, WATER-PROOF INSULATION

DOUBLE INSULATING GLASS

SINGLE GLASS

16" CONC. HEAT-STORAGE WALL

SLAB NOT REQ'D AD-JACENT TO TROMBE WALL.

PERIMETER RIGID INSUL'N

TROMBE WALL WITH WINDOW OPENING

White Rock, NM

Ken Brooks, Arch. Res. Cons./Rocky Mountain Sun Power
Santa Fe, NM

HEATED AREA: 1,063 FT²

NUMBER OF DEGREE DAYS: 5,853

NET THERMAL LOAD: 44.9 10⁶BTU/YR

AUXILIARY ENERGY: 5.71 BTU/DD/FT²

YEARLY SOLAR FRACTION: 54%

Private client.

CONTEXT
This one-bedroom residence is designed for a site in northern New Mexico which has a compelling view to the northeast of the Sangre de Christo Mountains and the Rio Grande River Gorge. Near Los Alamos, at an altitude of 6,300 feet, this site has a 6,000 degree day climate, which requires a considerable amount of heating. However, the building rests on a southern slope of the Jemez Mountains which further enhances the solar benefits of this clear, sunny region. Summers here are hot and dry with a large daily temperature swing. Natural cooling has traditionally been achieved in this region by building with massive walls which absorb heat during the day and release it to the outside at night. In this home design, a massive southern adobe wall and a concrete floor provide both natural solar heating in the winter and cooling in the summer.

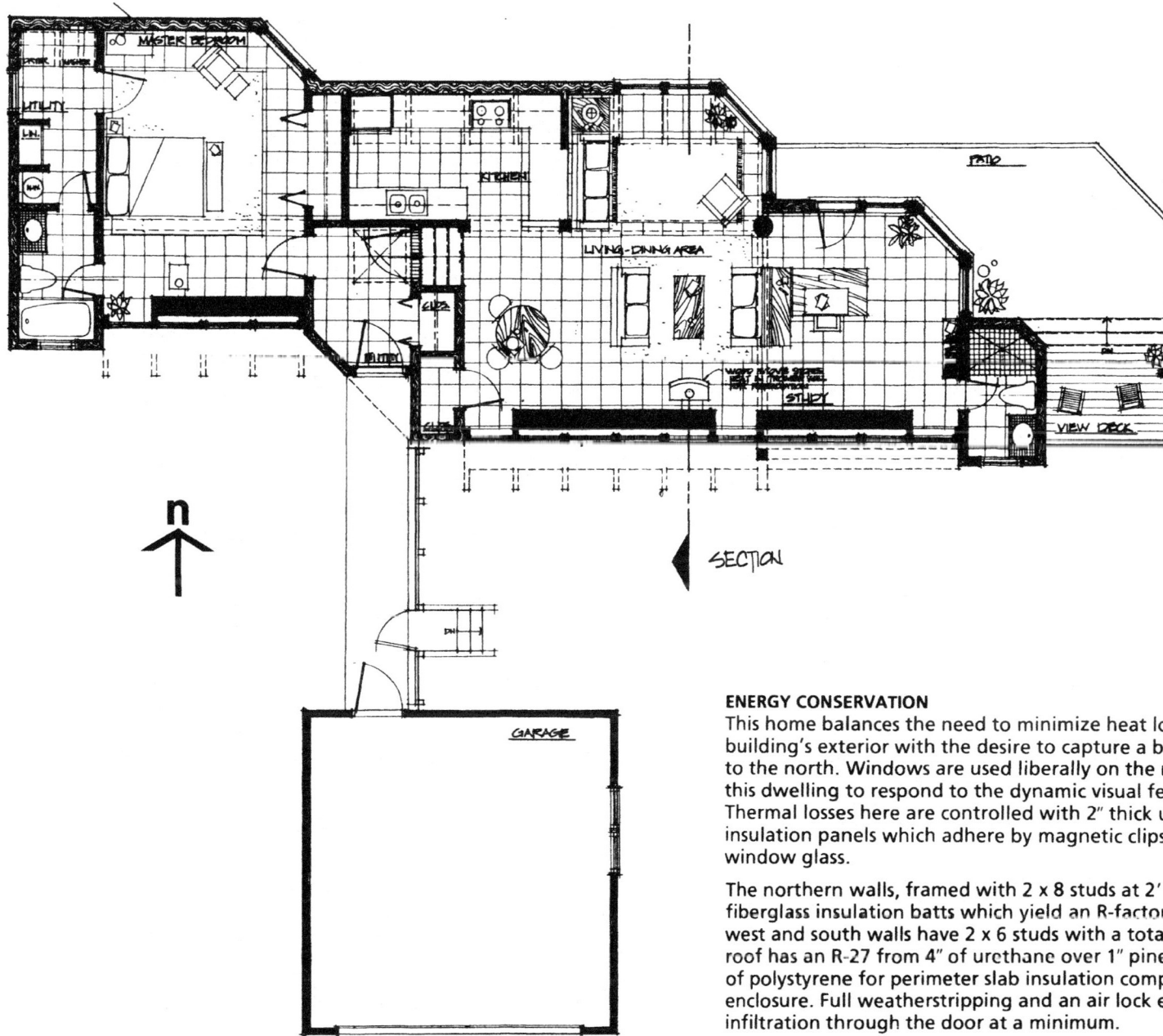

n (north arrow)

SECTION

Labels within floor plan:
- MASTER BEDROOM
- UTILITY
- LIN.
- KITCHEN
- LIVING-DINING AREA
- STUDY
- CLO.
- PATIO
- VIEW DECK
- GARAGE

ENERGY CONSERVATION

This home balances the need to minimize heat losses through the building's exterior with the desire to capture a breathtaking view to the north. Windows are used liberally on the northeast side of this dwelling to respond to the dynamic visual features of this site. Thermal losses here are controlled with 2" thick urethane, night insulation panels which adhere by magnetic clips to double-pane window glass.

The northern walls, framed with 2 x 8 studs at 2' on center have fiberglass insulation batts which yield an R-factor of 27. The east, west and south walls have 2 x 6 studs with a total R of 22. The flat roof has an R-27 from 4" of urethane over 1" pine deck. Two inches of polystyrene for perimeter slab insulation completes the thermal enclosure. Full weatherstripping and an air lock entry keep winter infiltration through the door at a minimum.

An adjustable wood awning shades the south-facing glass during the summer months but retracts to allow full solar exposure during the winter. The need for a sunshade which varies in length from season to season is overlooked in most south wall shading schemes. Although outside temperatures are vastly different, sun angles are identical on March 21 and September 21. Protecting southern glazing with a fixed overhang in August and September when cooling is still a concern will shade the solar gain which is needed for passive heating in March and April. The hinged awning shown here in the building section is an excellent solution to this problem.

HEATING
This design uses both a direct and indirect solar gain to provide 54% of the annual heating load. Roughly two-thirds of this gain is absorbed in a 14" thick adobe mass wall or "Trombe Wall". With an insulating glass wall 7" off the surface of this adobe wall, the entering solar energy is absorbed by the south face of this massive wall and trapped in this cavity. This heat then passes slowly through the wall mass during the day and is radiated into the home from this wall at night. Vents in this wall also allow heat to convect into the house during the day.

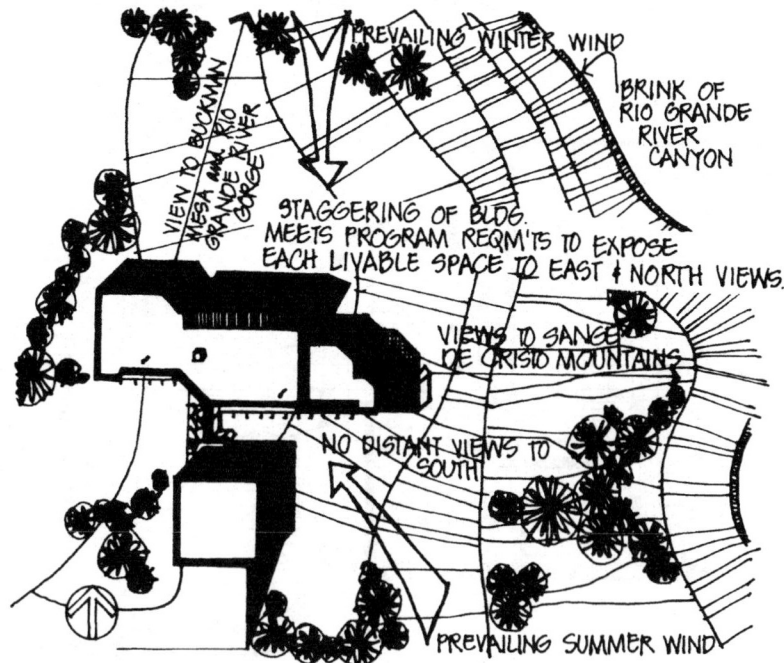

Light enters the house directly through southern windows and the clerestory light scoop toward the rear. This scoop helps to distribute this direct gain sunlight over the entire 4" concrete floor slab, allowing an even transfer of heat. Allowing a 12°F. temperature swing in the living space releases this heat at night.

Between the concrete floor slab and Trombe wall, there are 45,000 lbs. of masonry in this home which store 9,000 BTU's of heat for each 1°F. rise in temperature.

This balances with the total south glazing area of 320 sq. ft., most of which is Trombe Wall, since this computes to approximately 30 BTU's of storage for each square foot of glass. Two wood stoves, a large one in the living area and a smaller one in the bedroom, provide backup heating as needed.

COOLING
With an average daily temperature in July of 70°F. and an average maximum temperature of 82°F., cooling is not a serious problem in this climate. What cooling that is needed during the summer is provided by including ventilation through the house at night to cool the floor and wall mass when temperatures are below the comfort range. With vents closed during the heat of the day, this cool mass absorbs heat within the house. Prevailing summer breezes are from the south, and vents on this wall draw air into the house near the floor. Operating sashes on the clerestory windows open to act as high exhaust vents. The air flow induced by this low-intake, high-exhaust ventilation approach is generally effective, even in the absence of an outside breeze.

CONCLUSION
This design responds well to the features of a specific site to create an exciting living area for a single person or a couple without children. With only one bedroom and a character peculiar to homes in the Southwest, this plan is not widely marketable as shown. Modifying this plan to include another bedroom where the study is presently and reducing the number of angles and offsets on the exterior can turn it into a small, economical home which can be built anywhere. All exterior walls shown are light-framed with a stucco veneer which can be changed to wood siding or any exterior treatment popular in a given region.

Thermally this house can be improved by adding roof insulation, reducing north glazing, and by enlarging the clerestory to shower additional solar energy on a northern masonry storage wall. Not only will the fraction of solar heating be improved, but with a radiant storage mass on both the north and south walls, the heat in this space will be more even.

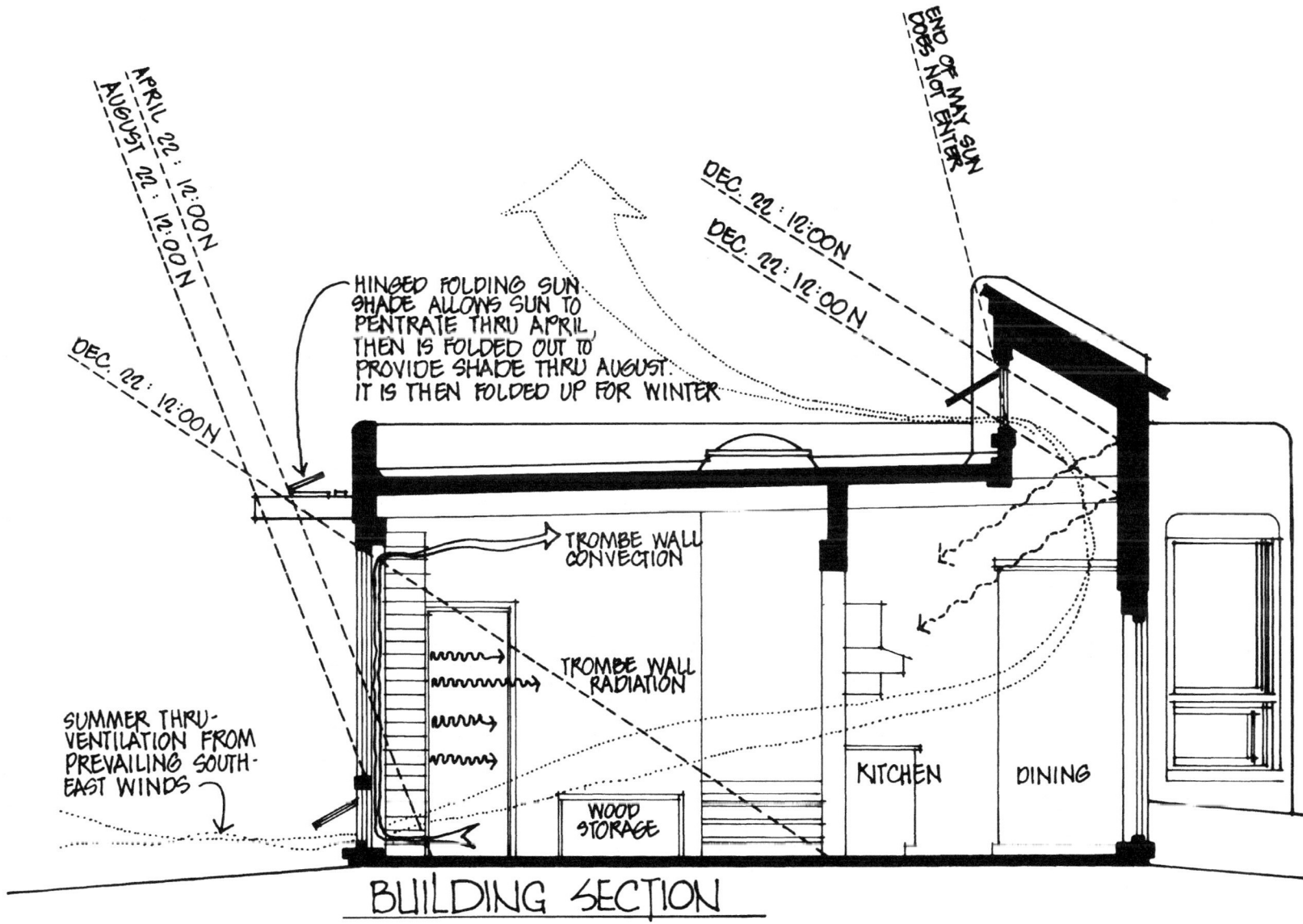

APRIL 22 : 12:00N
AUGUST 22 : 12:00N

DEC. 22 : 12:00N

END OF MAY SUN DOES NOT ENTER

DEC. 22 : 12:00N
DEC. 22 : 12:00N

HINGED FOLDING SUN-
SHADE ALLOWS SUN TO
PENTRATE THRU APRIL,
THEN IS FOLDED OUT TO
PROVIDE SHADE THRU AUGUST.
IT IS THEN FOLDED UP FOR WINTER.

TROMBE WALL
CONVECTION

TROMBE WALL
RADIATION

SUMMER THRU-
VENTILATION FROM
PREVAILING SOUTH-
EAST WINDS

WOOD
STORAGE

KITCHEN

DINING

BUILDING SECTION

99

Carbondale, CO

Ron Shore, Thermal Technology Corp.
Snowmass, CO

HEATED AREA: 1,220 FT²

NUMBER OF DEGREE DAYS: 7,339

NET THERMAL LOAD: 37.3 10⁶BTU/YR

AUXILIARY ENERGY: 2.9 BTU/DD/FT²

YEARLY SOLAR FRACTION: 88%

Open market sale.

CONTEXT
This small, one-story home designed for the cold 7339 degree day climate of the Roaring Fork Valley in central Colorado is an elegant integration of a number of passive design principles. Each of the three bedrooms in this scheme, which is laid out on an east-west axis, is permitted access to south-facing windows and part of a Trombe wall. Less frequently used spaces, such as the mechanical room, bathrooms, and work areas, are relegated to the north part of the plan to act as buffer zones.

While this scheme is oriented 15° west of south, this deviation from the optimum only modestly affects the solar performance of the house. This is yet another reminder that passive solar design is not limited to a small range of solutions and sites.

The simple foundation design of this residence allows a conventional 2 foot on center roof framing system which is broken only to allow natural light and heat into the remote rooms. These same roof monitors provide an outlet·for ventilation purposes in the summer.

A greenhouse/solarium space is the central organizing feature in the design. The kitchen, dining area and living room spaces open off this slightly sunken solarium area. The use of this open plan gives a more spacious feeling to this otherwise modest home. Low water columns help store heat in this portion of the house which is solar heated primarily by large direct gain windows. The excellent view to the south at this site is also captured through these large picture windows. People sitting in the slightly elevated living room area are allowed a view out into the scenery beyond.

This home, designed for the mass market, should appeal to the tastes and pocketbooks of young couples with one or two children. It is a realistic and buildable home which is compatible with other development in the area. It provides further proof that the affordable solar home is available now.

CONSERVATION

To protect against the extreme conditions that prevail in this popular ski resort area, many conservation features are given a top priority. Double glazed window units are employed every where, including the standard glazed units which are part of the Trombe wall construction detail. Two by twelve 2'-0" on center framing in the ceiling allows the use of a full 12" of insulation which provides a ceiling R-factor of 42. An unheated vestibule entrance punctuates the south elevation and cuts down on cold air infiltration rates. Complementing the open south elevation, the north side of the home is earth-bermed, as are parts of the east and west walls.

Exterior foundation insulation surrounds the entire perimeter of the home and insulates the thermal mass within the interior space. Automatically operated insulating curtains close off the Trombe wall areas from the cold glazing surfaces during the nighttime, and retain the heat in the home.

An often overlooked source of energy conservation in this design is the excellent use of passive natural daylight to displace the use of electric lighting fixtures. The roof monitors, which are equipped with moveable insulating panels in the evening, provide light to spaces which are too far from windows to be naturally lit. Again, the small auxiliary energy demand allows the use of baseboard electric as a backup system. Little mechanical space is needed; this allows the developer to construct fewer total square feet.

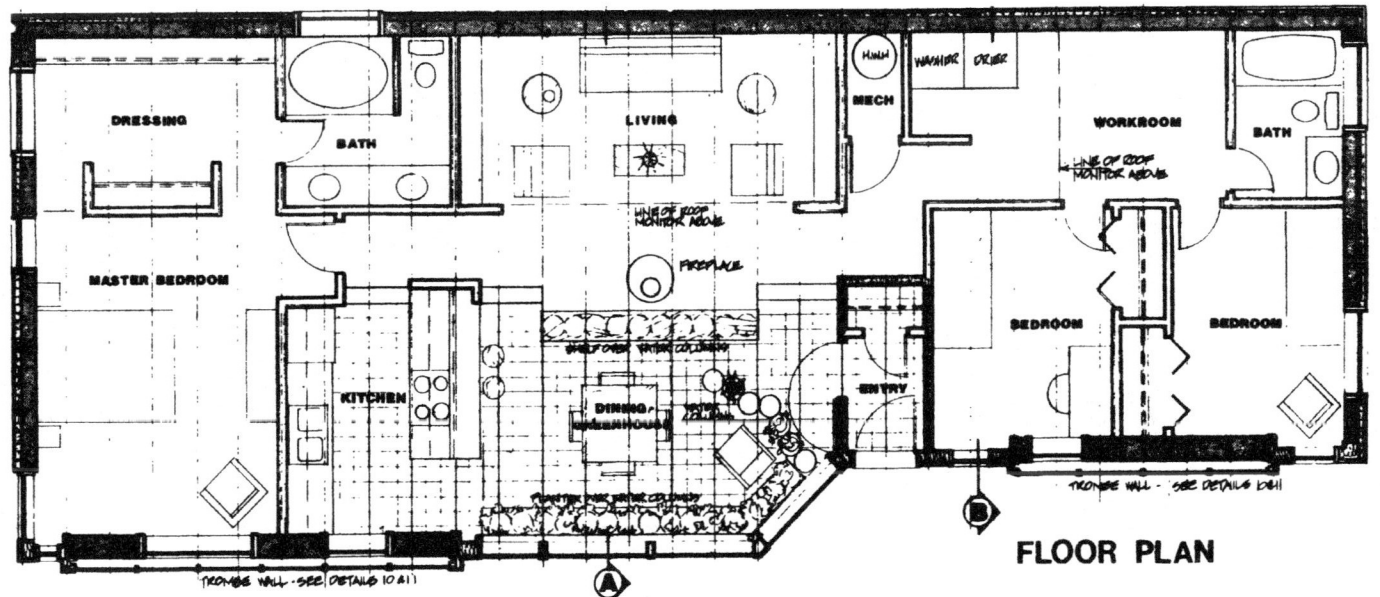

FLOOR PLAN

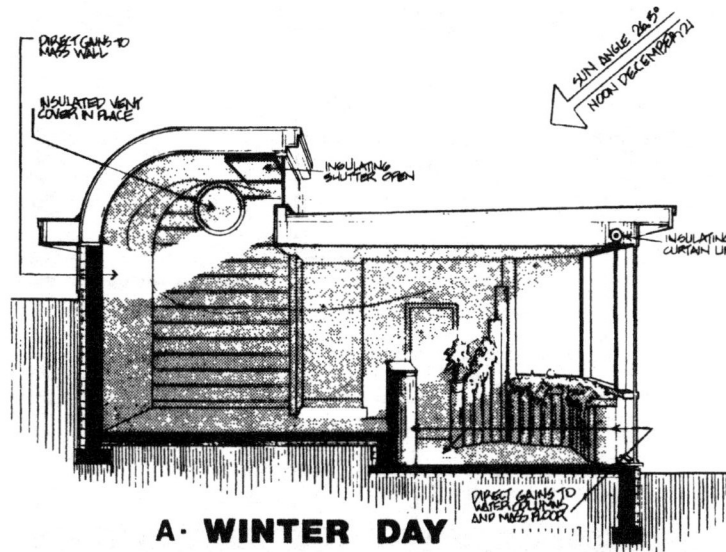

DIRECT GAINS TO MASS WALL

INSULATED VENT COVER IN PLACE

INSULATING SHUTTER OPEN

SUN ANGLE 26.5°
NOON DECEMBER 21

INSULATING CURTAIN UP

DIRECT GAINS TO WATER COLUMN AND MASS FLOOR

A· WINTER DAY

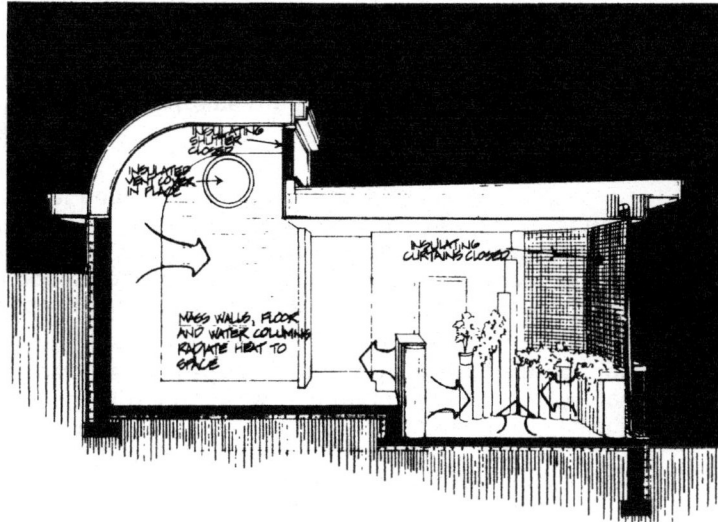

INSULATING SHUTTER CLOSED

INSULATED VENT COVER IN PLACE

INSULATING CURTAINS CLOSED

MASS WALLS, FLOOR AND WATER COLUMNS RADIATE HEAT TO SPACE

A· WINTER NIGHT

HEATING

A total of about 375 square feet of south-facing glazing in both Trombe walls and direct gain windows allows solar energy to penetrate into the house. This glazed area represents over 25% of the total heated floor area of the home. In this very well insulated building, this solar collection area is calculated to provide between 80% and 90% solar heating contribution in the Colorado climate for which this home is designed.

The Trombe wall operates in a conventional fashion with top and bottom vents to allow daytime convective heating. The natural convective air flow rates in this system are only modest. It may be preferable to locate the inlet plenum lower on the wall than indicated to provide a greater height differential between inlet and outlet. The use of nighttime insulating curtain in the Trombe wall eliminates the need for reverse flow backdraft dampers on the vents. The self-inflating insulating curtain in this design is detailed to fit nicely within the roof framing and to be out of the way of incoming radiation during the daytime. Tracks are provided at the Trombe wall jamb to ensure a tight seal which is essential to the performance of any moveable insulating device. Similarly, a weighted pole is used to ensure a tight seal at the bottom of the curtain. If room air is allowed to infiltrate between the insulation and the cold window, the performance of the device will be drastically reduced.

When sunlight is no longer available to drive the convective loop, the wall radiates energy from its interior surface. This radiant distribution warms the occupants, the furniture and the walls of the rooms in contact with the Trombe wall. In this design, as in all Trombe wall designs, it is very important that the primary rooms are allowed access to a segment of the Trombe wall.

In the solarium/direct gain portion of the design, additional thermal storage is provided by water drums and an insulated four-inch-thick concrete slab on grade. This area is insulated from the glass by an inflating curtain at night. As in any direct gain system, heat is distributed by the direct radiation and natural convection of energy off the warmed surfaces. Concern must be shown for the problems of excessive glare and the fading of fabrics; rugs which act as an insulating layer must not cover the thermal mass.

In the north walls, eight-inch structural concrete block provides some thermal storage for light entering through the clerestory window. This mass is protected from outside temperatures by a continuous blanket of exterior insulation. East and west masonry exterior walls provide some additional mass to further moderate interior room air temperatures.

Approximately 32 BTU's of directly illuminated solar storage per degree Fahrenheit is provided on an average for each square foot of glazing. While it has been shown that a computer simulation can predict the daily temperature swings that would be experienced in this house without auxiliary heating, previous experience has shown that the house should probably have swings no greater than 10°-15°F.

One of the strengths in the thermal approach of this design is the appropriate matching of the type of passive solar system and the use of the space which each system heats. Rooms which require more constant temperatures are located behind Trombe walls which will create more constant interior temperature environments.

COOLING

Because the climate of the Roaring Fork Valley is not characterized by high temperatures, the summer cooling load for the project is minimal. In addition, the large amount of thermal mass in the structure and the below-grade construction combine to create an inner environment resistant to sharp exterior temperature swings.

Summer solar direct gains have been minimized with overhangs on a portion of the south windows and on all the skylight monitors. Energy gains optimized in the winter by moveable insulation for Trombe walls can also be negated in the summer with the technique of reversing the insulating modes. (This technique has actually proved to be unnecessary in other homes in the Roaring Fork Valley having similar systems.)

Any daily gains in the heat storage mass of the building can be "dumped" at night by ventilating the building with the Valley's typical cool summer night air. Operable windows and skylight vents are designed to enhance natural and induced ventilation. To encourage cross ventilation, an attempt was made to provide each space with at least two openings, an outlet and an inlet.

The greenhouse space is provided with additional operable awning windows to alleviate any potential overheating problems in that space. In this cool climate, the use of wind ventilation and temperature-induced ventilation during periods of low breezes should keep this home and its inhabitants comfortable throughout the summer.

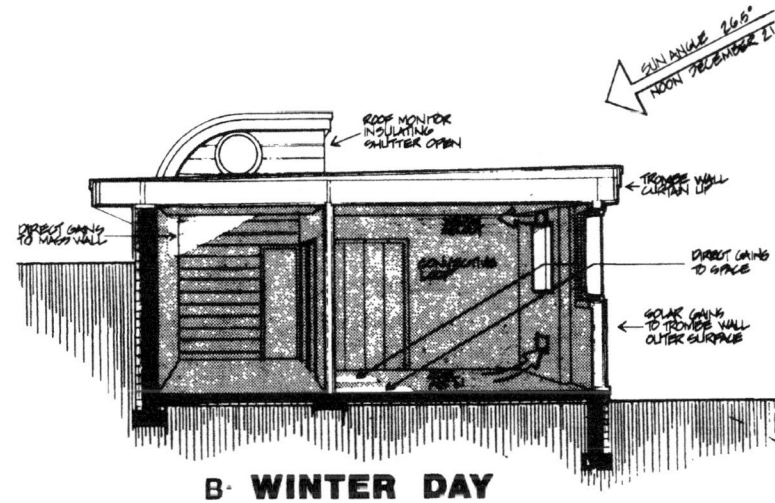

B· **WINTER DAY**

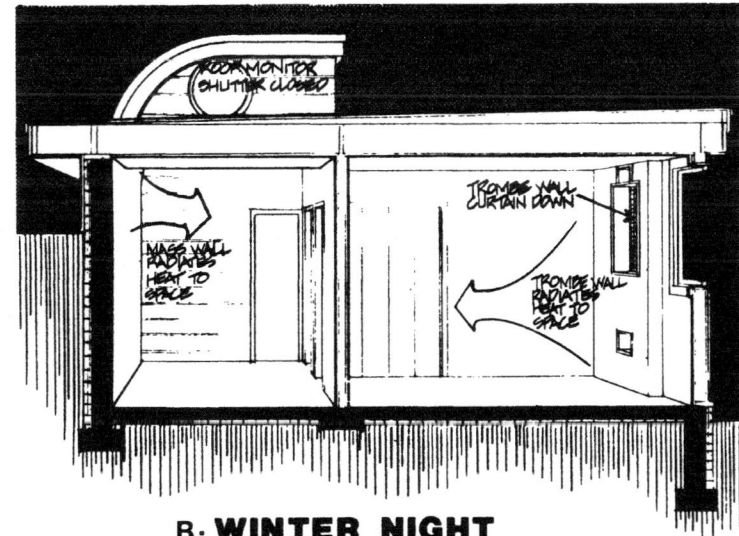

B· **WINTER NIGHT**

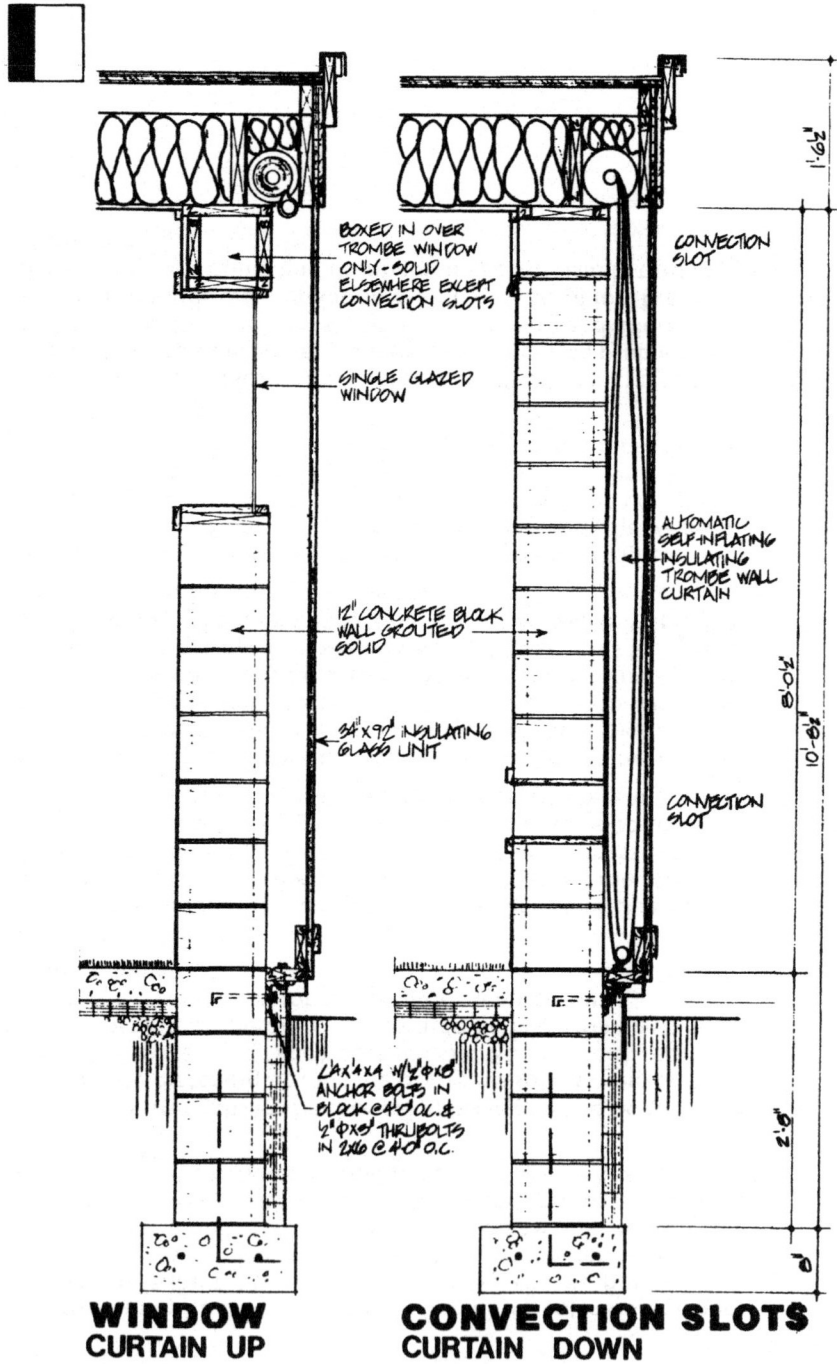

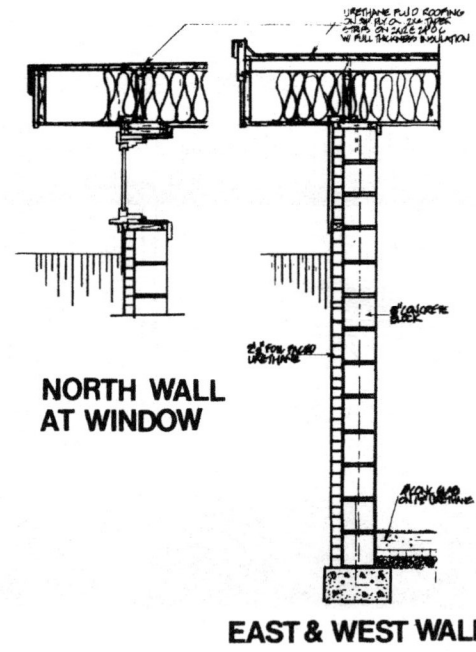

BOXED IN OVER
TROMBE WINDOW
ONLY - SOLID
ELSEWHERE EXCEPT
CONVECTION SLOTS

CONVECTION
SLOT

SINGLE GLAZED
WINDOW

AUTOMATIC
SELF-INFLATING
INSULATING
TROMBE WALL
CURTAIN

12" CONCRETE BLOCK
WALL GROUTED
SOLID

34"x92" INSULATING
GLASS UNIT

CONVECTION
SLOT

L 6"x4"x4 w/ ½"ϕx8"
ANCHOR BOLTS IN
BLOCK @ 4'-0" O.C. &
½"ϕx8" THRU BOLTS
IN 2x16 @ 4'-0" O.C.

WINDOW
CURTAIN UP

CONVECTION SLOTS
CURTAIN DOWN

URETHANE FLUID ROOFING
3" 16 PLY O.C. 2x4 TIMBER
CHIPS ON 2x4 @ 24" O.C.
W/ FULL BATT INSULATION

8" CONCRETE
BLOCK

2 ½" FOIL FACED
URETHANE

2½"x1½2"
CUT FOR URETHANE

**NORTH WALL
AT WINDOW**

EAST & WEST WALL

CONCLUSION

This scheme is representative of a whole range of new residential design solutions which offer their occupants an alternative to conventional housing styles and performance. While the inevitable large areas of south-facing collector glazing in passive design give the residence a modern appearance, it belies the essentially "conservative" nature of this compact and efficient little "machine for living."

It is a home which requires a modest amount of user participation in the operation and maintenance of its passive solar system, but it essentially relies on extensive thermal mass and simple solar collection to provide the bulk of the heating load for the residence. The design extracts some of the new ideas embodied in passive solar design and takes them to their logical conclusion. For instance, once a concrete block wall is dictated by thermal mass requirements, it is put to double use by retaining the earth berming to the north. Once a Trombe wall is dictated, it too is expected to act as structure.

It is a tough, durable home that should appeal to the rugged young homebuyer in today's market.

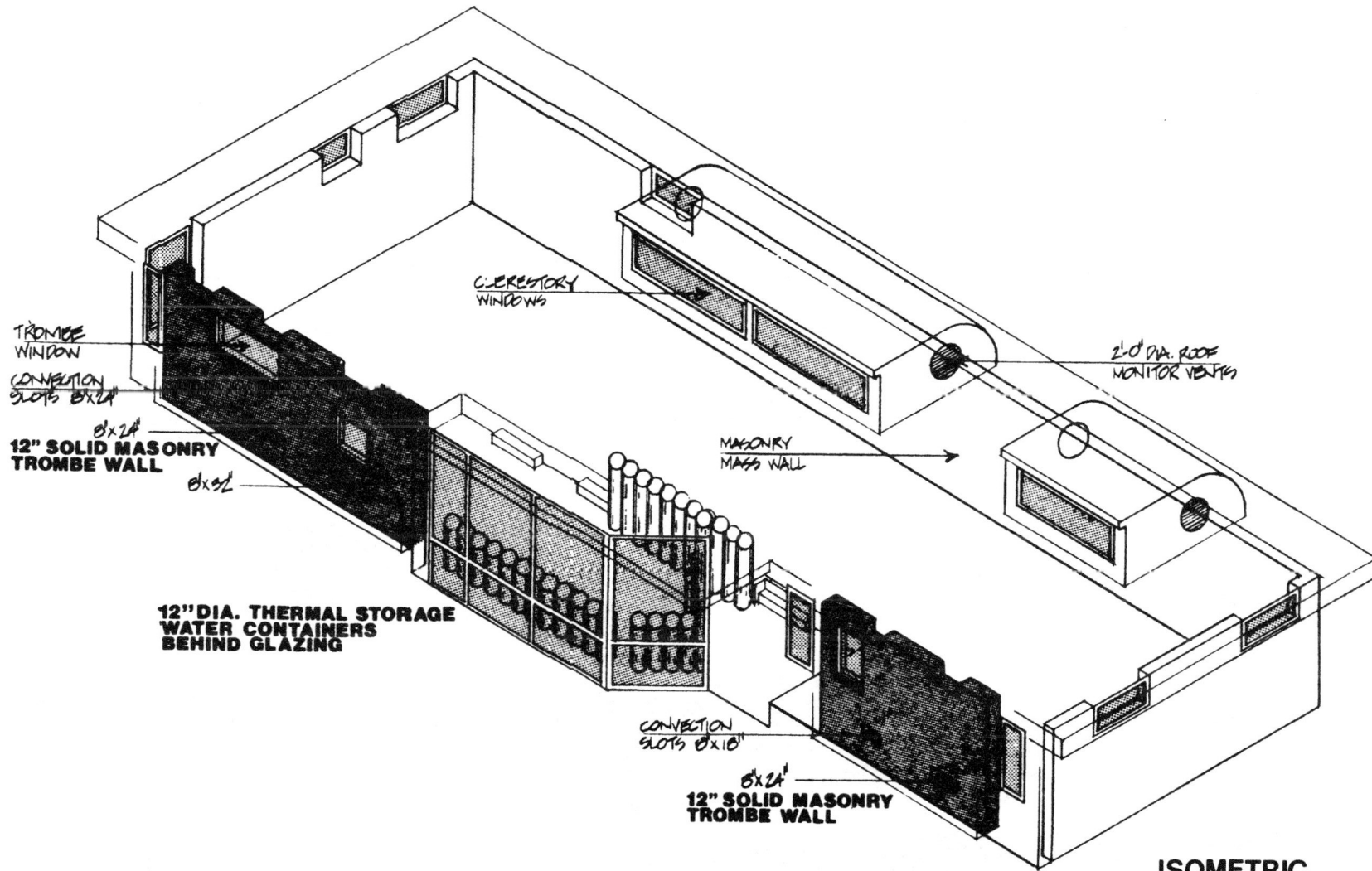

TROMBE
WINDOW

CONVECTION
SLOTS 8"x24"

8"x24"

**12" SOLID MASONRY
TROMBE WALL**

8"x32"

**12"DIA. THERMAL STORAGE
WATER CONTAINERS
BEHIND GLAZING**

CLERESTORY
WINDOWS

MASONRY
MASS WALL

2'-0" DIA. ROOF
MONITOR VENTS

CONVECTION
SLOTS 8"x18"

8"x24"

**12" SOLID MASONRY
TROMBE WALL**

ISOMETRIC

105

Eugene, OR

David Nofferi and Randall Shafer
Eugene, OR

HEATED AREA: 1,492 FT²

NUMBER OF DEGREE DAYS: 4,599

NET THERMAL LOAD: 21.8 10⁶BTU/YR

AUXILIARY ENERGY: 2.87 BTU/DD/FT²

YEARLY SOLAR FRACTION: 87%

CONTEXT

The "Sunstead" home is the result of a design process combining solar energy strategies with continuous market feasibility and site compatibility information.

Located in the Willamette Valley in Eugene, Oregon, this home is subject to a 4,600 degree day heating season extending from late September through early May. Long periods of winter cloudiness intensify the need for a passive system that charges quickly, in harmony with a floor plan and building envelope which take maximum advantage of natural light and energy conservation opportunities.

As a developer's home, the three-bedroom, 1,500 sq. ft. design can adapt to a number of available sites. The presence of a high water table influenced the selection of a slab on grade, wood frame building system which uses locally available materials. The design reflects the shed-roof style of the surrounding wood frame homes and is in keeping with local construction practices.

CONSERVATION

The home's wall construction consists of 2 x 6 studs spaced 24" on center to accommodate R-19 batt insulation. A 6 mil polyethelene

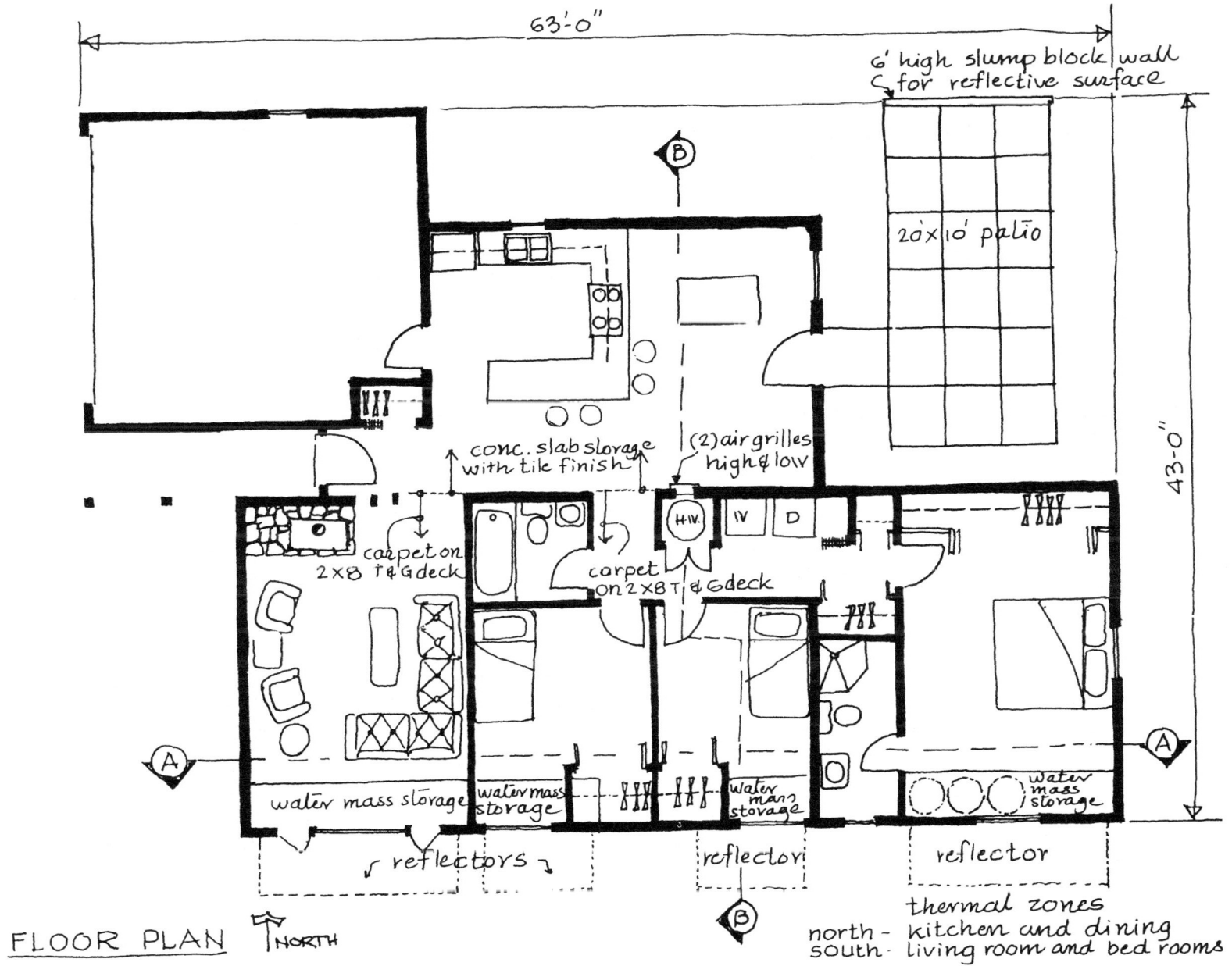

63'-0"

6' high slump block wall
for reflective surface

20'x10' patio

B

conc. slab storage
with tile finish

(2) air grilles
high & low

43'-0"

carpet on
2x8 t&G deck

H.W.   W   D

carpet
on 2x8 T & G deck

A   A

water mass storage   water mass storage   water mass storage   water mass storage

reflectors   reflector   reflector

B

FLOOR PLAN   ↑ NORTH

thermal zones
north - kitchen and dining
south - living room and bed rooms

107

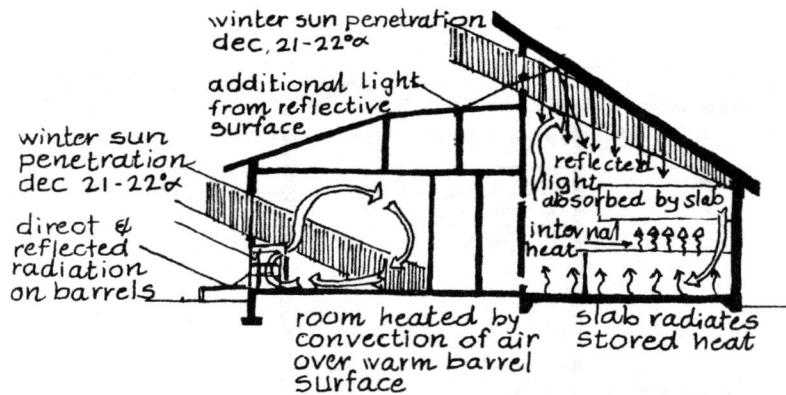

winter sun penetration dec, 21·22°α

additional light from reflective surface

winter sun penetration dec 21·22°α

direct & reflected radiation on barrels

reflected light absorbed by slab

internal heat

room heated by convection of air over warm barrel surface

slab radiates stored heat

WINTER ENERGY FLOW

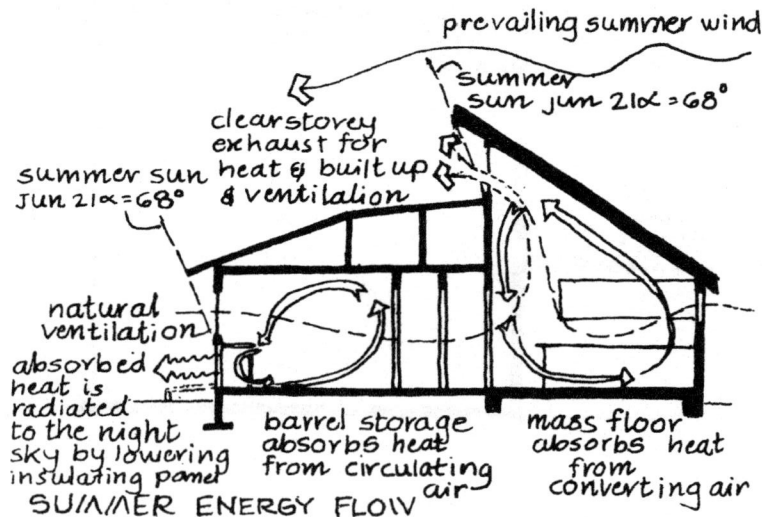

prevailing summer wind

summer sun jun 21α=68°

clearstorey exhaust for heat & built up & ventilation

summer sun jun 21α=68°

natural ventilation

absorbed heat is radiated to the night sky by lowering insulating panel

barrel storage absorbs heat from circulating air

mass floor absorbs heat from convecting air

SUMMER ENERGY FLOW

film provides a continuous vapor barrier and aids in reducing infiltration. One half inch plywood sheathing and lapped cedar siding complete the exterior and ensure an appearance much like that of other homes in the area. The interior finish is 1/2" gypsum board throughout. The roof is of similar construction but uses R-38 batt insulation in the ceiling cavity above the bedroom and living areas, and R-30 batt insulation in the sloped roof structure above the kitchen. All windows used in the "Sunstead" are stock wood-frame insulated glass casements with an R-factor of 1.61. The clerestory windows are double glazed hopper type casements which can be opened for natural ventilation. A hinged moveable insulating panel system of 1" urethane framed in wood and covered with a decorative cloth provides a winter insulating value of R-8 with the double clerestory glazing. A draw cord has been included to manually control the insulating panels, allowing natural light during the daytime while preventing heat loss at night.

Other energy conservation features include wind-sheltered entrances and deep overhangs for summer shading of the south-facing windows. The garage may be attached in numerous ways to provide wind protection as dictated by a particular site and micro-climate condition. The design also includes three flat plate collector panels for solar domestic hot water heating.

HEATING
The layout of the home combines a functional arrangement of the living spaces with a logical introduction of two different passive heating concepts—direct solar gain and indirect solar gain.

The kitchen and dining area constitutes the north thermal zone. By locating this space to the north, the design takes advantage of the high internal heat gains generated in this area from cooking and mealtime congregations of people to offset the proportionately higher heat losses usually associated with north-facing rooms. Solar energy which is introduced directly to this zone through the clerestory reflects off the sloped ceiling providing natural light and heat. The floor, which is a 4" concrete slab covered with half thickness brick (1"), serves as storage for this heat.

The south zone is composed of the bedrooms, baths and living area, which form a rectangle oriented east-west. The indirect passive heating system, consisting of eighteen 50-gallon water drums behind standard double insulated storm door glazing, is located in this zone. The number of drums is distributed according to room size on the floor along the south wall. The reflective insulating panels outside are manually opened from within the rooms (via ratcheted cranks) to expose and reflect solar energy onto the barrels and manually closed to trap the captured heat at night. Room air convects over the solar heated barrels, entering the cabinet-like enclosure through a continuous slot at floor level, and exiting through dampered diffusers near the top of the cabinet.

Additional heat is provided through the use of a wood stove in the living area and a radiant electric heating system in the ceilings of the home.

COOLING

During the summer, the winter operating sequence of the insulating panels is reserved. During the day the panels are closed so that no solar energy can strike the barrels. The barrels remain at room temperature and are free to absorb excess heat from the interior. At night the panels are dropped, allowing the barrels to radiate their heat to the night air. Nighttime ventilation through the home further cools the barrels. Ideally, the barrels are cooled sufficiently during the evening to effectively delay or eliminate heat buildup in the interior on the following day.

CONCLUSION

The advantages of this home arrangement and passive system design are numerous. The use of water as an indirect gain storage medium provides the response necessary in the cloudy Eugene environment. The water barrels supply the maximum amount of storage possible while still freeing a substantial amount of south-facing wall area for windows. The mixture of direct and indirect gain passive systems provides some degree of daily control and an equitable distribution of stored energy during the evening hours. The use of water storage cabinets can be easily applied to alternate plans, and maintenance and freeze protection can be easily handled.

In order for this design to be completely successful owner participation is necessary. Inside, the moveable panels must be opened and closed daily in response to weather conditions and seasonal space conditioning needs. Outside, some attention will be necessary to free the panels after heavy snows. Because of the flexible nature of the actuating cable, the panels will be subject to movement and racking in high winds. Their accessibility could also make the system susceptible to vandalism.

The home as designed is a simple scheme showing highly marketable features and easy repeatability. The combination of the two distinct passive systems is effective and controllable. Together they logically service the home's north and south thermal zones and provide the multiple benefits of passive heating, nighttime cooling, natural ventilation and effective daylighting. The passive concepts which are implemented can be readily adapted to any number of house plans without sacrificing any of the attributes inherent in the "Sunstead".

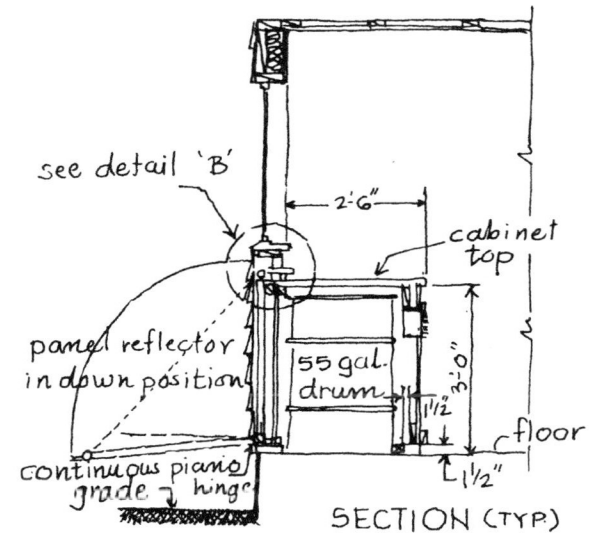

SECTION (TYP.)

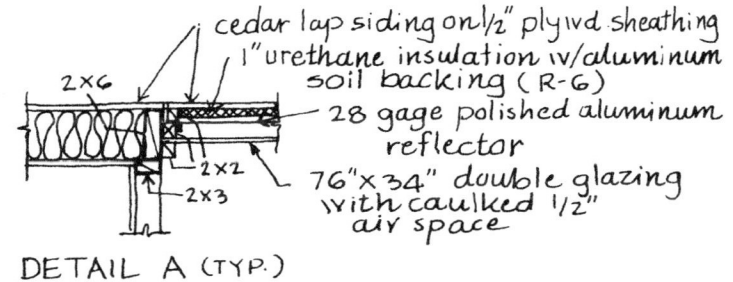

DETAIL A (TYP.)

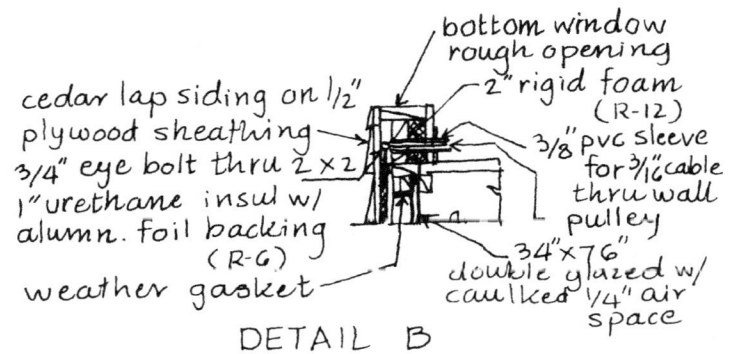

DETAIL B

109

INDIRECT SOLAR GAIN

The following project pages are devoted to a brief description of the other INDIRECT SOLAR GAIN homes which were selected for awards. Each project is shown in perspective and accompanied by either a plan or a section. The projected information extracted from the grant application is as follows:

Location of the home

Designer's name and firm
City and state of the designer*

HEATED AREA in square feet

NUMBER OF HEATING DEGREE DAYS

NET THERMAL LOAD in millions of British Thermal Units per year

AUXILIARY heating load in British Thermal Units per degree day per square foot

YEARLY SOLAR FRACTION: the percentage of heating energy provided by solar

COLLECTOR: Description and number of square feet

STORAGE: Description and Capacity in British Thermal Units per degree Fahrenheit

CONTROLS: Description

BACKUP: Type and capacity in British Thermal Units per hour

*Space limitations did not permit printing complete address. If you would like to contact the designer or builder concerning any of these projects, simply contact the National Solar Heating and Cooling Information Center (PROFESSIONALS FILE) by calling 800-523-2929 or 800-462-4983 (if calling from Pennsylvania), or by writing P.O. Box 1607, Rockville, MD 20850.

This new 1-story, 3-bedroom detached home has a stucco finish and is priced in the $50,000 range. There is no existing vegetation on this flat lot.

Open market sale.

Tucson, AZ

Jack Cohen, Goldblatt, Cohen and Aros
Tucson, AZ

HEATED AREA: 1,595 FT²

NUMBER OF DEGREE DAYS: 1,855

NET THERMAL LOAD: 7.42 10⁶BTU/YR

AUXILIARY ENERGY: 3.92 BTU/DD/FT²

YEARLY SOLAR FRACTION: 94%

COLLECTOR: Venetian blinds over south-facing window Area: 140 FT²

STORAGE: Concrete, sand filled, masonry walls Capacity: 26,987 BTU/°F

CONTROLS: Operable blinds, fan speed controls

BACKUP: 20,300 BTUH electric baseboard; 250W heat lamps; fireplace

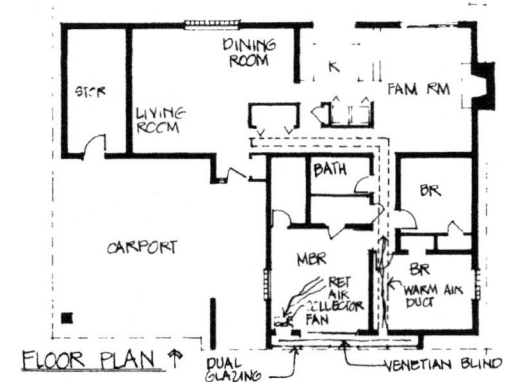

FLOOR PLAN

Devonshire, CA

Dick Munday
Tahoe City, CA

HEATED AREA: 1,790 FT²

NUMBER OF DEGREE DAYS: 8,208

NET THERMAL LOAD: 100 10⁶BTU/YR

AUXILIARY ENERGY: 6.15 BTU/DD/FT²

YEARLY SOLAR FRACTION: 34%

COLLECTOR: South-facing windows
Area: 124 FT²

STORAGE: Solar mass wall
Capacity: 3,089 BTU/°F

CONTROLS: Insulating panels, operable
windows

This new 2-story, 3-bedroom detached house on its relatively flat site is in the $70,000 price range. When the light vegetation around the house matures, it will provide winter protection.

Private client.

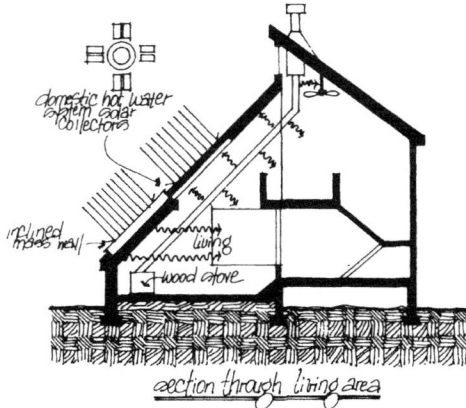

section through living area

Fair Oaks, CA

Tom Carver, Sierra Engineering
Lodi, CA

HEATED AREA: 2,300 FT²

NUMBER OF DEGREE DAYS: 2,374

NET THERMAL LOAD: 26.5 10⁶BTU/YR

AUXILIARY ENERGY: 3.28 BTU/DD/FT²

YEARLY SOLAR FRACTION: 88%

COLLECTOR: South-facing windows and
storage mass wall Area: 581 FT²

STORAGE: Pre-cast concrete floor and walls
Capacity: 10,173 BTU/°F

CONTROLS: Operable awning, southside
overhang, and insulation shutters;
ridge roof vents and soffit vents

BACKUP: 80,000 BTUH natural gas furnace

This 4-bedroom ranch-style home has a decorative plywood exterior with a rough lumber trim and a patio overlooking a wooded canyon. The garage was placed on the west not only to shade the house from the summer sun, but also to provide a sound barrier from a major roadway. The price is in the $125,000 range.

Open market sale.

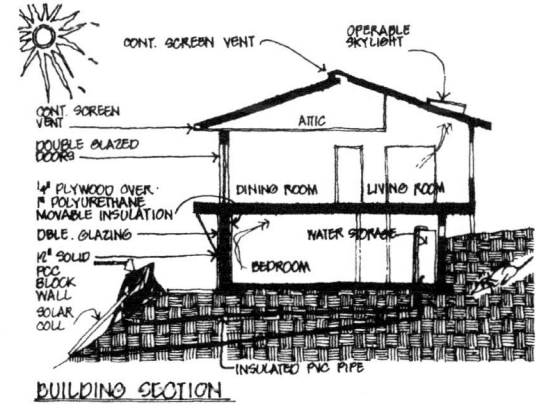

BUILDING SECTION

This 2-story, 3-bedroom wood frame house on its relatively flat lot is priced in the $80,000 range. It has wood siding and white reflective stones in front of the solar mass wall.

Open market sale.

Boulder, CO

Paul Shipee, Colorado Sunworks
Boulder, CO

HEATED AREA: 1,528 FT²

NUMBER OF DEGREE DAYS: 5,540

NET THERMAL LOAD: 38.8 10^6BTU/YR

AUXILIARY ENERGY: 2.75 BTU/DD/FT²

YEARLY SOLAR FRACTION: 92%

COLLECTOR: South-facing windows, solar mass wall Area: 467 FT²

STORAGE: Solar mass wall, concrete floor Capacity: 13,704 BTU/°F

CONTROLS: Insulated curtains, solar mass wall vents, operable windows and doors

BACKUP: 35,000 BTUH gas hot water baseboard; wood stove

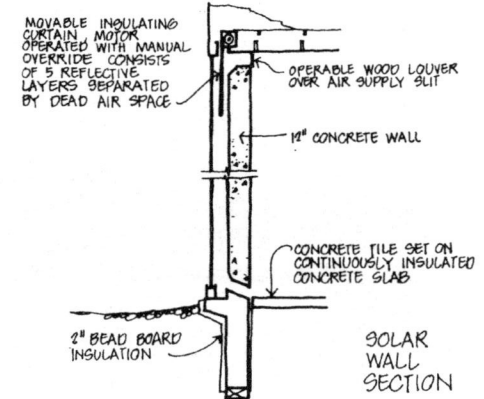

MOVABLE INSULATING CURTAIN MOTOR OPERATED WITH MANUAL OVERRIDE CONSISTS OF 5 REFLECTIVE LAYERS SEPARATED BY DEAD AIR SPACE

OPERABLE WOOD LOUVER OVER AIR SUPPLY SLIT

12" CONCRETE WALL

CONCRETE TILE SET ON CONTINUOUSLY INSULATED CONCRETE SLAB

2" BEAD BOARD INSULATION

SOLAR WALL SECTION

This new, 2-story, 3-bedroom house has natural cedar wood siding and is priced in the $92,500 range. The house is slightly set into the site with berming on the north side, and harsh western winds are also buffered by the garage. Deciduous trees on the southeast and southwest sides shade the house in the summer.

Open market sale.

Boulder, CO

Deidre McCrystal, McCrystal Design
Denver, CO

HEATED AREA: 1,716 FT²

NUMBER OF DEGREE DAYS: 6,051

NET THERMAL LOAD: 123 10^6BTU/YR

AUXILIARY ENERGY: 4.86 BTU/DD/FT²

YEARLY SOLAR FRACTION: 72%

COLLECTOR: South-facing greenhouse, heat storage mass wall, rock storage Area: 643 FT²

STORAGE: Concrete wall and floors, rockbed Capacity: 28,707 BTU/°F

CONTROLS: Operable skylights, vents, solarium fan, louvers

BACKUP: 50,000 BTUH gas furnace; 20,000 BTUH gas heater; 35,000-50,000 BTUH wood stove

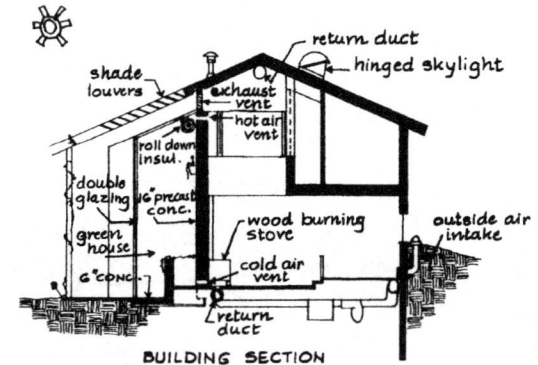

return duct
hinged skylight
shade louvers
exhaust vent
hot air vent
roll down insul.
double glazing
16" precast conc.
green house
wood burning stove
cold air vent
outside air intake
6" conc.
return duct

BUILDING SECTION

This existing 1-story, 3-bedroom detached house has been retrofitted with a solarium and solar mass wall to collect solar energy. The flat site features mature spruce trees and hedges to provide wind protection.

Retrofit.

Boulder, CO

Doug Graybeal, J. Welch and D. Graybeal
Boulder, CO and Aspen, CO

HEATED AREA: 2,210 FT²

NUMBER OF DEGREE DAYS: 5,275

NET THERMAL LOAD: 113 10⁶BTU/YR

AUXILIARY ENERGY: 7.8 BTU/DD/FT²

YEARLY SOLAR FRACTION: 80%

COLLECTOR: Solar mass wall, solarium
Area: 238 FT²

STORAGE: Solar mass wall, flagstone, solarium
floor Capacity: 6,458 BTU/°F

CONTROLS: Fans, vents, operable solarium
door

BACKUP: Gas furnace

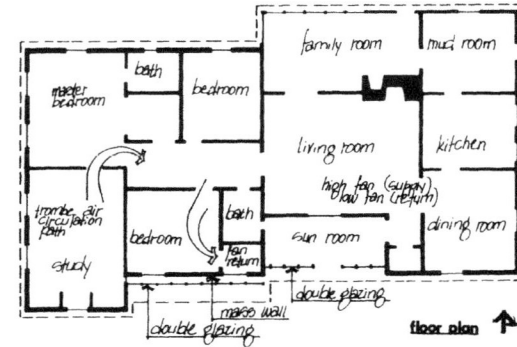

Boulder, CO

Craig Christensen, Rohde & Christensen
Boulder, CO

HEATED AREA: 1,400 FT²

NUMBER OF DEGREE DAYS: 5,368

NET THERMAL LOAD: 52.2 10⁶BTU/YR

AUXILIARY ENERGY: 4.80 BTU/DD/FT²

YEARLY SOLAR FRACTION: 61.4%

COLLECTOR: South-facing glass doors and
windows, solar mass wall
Area: 310 FT²

STORAGE: Concrete floor and solar mass wall
Capacity: 5,740 BTU/°F

CONTROLS: Automatic motorized insulating
curtain, operable air registers

BACKUP: 50,000 BTUH electric baseboard
heater

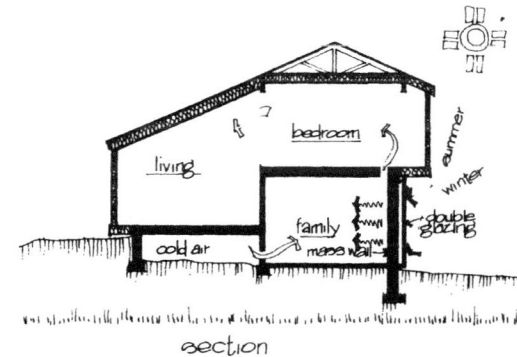

This new 2-story, 3-bedroom house is priced in the $60,000 range. Earth berming on the east, north, and west sides of the house insulate it, while a main entrance on the east side reduces winter wind infiltration.

Open market sale.

This new 1-story, 3-bedroom detached house is priced in the $75,000 range. Trees on this flat site block the summer afternoon sun. Winter winds are buffered by houses, fences, and high-density vegetation to the north.

Open market sale.

Carbondale, CO

Doug Davis, Sunshine Design
Carbondale, CO

HEATED AREA: 1,420 FT²

NUMBER OF DEGREE DAYS: 7,340

NET THERMAL LOAD: 48.1 10⁶BTU/YR

AUXILIARY ENERGY: 3.31 BTU/DD/FT²

YEARLY SOLAR FRACTION: 69%

COLLECTOR: Clerestory and skylights, and solarium glazing
Area: 336.25 FT²

STORAGE: Masonry mass walls, brick-earth mass planter
Capacity: 8,425.6 BTU/°F

CONTROLS: Automatic insulating curtains, manual insulating panels, operable windows

BACKUP: 39,660 BTUH electric baseboard; 20,000 BTUH woodburning fireplace

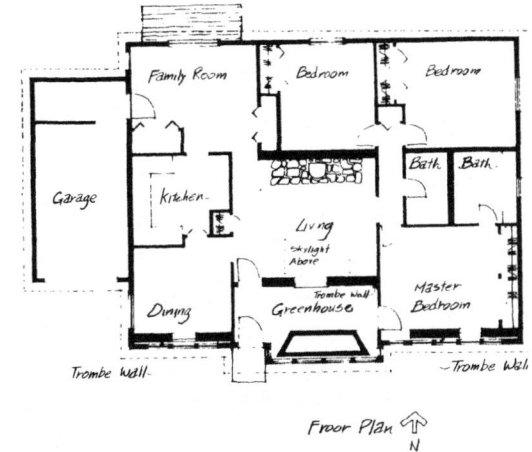

Froor Plan N

This new, 2-story residence is one of eight that will be built on a 2-acre subdivision designed primarily for moderate income families. The garage is located on the north side of the house and acts as a buffer against cold northwest winter winds. The house is in the $41,500 price range.

Open market sale.

Loveland, CO

Darrel Smith
Loveland, CO

HEATED AREA: 1,146 FT²

NUMBER OF DEGREE DAYS: 6,202

NET THERMAL LOAD: 74.1 10⁶BTU/YR

AUXILIARY ENERGY: 5.58 BTU/DD/FT²

YEARLY SOLAR FRACTION: 61%

COLLECTOR: South-facing windows and heat storage mass wall Area: 509 FT²

STORAGE: Concrete storage mass wall, concrete floor
Capacity: 21,061 BTU/°F

CONTROLS: Operable dampers, shades, and windows

BACKUP: 50,000 BTUH wood stove; 46,000 BTUH electric furnace

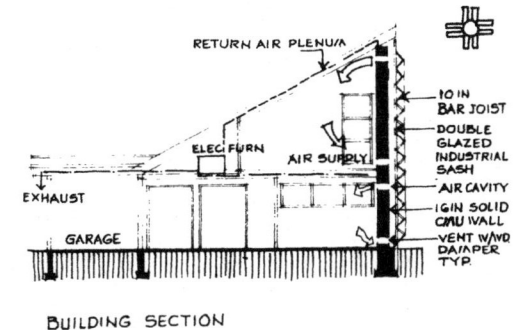

BUILDING SECTION

Located in the Rocky Mountain foothills, this 2-story, 3-bedroom wood frame house is priced in the $85,000 range. Pines to the west form a natural windbreak.

Open market sale.

Lyons, CO

Bob King, Allen-King Builders
Boulder, CO

HEATED AREA: 1,669 FT²

NUMBER OF DEGREE DAYS: 5,446

NET THERMAL LOAD: 42.4 10⁶BTU/YR

AUXILIARY ENERGY: 2.75 BTU/DD/FT²

YEARLY SOLAR FRACTION: 81%

COLLECTOR: Solar mass wall, south-facing glass
Area: 454 FT²

STORAGE: Concrete, brick and gravel floors
and walls Capacity: 16,959 BTU/°F

CONTROLS: Operable shades, thermal chimney
outlets and window insulation

BACKUP: 30,716 BTUH electric resistance
baseboard; wood stove/fireplace

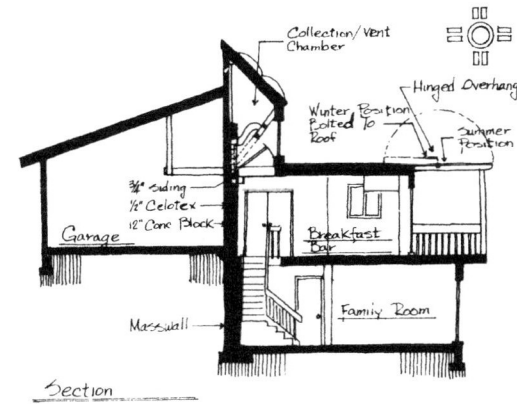

This new raised ranch, 3-bedroom house is in the $85,000 price range. In addition to an air-lock entrance on the south side of the house, earth berming and a northern wooded area reduce winter wind infiltration.

Open market sale.

Shelton, CT

Carl Mezoff, Wormser Scientific Corporation
Stamford, CT

HEATED AREA: 1,966 FT²

NUMBER OF DEGREE DAYS: 5,102

NET THERMAL LOAD: 37.9 10⁶BTU/YR

AUXILIARY ENERGY: 2.02 BTU/DD/FT²

YEARLY SOLAR FRACTION: 85%

COLLECTOR: Southeast-facing solarium,
clerestory windows Area: 504 FT²

STORAGE: Water drum, water filled glass
block, phase change tile, concrete
wall, brick Capacity: 58,956 BTU/°F

CONTROLS: Manually open and close windows

BACKUP: 24,000 BTUH electric heat pump;
20,000 BTUH wood stove; 15,000
BTUH wood fireplace

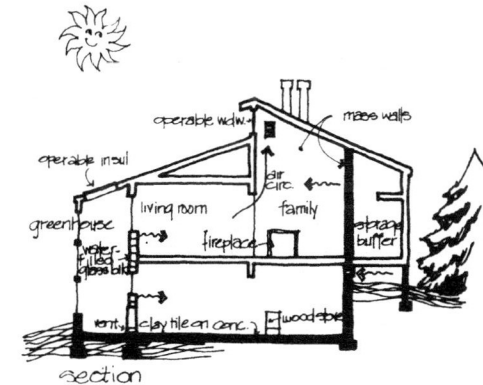

INDIRECT SOLAR GAIN

This new 1-story, 3-bedroom detached home with brick veneer is in the $45,000 range. Evergreens on the northwest on this flat site will protect the home from winter winds. Trees and vines on east and west will provide summer shading.

Open market sale.

Shenandoah, GA

Preston Stevens, Jr., Stevens and Wilkinson
Atlanta, GA

HEATED AREA: 1,184 FT²

NUMBER OF DEGREE DAYS: 2,800

NET THERMAL LOAD: 13.6 10^6BTU/YR

AUXILIARY ENERGY: 3.27 BTU/DD/FT²

YEARLY SOLAR FRACTION: 61%

COLLECTOR: Double-glazed solar mass wall with reflector Area: 132 FT²

STORAGE: Concrete wall with steel fiber Capacity: 3,649 BTU/°F

CONTROLS: Thermal sensor controlled fans and manual dampers

BACKUP: 25,000 BTUH gas furnace

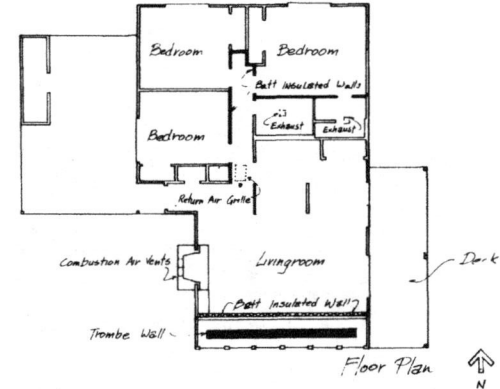

Floor Plan

This existing 2-story, 4-bedroom attached house has been retrofitted with a solarium to collect the sun's heat. Trees shade the west side of the house, protecting it from overheating during the summer.

Retrofit.

Bonner Springs, KS

C. Eugene Moeller
Bonner Springs, KS

HEATED AREA: 2,826 FT²

NUMBER OF DEGREE DAYS: 4,711

NET THERMAL LOAD: 71.3 10^6BTU/YR

AUXILIARY ENERGY: 3.5 BTU/DD/FT²

YEARLY SOLAR FRACTION: 43%

COLLECTOR: Triple glazed, south-facing windows Area: 322 FT²

STORAGE: Water drums Capacity: 13,838 BTU/°F

CONTROLS: Vents, fan

BACKUP: 84,000 BTUH propane forced air furnace; fireplaces

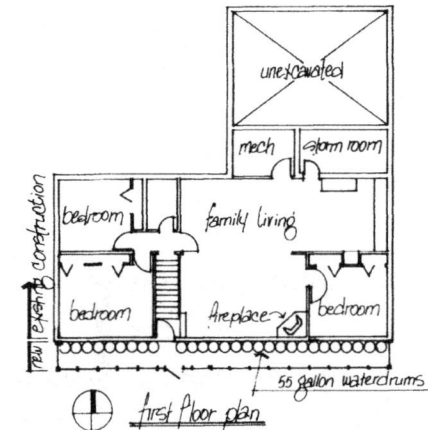

first floor plan

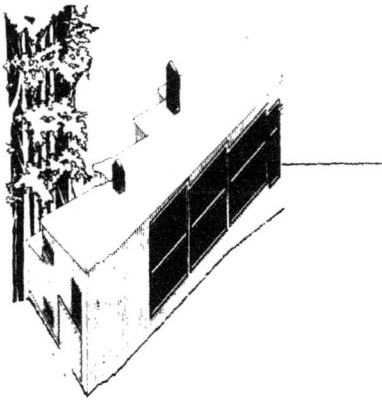

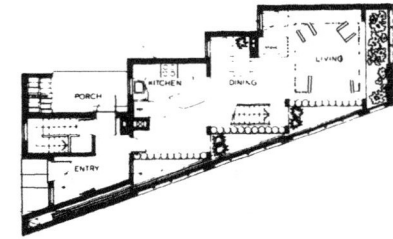

Groton, MA

Gifford Pierce, Beckman, Blydenburgh and Associates
Groton, MA

HEATED AREA: 1,820 FT²

NUMBER OF DEGREE DAYS: 6,424

NET THERMAL LOAD: 76.2 10⁶BTU/YR

AUXILIARY ENERGY: 3.77 BTU/DD/FT²

YEARLY SOLAR FRACTION: 76%

COLLECTOR: South-facing windows and water storage wall Area: 569 FT²

STORAGE: Water storage mass wall Capacity: 23,267 BTU/°F

CONTROLS: Operable vents and windows; moveable insulation

BACKUP: 40,000 BTUH wood stoves; 25,000 BTUH electric heater

FIRST FLOOR

This new 2-story house is priced in the $48,000 range. Winter wind infiltration is reduced by a recessed main entrance, and cold breezes are deflected by trees to the north and west.

Private client.

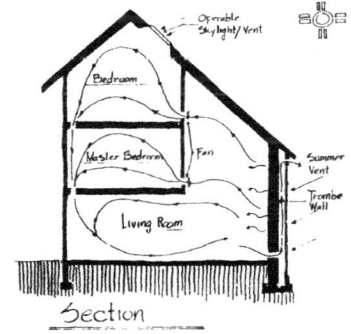

Georgetown, ME

Richard Zamore, Suntech Homes, Inc.
Brunswick, ME

HEATED AREA: 1,976 FT²

NUMBER OF DEGREE DAYS: 7,246

NET THERMAL LOAD: 53.2 10⁶BTU/YR

AUXILIARY ENERGY: 2.9 BTU/DD/FT²

YEARLY SOLAR FRACTION: 75%

COLLECTOR: Heat storage mass wall Area: 405 FT²

STORAGE: Concrete wall Capacity: 16,420 BTU/°F

CONTROLS: Operable insulating curtain, skylight insulation, and vents

BACKUP: 40,000 BTUH wood stove; 30,000 BTUH electric baseboard heating

This new 3-story, 3-bedroom house is designed to conform to the traditional New England saltbox architectural style. Priced in the $74,500 range, the house is protected from winter winds by tall pine trees to the north and shaded during the summer by maple trees to the south. The heat storage mass wall is located in a spot where it would not be the focal point of the design.

Open market sale.

Grand Rapids, MI

Thomas and Jill Newhouse
Grandville, MI

HEATED AREA: 1,185 FT²

NUMBER OF DEGREE DAYS: 7,279

NET THERMAL LOAD: 41.6 10⁶BTU/YR

AUXILIARY ENERGY: 2.08 BTU/DD/FT²

YEARLY SOLAR FRACTION: 69%

COLLECTOR: Solar mass wall, south-facing
windows Area: 396 FT²

STORAGE: Solar mass wall
Capacity: 6,392 BTU/°F

CONTROLS: Thermosyphon doors, registers,
thermal shutters, operable
windows

BACKUP: 72,000 BTUH wood stoves; 31,500
BTUH electric baseboard

This 2-bedroom, 1-story house of concrete
and cedar is in the $60,000 range. This low-
profile structure with an earth-covered roof is
built into a natural ridge. Large trees for
summer shade are located on the south and
west sides of this rural site.

Private client.

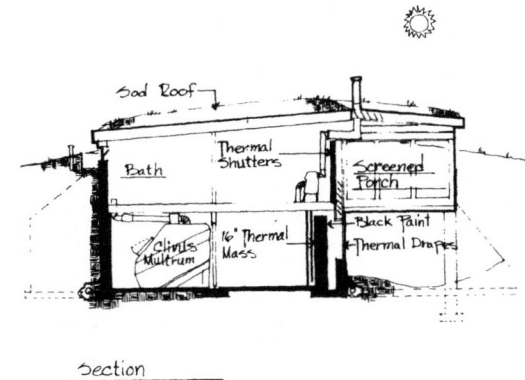

Section

Three Rivers, MI

B. Monroe/R. Pryor
Jones, MI

HEATED AREA: 1,770 FT²

NUMBER OF DEGREE DAYS: 6,782

NET THERMAL LOAD: 89.1 10⁶BTU/YR

AUXILIARY ENERGY: 6.77 BTU/DD/FT²

YEARLY SOLAR FRACTION: 48%

COLLECTOR: South-facing windows, solarium
Area: 596 FT²

STORAGE: Brick floor, concrete floor, gravel
floor, chimney
Capacity: 24,130 BTU/°F

CONTROLS: Operable shutters, shades, chimney
dampers, vents, automatic fan and
skylights

BACKUP: 60,000 BTUH furnace;
wood-burning heater

This 2-story house is designed for a family of
four to five people and is priced in the
$75,000 range. Dense woods and earth berm-
ing on all sides of the house minimize winter
wind infiltration.

Open market sale,

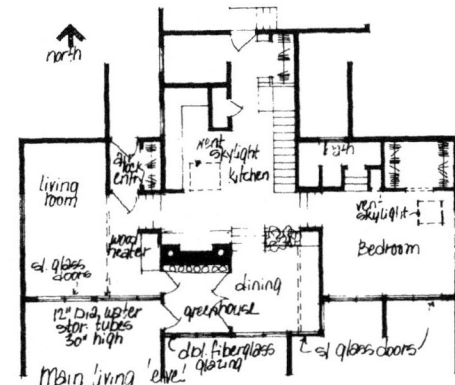

This 2-story, 4-bedroom wood frame house is located on a wooded 3-acre tract. It is priced in the $65,000 range.

Private client.

Columbia, MO

Jeffrey Barger, Sol-Terra Design
Columbia, MO

HEATED AREA: 2,136 FT²

NUMBER OF DEGREE DAYS: 5,046

NET THERMAL LOAD: 57.4 10^6BTU/YR

AUXILIARY ENERGY: 2.5 BTU/DD/FT²

YEARLY SOLAR FRACTION: 55%

COLLECTOR: South-facing windows, solar mass wall Area: 397 FT²

STORAGE: Concrete wall and floor Capacity: 9,692 BTU/°F

CONTROLS: Automatic fan

BACKUP: 30,000 BTUH electric heat pump; 100,000 BTUH wood stoves

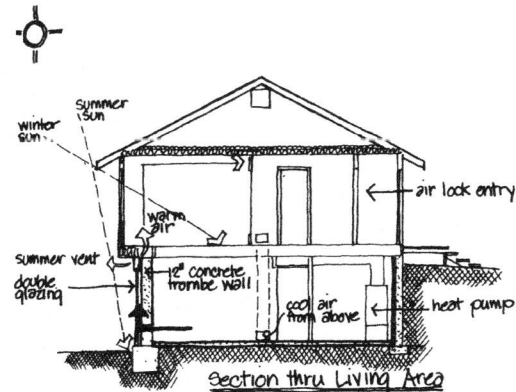

Section thru Living Area

This is a 1-story, 3-bedroom wood frame house in the $45,000 range. Located on level terrain, the home is shaded in summer by trees on its south side.

Open market sale.

Columbia, MO

Nicholas Peckham and Bradley Wright
Columbia, MO

HEATED AREA: 1,024 FT²

NUMBER OF DEGREE DAYS: 5,083

NET THERMAL LOAD: 18.6 10^6BTU/YR

AUXILIARY ENERGY: 4.12 BTU/DD/FT²

YEARLY SOLAR FRACTION: 86%

COLLECTOR: Solar mass wall, south-facing windows Area: 333 FT²

STORAGE: Water solar mass wall, concrete floor Capacity: 16,187 BTU/°F

CONTROLS: Operable windows, insulated drapes, and vent fans

BACKUP: 20,000 BTUH electric furnace

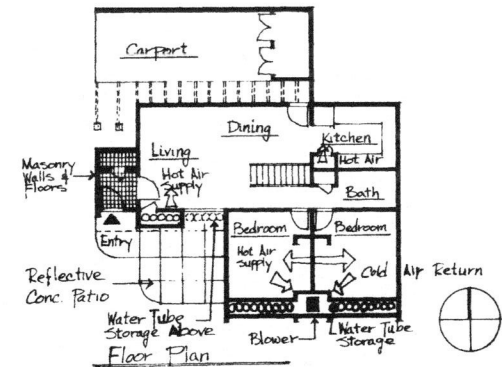

Floor Plan

119

This 3-bedroom underground structure with walkout basement is priced in the $85,000 range. Located on the south slope of a 7-acre, heavily wooded area, English ivy on the roof and earth berms shade the roof in summer and form an insulating layer of air in winter.

Private client.

Glencoe, MO

Terry A. Hoffman, Hunter Hunter Associates-Architects
St. Louis, MO

HEATED AREA: 3,200 FT²

NUMBER OF DEGREE DAYS: 4,705

NET THERMAL LOAD: 40.4 10⁶BTU/YR

AUXILIARY ENERGY: 1.84 BTU/DD/FT²

YEARLY SOLAR FRACTION: 87%

COLLECTOR: South-facing windows, solar mass wall Area: 910 FT²

STORAGE: Concrete and brick walls, concrete roof, concrete floor Capacity: 22,101 BTU/°F

CONTROLS: Dampers, moveable insulation, exhaust fan, vents

BACKUP: 75,000 and 50,000 BTUH fireplaces

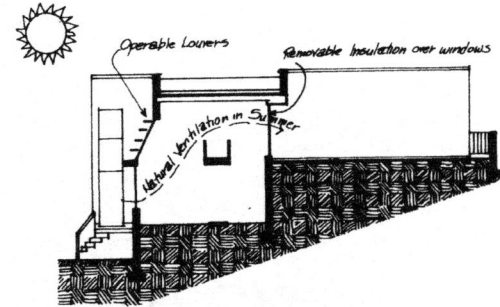

Section Through Greenhouse

This ranch-style, 3-bedroom house is in the $18,000 price range. This retrofit design proposal involves one in a group of 30 moderate-income tract houses; the design could be applied directly to half of the tract houses, and to the others with slight modifications.

Retrofit.

Joplin, MO

G. Herbert Gill
Appleton City, MO

HEATED AREA: 810 FT²

NUMBER OF DEGREE DAYS: 4,088

NET THERMAL LOAD: 37.8 10⁶BTU/YR

AUXILIARY ENERGY: 8.26 BTU/DD/FT²

YEARLY SOLAR FRACTION: 44%

COLLECTOR: South-facing greenhouse, water wall Area: 253 FT²

STORAGE: Water storage mass wall (320 2-gallon jugs; 500 1-gallon cans), greenhouse floor Capacity: 6,806 BTU/°F

CONTROLS: Window fans

BACKUP: 75,000 BTUH gas forced-air furnace

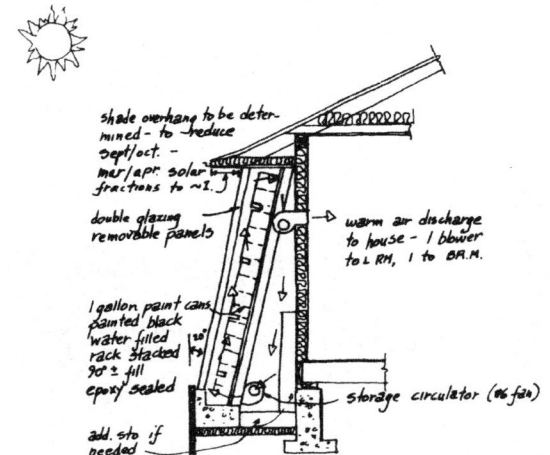

Section through Greenhouse addition

This 2-story house with masonry walls has 3 bedrooms plus 2 offices. Situated on a south-facing slope, it is surrounded by pastures on three sides, and a dense treeline to the north forms a natural windbreak. The home is in the $65,000 range.

Private client.

Chester, NH

Charles Pearson, B. V. Pearson Associates
Derry, NH

HEATED AREA: 2,196 FT²

NUMBER OF DEGREE DAYS: 7,246

NET THERMAL LOAD: 45.7 10⁶BTU/YR

AUXILIARY ENERGY: 3.26 BTU/DD/FT²

YEARLY SOLAR FRACTION: 82%

COLLECTOR: Solar mass wall, clerestory
                    Area: 576 FT²

STORAGE: Sand or concrete mass wall,
                    concrete floor slab
                    Capacity: 23,624 BTU/°F

CONTROLS: Fan, insulating shutters, vents

BACKUP: 60,000 BTUH and 50,000 BTUH wood
                    stoves; 30,000 BTUH wood/coal stove

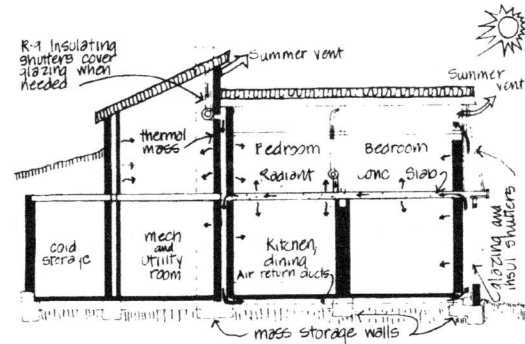

Section

This 2-story, wood frame house is priced in the $69,000 range. Evergreen trees to the northeast and northwest of the home provide winter wind protection, and tall deciduous trees to the south shade the house in the summer.

Open market sale.

Hampton, NH

Peter B. Olney, AIA
Hampton, NH

HEATED AREA: 1,480 FT²

NUMBER OF DEGREE DAYS: 7,246

NET THERMAL LOAD: 44.1 10⁶BTU/YR

AUXILIARY ENERGY: 3.85 BTU/DD/FT²

YEARLY SOLAR FRACTION: 66%

COLLECTOR: Southeast-facing solarium
                    Area: 424 FT²

STORAGE: Concrete wall and floor
                    Capacity: 27,673 BTU/°F

CONTROLS: Operable vents and exterior blind;
                    automatic furnace fans, clock and
                    thermostats

BACKUP: 60,000 BTUH gas forced-air furnace;
                    2 10,000 BTUH wood stoves

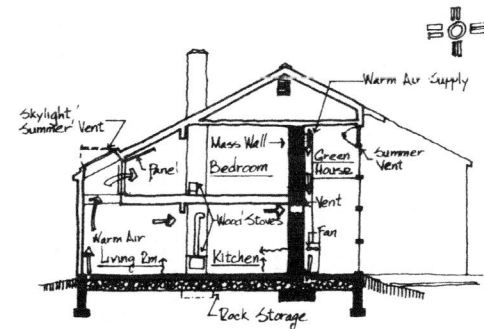

Section - Winter

121

This ranch-style house is priced in the $45,000 range and is intended to be a starter house for a young family or a retirement home for an older couple. The main entry is surrounded by a vestibule which prevents winter wind in-filtration.

Open market sale.

Plymouth, NH

Richard Holt, Evog Associates, Inc.
Hebron, NH

HEATED AREA: 1,212 FT²

NUMBER OF DEGREE DAYS: 8,177

NET THERMAL LOAD: 60.1 10⁶BTU/YR

AUXILIARY ENERGY: 3.80 BTU/DD/FT²

YEARLY SOLAR FRACTION: 37%

COLLECTOR: South-facing windows and heat storage mass wall Area: 194 FT²

STORAGE: Brick heat storage mass wall Capacity: 2,216 BTU/°F

CONTROLS: Operable windows and storage mass wall registers; automatic back-draft dampers

BACKUP: 84,000 BTUH oil furnace

FLOOR PLAN

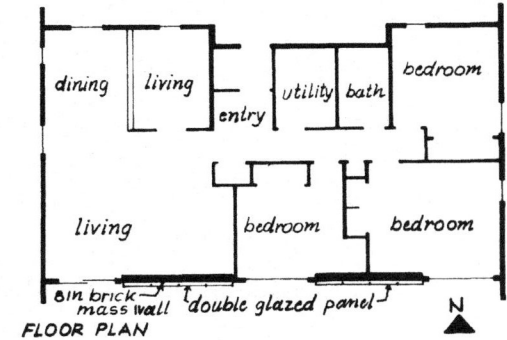

This modern residence is designed to meet the needs and lifestyle of most one- to three-child families. This house is built with standard construction materials and techniques with an emphasis on natural wood. The price is in the $78,000 range. The house is stretched out east-to-west to maximize solar exposure, and it is set into the side of a hill to minimize heat losses.

Private client.

Andover, NJ

Doug Kelbaugh, Environmint Partnership
Princeton, NJ

HEATED AREA: 1,681 FT²

NUMBER OF DEGREE DAYS: 5,696

NET THERMAL LOAD: 60.5 10⁶BTU/YR

AUXILIARY ENERGY: 2.4 BTU/DD/FT²

YEARLY SOLAR FRACTION: 83%

COLLECTOR: South-facing windows, solar mass wall, water drums, roof aperture, solarium window Area: 651 FT²

STORAGE: Water barrels, concrete walls and floor, water storage tank in skylight, gravel rock bed Capacity: 29,917 BTU/°F

CONTROLS: Operable shutters and curtains; automatic fan

BACKUP: 72,000 BTUH oil/wood central furnace; 25,000 BTUH wood fireplace

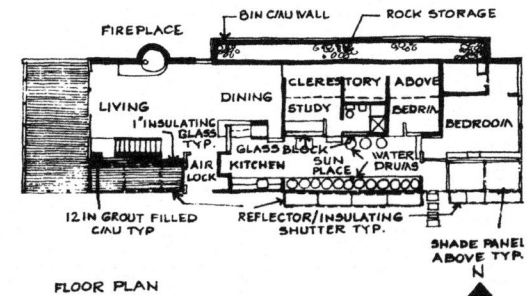

FLOOR PLAN

Flemington, NJ

Tom Wilson, Star Route Studios
Upper Black Eddy, PA

HEATED AREA: 1,148 FT²

NUMBER OF DEGREE DAYS: 5,733

NET THERMAL LOAD: 61.5 10⁶BTU/YR

AUXILIARY ENERGY: 5.5 BTU/DD/FT²

YEARLY SOLAR FRACTION: 57%

COLLECTOR: Stone storage mass wall
Area: 425 FT²

STORAGE: Stone storage mass wall
Capacity: 13,638 BTU/°F

CONTROLS: Operable vents

BACKUP: Electric baseboard

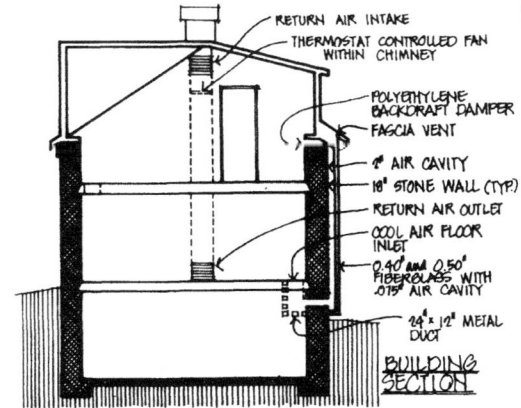

This project involves creating a masonry Trombe wall from an existing 18" stone wall by painting it black and covering it with dual level fiberglass-reinforced plastic glazing. Vents provide daytime distribution; the applicant plans to remove the existing interior insulation to permit direct radiation from the masonry wall. The retrofit project will cost less than $1.75/FT² of wall surface.

Retrofit.

Stillwater, NJ

William Collins
Cambridge, MA

HEATED AREA: 1,500 FT²

NUMBER OF DEGREE DAYS: 5,810

NET THERMAL LOAD: 58.2 10⁶BTU/YR

AUXILIARY ENERGY: 7.15 BTU/DD/FT²

YEARLY SOLAR FRACTION: 100%

COLLECTOR: Windows, solarium, skylight, and
storage mass wall Area: 510 FT²

STORAGE: Concrete and stone floor, storage
mass wall, rock bin, concrete and
stone central mass
Capacity: 16,564 BTU/°F

CONTROLS: Operable shutters, vents, windows,
and doors

BACKUP: 50,000 BTUH gas floor convection
furnace; fireplace; wood stove

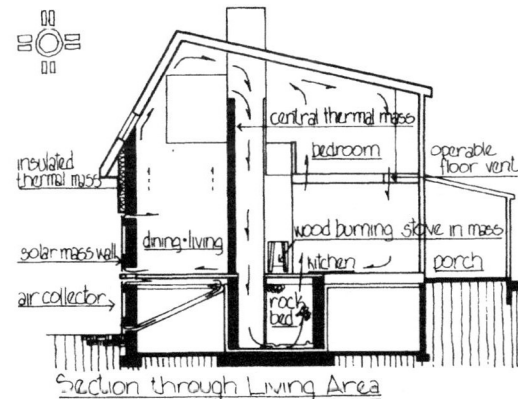

This 2-story, 3-bedroom house is constructed with concrete and on-site glacial rubble. Priced in the $55,000 range, the house is designed with centralized heating elements which distribute heat efficiently. An airlock entry reduces winter wind infiltration.

Private client.

123

INDIRECT SOLAR GAIN

This 3-story, 2-bedroom and loft wood frame house is priced in the $60,000 range. Located in a rural area, it is surrounded by heavy woods on the east, north and west.

Private client.

Upper Freehold, NJ

Doug Kelbaugh, Environmint Partnership
Princeton, NJ

HEATED AREA: 1,655 FT²

NUMBER OF DEGREE DAYS: 4,911

NET THERMAL LOAD: 59.4 10^6BTU/YR

AUXILIARY ENERGY: 2.11 BTU/DD/FT²

YEARLY SOLAR FRACTION: 85%

COLLECTOR: South-facing windows, solar mass wall Area: 613.5 FT²

STORAGE: Water drums, solar mass wall, concrete floor, triple sheetrock ceiling Capacity: 18,707 BTU/°F

CONTROLS: Thermal curtains and shades, operable windows, solar mass wall damper

BACKUP: 72,000 BTUH oil/wood furnace; 30,000 BTUH wood stove

FLOOR PLAN

A solar mass wall and solarium are retrofitted onto this 1-story, 4-bedroom adobe house. Trees on the north side of this lot provide a natural windbreak.

Retrofit.

Pojoaque, NM

W. A. Scott
Pojoaque, NM

HEATED AREA: 1,700 FT²

NUMBER OF DEGREE DAYS: 6,170

NET THERMAL LOAD: 199 10^6BTU/YR

AUXILIARY ENERGY: 12.3 BTU/DD/FT²

YEARLY SOLAR FRACTION: 99%

COLLECTOR: Solar mass wall, solarium Area: 348 FT²

STORAGE: Solar mass wall, water drums, hot tub Capacity: 16,964 BTU/°F

CONTROLS: Vents, shades, operable door

BACKUP: 120,000 BTUH natural gas furnace; fireplace

SECTION · GREENHOUSE ADDITION

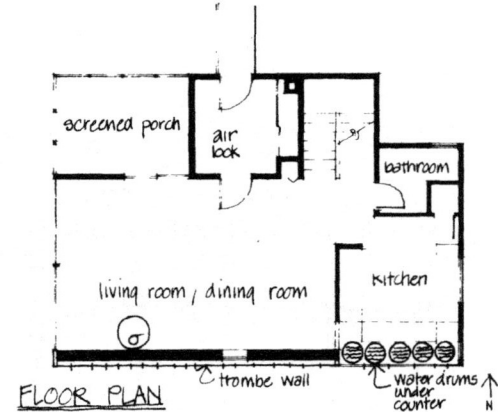

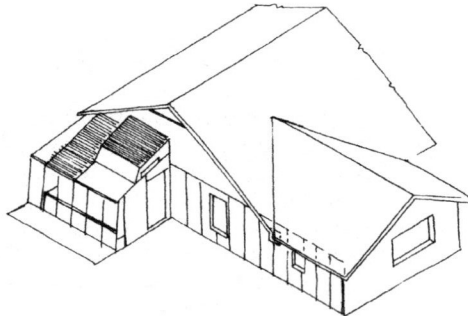

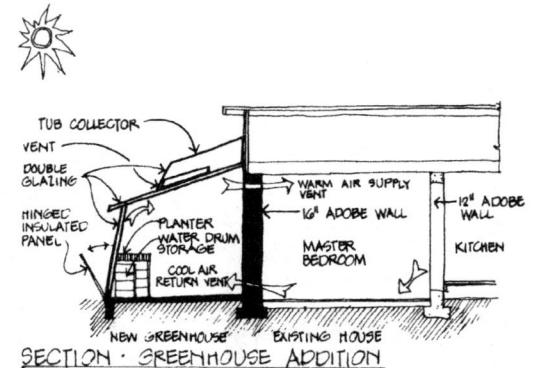

This new house is designed to require minimal non-solar heating, since utilities are inaccessible to most of the Ramah Navaho community. The house is priced in the $55,000 range. A wind-deflecting airlock entrance helps to conserve heat in the winter.

Private client.

Ramah, NM

Buck Rogers, Interface Design Group
Santa Fe, NM

HEATED AREA: 1,505 FT²

NUMBER OF DEGREE DAYS: 6,576

NET THERMAL LOAD: 55.4 10⁶BTU/YR

AUXILIARY ENERGY: 2.48 BTU/DD/FT²

YEARLY SOLAR FRACTION: 75%

COLLECTOR: South-facing windows, storage mass wall, and greenhouse Area: 382 FT²

STORAGE: Adobe walls, rock storage, masonry/earth over pumic floors Capacity: 89,187 BTU/°F

CONTROLS: Operable insulating drapes, vents, air intakes, doors, and windows; automatic fan

BACKUP: 37,000 BTUH wood fireplace; variable capacity electric baseboard

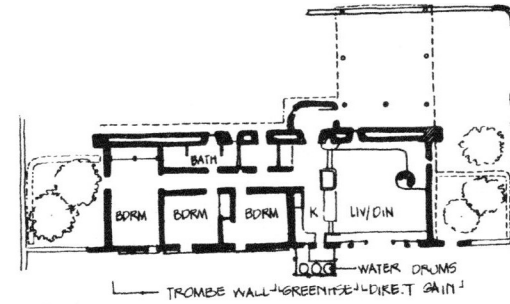

FLOOR PLAN

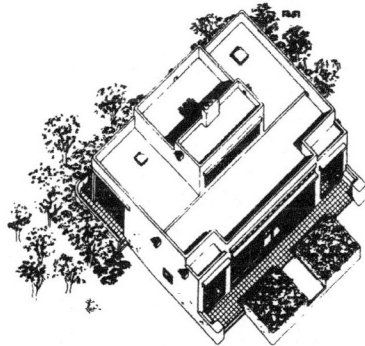

Built 100 feet below the ridge line on a south-facing slope, this 1-story, 2-bedroom wood frame house is a blend of pueblo and contemporary design. A tree line of dense pinõn cover shields the home from winter winds. It is priced in the $150,000 range.

Private client.

Santa Fe, NM

Robert Peters, Addy Associates
Albuquerque, NM

HEATED AREA: 1,795 FT²

NUMBER OF DEGREE DAYS: 5,586

NET THERMAL LOAD: 71.3 10⁶BTU/YR

AUXILIARY ENERGY: 4.27 BTU/DD/FT²

YEARLY SOLAR FRACTION: 68%

COLLECTOR: South-facing walls and clerestories, solar mass wall, and solarium Area: 498 FT²

STORAGE: Solar mass wall, adobe wall, concrete solarium floor, concrete fireplace Capacity: 12,902 BTU/°F

CONTROLS: Operable vents and shutters

BACKUP: 40,000 BTUH electric radiant cable in floor

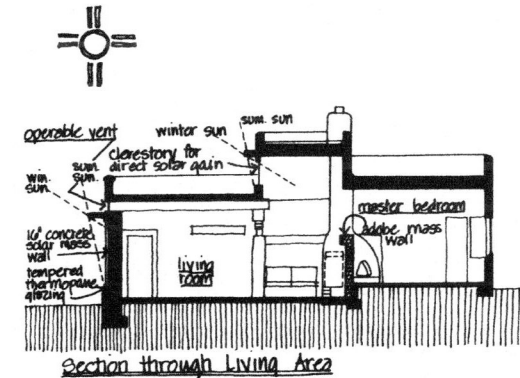

Section through Living Area

This new wood frame house conforms to the historic architectural style of Santa Fe. The house has good solar access and is set into the south-facing slope of an east-west ridge. Winter winds are redirected by vegetation, earth berms, and a garage and guest house located to the north of the main building. The price is in the $120,000 range.

Private client.

Santa Fe, NM

Mark Jones
Santa Fe, NM

HEATED AREA: 2,300 FT²

NUMBER OF DEGREE DAYS: 6,016

NET THERMAL LOAD: 5.8 10⁶BTU/YR

AUXILIARY ENERGY: 1.24 BTU/DD/FT²

YEARLY SOLAR FRACTION: 78%

COLLECTOR: South-facing solar mass walls, solarium, and windows
    Area: 608 FT²

STORAGE: Concrete solar mass walls and floor, rockbed Capacity: 23,103 BTU/°F

CONTROLS: Operable insulation, windows, vents, and shades; automatic fan

BACKUP: 20,000 BTUH electric baseboards; 10,000 BTUH electric baseboards; wood fireplace; heat lamp in bathroom

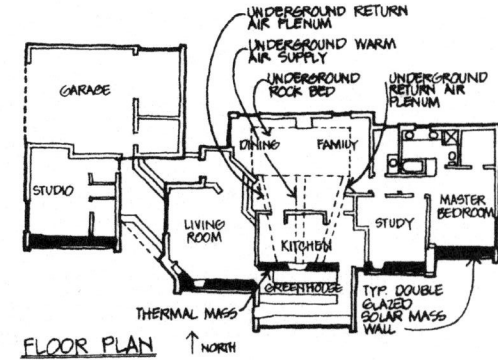

FLOOR PLAN   ↑NORTH

This 1-story, 2-bedroom pueblo house is of frame and block construction and is priced in the $75,000 range. The home, located below a ridge on a gentle southwest slope, is buffered by a line of trees.

Open market sale.

Santa Fe, NM

Susan Nichols, Communico
Santa Fe, NM

HEATED AREA: 1,162 FT²

NUMBER OF DEGREE DAYS: 5,586

NET THERMAL LOAD: 25.8 10⁶BTU/YR

AUXILIARY ENERGY: 1.99 BTU/DD/FT²

YEARLY SOLAR FRACTION: 99%

COLLECTOR: South-facing windows, solar mass wall, mass-backed solarium
    Area: 329 FT²

STORAGE: Solar mass wall, adobe wall, solarium brick floor
    Capacity: 8,167 BTU/°F

CONTROLS: Vents, skylight shutters, solarium door

BACKUP: 17,000 BTUH electric baseboard

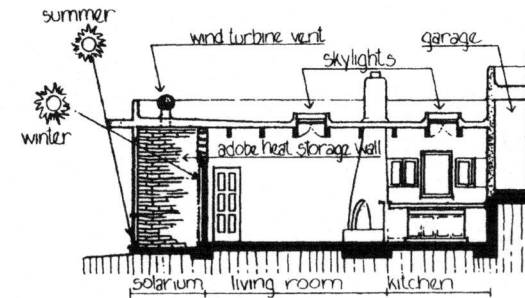

Section through Living Area

Bedford Heights, OH

William Kolar, SOA Energy Consortium
Arlington, VA

HEATED AREA: 1,530 FT²

NUMBER OF DEGREE DAYS: 5,993

NET THERMAL LOAD: 48.8 10⁶BTU/YR

AUXILIARY ENERGY: 6.0 BTU/DD/FT²

YEARLY SOLAR FRACTION: 49%

COLLECTOR: South-facing windows, water storage mass wall, east- and south-facing solarium, west-facing windows Area: 286 FT²

STORAGE: Concrete floor, masonry walls, water storage mass wall, rock storage Capacity: 25,309 BTU/°F

CONTROLS: Insulating curtains, vents, windows, shutters, and fan

BACKUP: 24,000 BTUH electric heater

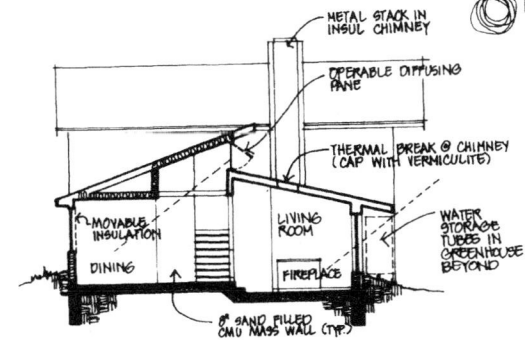

METAL STACK IN INSUL CHIMNEY
OPERABLE DIFFUSING PANE
THERMAL BREAK @ CHIMNEY (CAP WITH VERMICULITE)
MOVABLE INSULATION
LIVING ROOM
WATER STORAGE TUBES IN GREENHOUSE BEYOND
DINING
FIREPLACE
8" SAND FILLED CMU MASS WALL (TYP.)

SECTION AT CLERESTORY

This new, 3-bedroom, split-level house is priced in the $65,000 range. The use of earth berms, a south-facing airlock entrance, conifer trees, and a garage on the building's north side reduces winter heat loss.

Open market sale.

Cleveland, OH

Gregory Goss and Terry Sefchick
Cleveland Heights, OH

HEATED AREA: 1,328 FT²

NUMBER OF DEGREE DAYS: 6,351

NET THERMAL LOAD: 40.1 10⁶BTU/YR

AUXILIARY ENERGY: 9.91 BTU/DD/FT²

YEARLY SOLAR FRACTION: 90%

COLLECTOR: Solar mass wall, clerestory Area: 780 FT²

STORAGE: Concrete solar mass wall Capacity: 22,174 BTU/°F

CONTROLS: Operable windows and solar mass wall louvers; automatic insulating beadwall

BACKUP: 30,000 BTUH electric heat pump; wood or coal fireplace

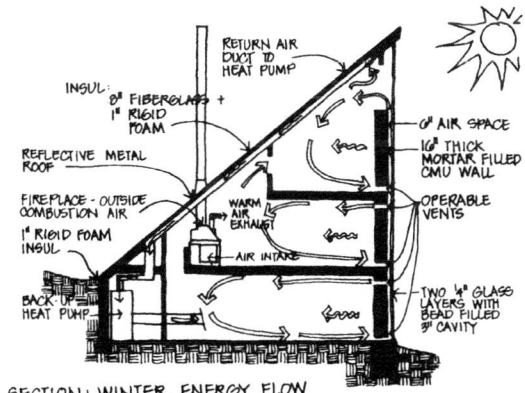

RETURN AIR DUCT TO HEAT PUMP
INSUL 8" FIBERGLASS + 1" RIGID FOAM
REFLECTIVE METAL ROOF
6" AIR SPACE
16" THICK MORTAR FILLED CMU WALL
FIREPLACE - OUTSIDE COMBUSTION AIR
1" RIGID FOAM INSUL
WARM AIR EXHAUST
AIR INTAKE
OPERABLE VENTS
BACK-UP HEAT PUMP
TWO ⅛" GLASS LAYERS WITH BEAD FILLED 9" CAVITY

SECTION: WINTER ENERGY FLOW

This new, 3-story, 3-bedroom detached house is in the $72,000 price range. The steep slope of the north-facing roof deflects winter winds, and shade trees surround the house for summer cooling.

Open market sale.

This 3-bedroom residence is a retrofit of a house built in the 1960's. The house is located in a region with fairly high insolation values and is oriented on its site with good solar exposure. However, proper reinsulation of the structure was necessary. A garage is located on the west end and protects against heat gain on west walls of conditioned space.

Retrofit.

Burns, OR

T. Kuntzman/D. Knokey
Eugene, OR

HEATED AREA: 1,170 FT^2

NUMBER OF DEGREE DAYS: 6,871

NET THERMAL LOAD: 49.6 10^{6}BTU/YR

AUXILIARY ENERGY: 3.80 BTU/DD/FT^2

YEARLY SOLAR FRACTION: 63%

COLLECTOR: South-facing windows, solar mass wall, and solarium Area: 261 FT^2

STORAGE: Concrete solar mass wall, clay tile floor Capacity: 9,645 BTU/°F

CONTROLS: Operable insulating shutters, windows, and vents, automatic fan

BACKUP: Electric baseboard

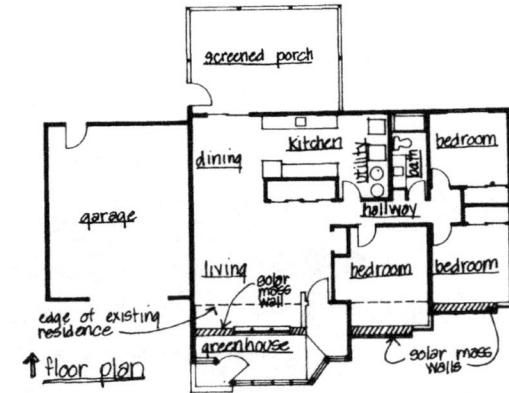

This 3-bedroom detached house is designed to conform to the architectural style of its mountain setting, having view windows toward the mountains to the southwest, decks for outdoor living, and the use of natural wood wherever possible. The price is in the $102,000 range (including land and design). Earth berming and a protected primary entrance reduce winter wind infiltration.

Open market sale.

Sisters, OR

James Emerson, Phase I
Sisters, OR

HEATED AREA: 1,819 FT^2

NUMBER OF DEGREE DAYS: 6,598

NET THERMAL LOAD: 66.4 10^{6}BTU/YR

AUXILIARY ENERGY: 3.23 BTU/DD/FT^2

YEARLY SOLAR FRACTION: 71%

COLLECTOR: South-facing windows and heat storage mass wall Area: 630 FT^2

STORAGE: Concrete solar mass wall Capacity: 13,014 BTU/°F

CONTROLS: Automatic solar mass wall fan, operable insulation, vents, and drapes

BACKUP: 51,195 BTUH electric furnace; 40,000 BTUH air-tight woodstove

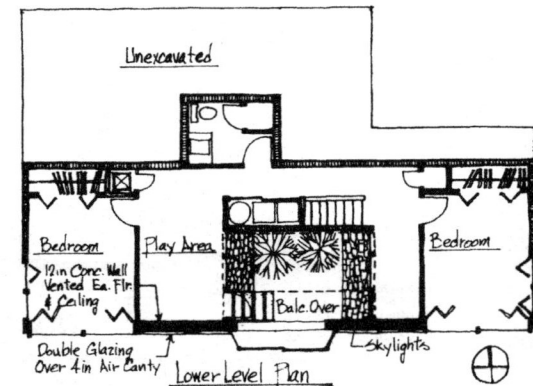

This new 2-story, 3-bedroom house has a plywood exterior. The house is in the $50,000 price range. The front face of the solarium is earth bermed, and the house is also protected from winter winds by the slope of the site and by trees on the north and west side of the house.

Private client.

Lakeville, PA

Charles Skowronski
Merrimac, MA

HEATED AREA: 1,697 FT²

NUMBER OF DEGREE DAYS: 6,202

NET THERMAL LOAD: 52.6 10⁶BTU/YR

AUXILIARY ENERGY: 2.27 BTU/DD/FT²

YEARLY SOLAR FRACTION: 71%

COLLECTOR: South-facing windows, greenhouse, water storage mass wall, skylights Area: 541 FT²

STORAGE: Water storage mass wall, concrete wall, rock and concrete floor Capacity: 21,465 BTU/°F

CONTROLS: Operable shutters, skylights, awnings, and dampers; Automatic thermostat and fans

BACKUP: 50,000 BTUH electric baseboard; 30,000 & 15,000 BTUH wood stoves

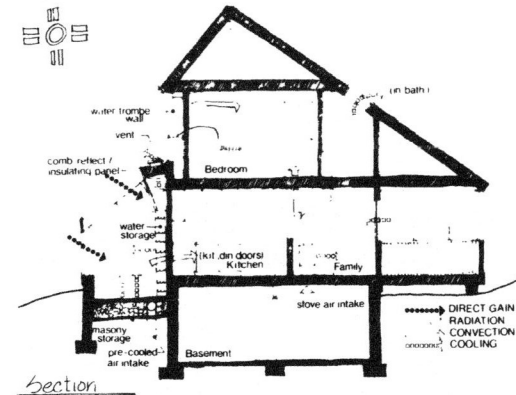

Section

This new, 2-story, 3-bedroom detached wood frame house is priced in the $50,000 range. Trees on the north side of the house will provide winter wind protection.

Private client.

Lititz, PA

J. Wylie Bradley, deVitry, Gilbert and Bradley
Lancaster, PA

HEATED AREA: 2,128 FT²

NUMBER OF DEGREE DAYS: 5,251

NET THERMAL LOAD: 68.1 10⁶BTU/YR

AUXILIARY ENERGY: 4.64 BTU/DD/FT²

YEARLY SOLAR FRACTION: 46%

COLLECTOR: Solar mass wall Area: 342 FT²

STORAGE: Brick wall Capacity: 13,591 BTU/°F

CONTROLS: Valves, canvas awnings

BACKUP: Wood stove

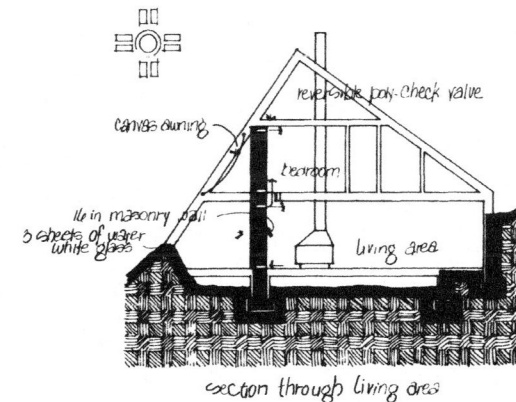

section through living area

This 2-story, 3-bedroom house is located on a slightly sloping site is priced in the $80,000 range. Earth berming, northern conifer trees, and a south entrance minimize winter heat losses, while deciduous trees on the south side shade the house in the summer.

Private client.

New Tripoli, PA

Shelter Design
New Tripoli, PA

HEATED AREA: 1,500 FT²

NUMBER OF DEGREE DAYS: 5,827

NET THERMAL LOAD: 57.9 10⁶BTU/YR

AUXILIARY ENERGY: 2.73 BTU/DD/FT²

YEARLY SOLAR FRACTION: 73%

COLLECTOR: South-facing windows and solarium Area: 496 FT²

STORAGE: Concrete slab, water tubes Capacity: 20,460 BTU/°F

CONTROLS: Operable windows and doors

BACKUP: 25,000 BTUH wood stove; 18,700 BTUH electric baseboard

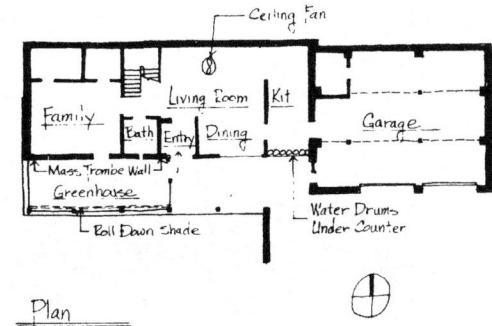

Plan

This new 2-story, 3-bedroom (with solarium) detached house is shaded by trees to its north and east and is in the $90,000 price range. The vegetation provides a natural windbreak for winter wind protection.

Private client.

Unionville, PA

Don Prowler, South Street Design
Philadelphia, PA

HEATED AREA: 2,040 FT²

NUMBER OF DEGREE DAYS: 5,101

NET THERMAL LOAD: 56.6 10⁶BTU/YR

AUXILIARY ENERGY: 3.1 BTU/DD/FT²

YEARLY SOLAR FRACTION: 76%

COLLECTOR: Solarium, solar mass wall, south-facing window Area: 850 FT²

STORAGE: Solar mass wall, water drums, block walls Capacity: 3,714.5 BTU/°F

CONTROLS: Insulated shutters, vents, damper, fan, solarium canvas shades

BACKUP: 40,000 BTUH oil furnace

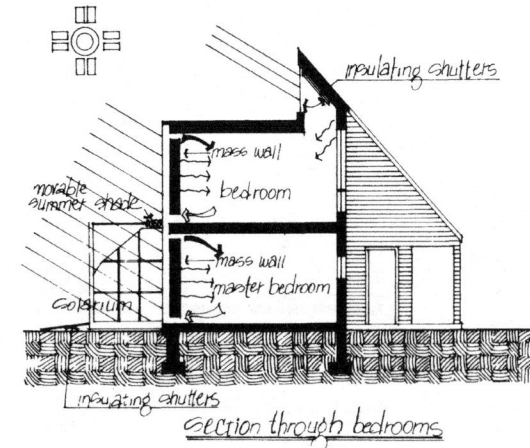

Section through bedrooms

Summerville, SC

Ralph F. McCay, Solar Engineering
Charleston, SC

HEATED AREA: 1,629 FT²

NUMBER OF DEGREE DAYS: 2,146

NET THERMAL LOAD: 15.3 10⁶BTU/YR

AUXILIARY ENERGY: 5.16 BTU/DD/FT²

YEARLY SOLAR FRACTION: 79%

COLLECTOR: Solarium Area: 281 FT²

STORAGE: Concrete floor, brick walls
Capacity: 7,965 BTU/°F

CONTROLS: Operable doors, air handler

BACKUP: 17,000 BTUH and 21,000 BTUH
electric heat pumps; 25,000 BTUH
fireplace

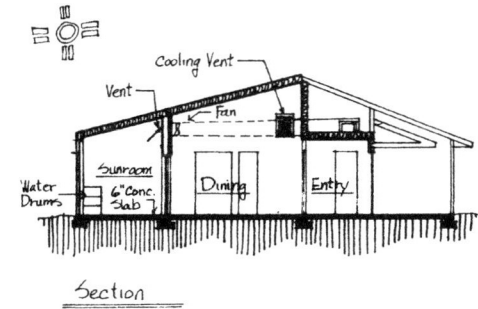

This 1-story, 3-bedroom wood frame house is in the $55,000 range. There is a pine forest 120 feet southwest of the home.

Open market sale.

Knoxville, TN

Bill Barth, Barth and Ransbottom
Knoxville, TN

HEATED AREA: 2,882 FT²

NUMBER OF DEGREE DAYS: 3,494

NET THERMAL LOAD: 35.9 10⁶BTU/YR

AUXILIARY ENERGY: 3.5 BTU/DD/FT²

YEARLY SOLAR FRACTION: 45%

COLLECTOR: Heat storage mass wall and
solarium Area: 620 FT²

STORAGE: Heat storage mass wall, masonry
solarium wall, solarium floor
Capacity: 22,502 BTU/°F

CONTROLS: Operable doors, windows, blowers,
wall vent panel, and dampers

BACKUP: 30,000 BTUH electric baseboard;
wood stove

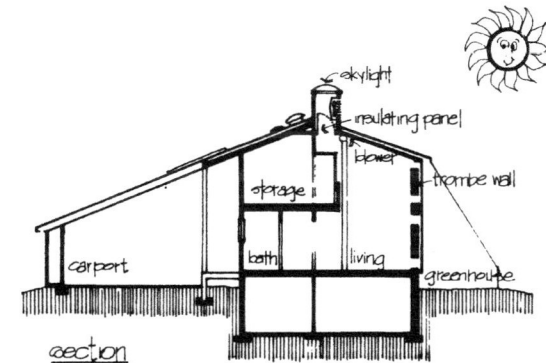

This 2-story, 3-bedroom house is wood-framed and is in the $75,000 price range. Located on a hilly, wooded undeveloped country site within commuting distance from Knoxville, the house is protected from winter winds by a carport roof and a storage area on its north side. Poplar trees will be planted on the house's south side for summer sun protection.

Private client.

This new 1-story, 2-bedroom detached house with its adobe exterior is a rustic design in the $90,000 price range. Junipers on the hillside site protect the house from northwest winds during the winter.

Open market sale.

El Paso, TX

Mack Caldwell, Phillip Mack Caldwell
El Paso, TX

HEATED AREA: 1,635 FT^2

NUMBER OF DEGREE DAYS: 2,700

NET THERMAL LOAD: 27.4 10^{6}BTU/YR

AUXILIARY ENERGY: 2.4 BTU/DD/FT^2

YEARLY SOLAR FRACTION: 97%

COLLECTOR: South-facing window
Area: 265 FT^2

STORAGE: Concrete floor, adobe walls, ramp and banco Capacity: 54,780 BTU/°F

CONTROLS: Operable shutters, exhaust fan

BACKUP: Two heatilator fireplaces 10-15,000 BTUH

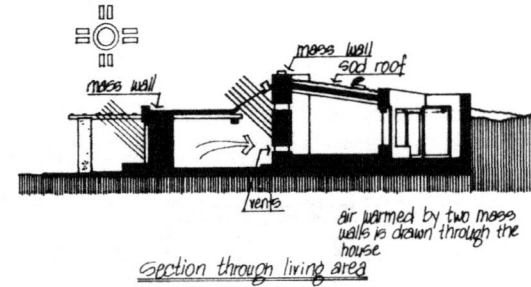

air warmed by two mass walls is drawn through the house

Section through living area

This new, 1-story with clerestory, 3-bedroom detached home is priced in the $42,500 range. Deciduous trees shade windows and the heat storage mass on the south side of the house in the summer; evergreen trees on the northeast side will redirect winter winds.

Open market sale.

Cedar City, UT

John W. Stewart, Solarama
Colorado City, AZ

HEATED AREA: 1,232 FT^2

NUMBER OF DEGREE DAYS: 5,200

NET THERMAL LOAD: 27.4 10^{6}BTU/YR

AUXILIARY ENERGY: N/A

YEARLY SOLAR FRACTION: 99%

COLLECTOR: South-facing solar mass wall and windows Area: 350 FT^2

STORAGE: Masonry solar mass wall, concrete floor Capacity: 4,215 BTU/°F

CONTROLS: Operable vents, blinds, shutters, and windows

BACKUP: Electric ceiling cable heat

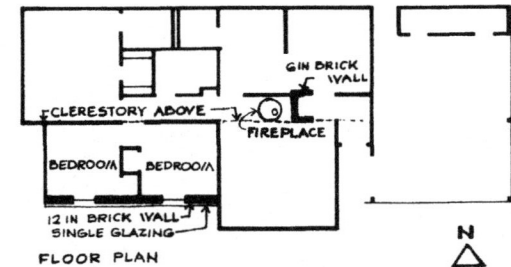

FLOOR PLAN

Kanab, UT

John W. Stewart, Solarama
Colorado City, AZ

HEATED AREA: 1,506 FT²

NUMBER OF DEGREE DAYS: 5,200

NET THERMAL LOAD: 16.3 10⁶BTU/YR

AUXILIARY ENERGY: 2.6 BTU/DD/FT²

YEARLY SOLAR FRACTION: 100%

COLLECTOR: South-facing solar mass wall,
solarium Area: 366 FT²

STORAGE: Concrete floor, solar mass wall
Capacity: 7,452 BTU/°F

CONTROLS: Solar mass wall vents, insulated
blinds, ground ducts, operable
solarium windows

BACKUP: Electric coil

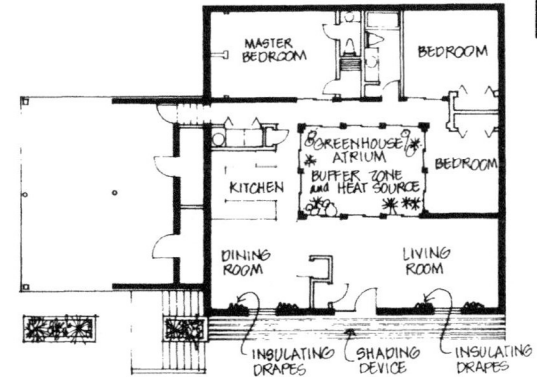

This 1-story, 3-bedroom underground house is priced in the $60,000 range. The south wall is exposed for heat collection, and a carport and sundeck are built above ground.

Open market sale.

Duffield, VA

Billy Born, Architectural Design Branch, TVA
Knoxville, TN

HEATED AREA: 1,016 FT²

NUMBER OF DEGREE DAYS: 4,121

NET THERMAL LOAD: 17.5 10⁶BTU/YR

AUXILIARY ENERGY: 2.66 BTU/DD/FT²

YEARLY SOLAR FRACTION: 79%

COLLECTOR: South-facing Trombe wall
Area: 294 FT²

STORAGE: Concrete block filled with concrete
Capacity: N/A

CONTROLS: Open exterior vents to Trombe
wall

BACKUP: 3,000 BTUH wood stove; 1-1/2-ton
cooling electric heat pump

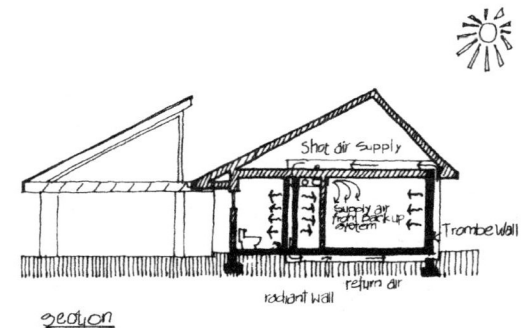

This new, 3-bedroom detached house has a wood siding and is priced in the $30,000 range. The plan is arranged to allow passive solar additions to be made to the east without disturbing the original house. Rooms on the west end of the house may be thermally isolated if the space is not needed.

Open market sale.

This new 2-1/2-story, 3-bedroom house has a traditional residential appearance. Its solar heating system is designed to require little active participation from the occupants. A sheltered entrance prevents winter wind infiltration.

Private client.

Norfolk, VA

Richard J. Fitts
Virginia Beach, VA

HEATED AREA: 1,402 FT²

NUMBER OF DEGREE DAYS: 3,384

NET THERMAL LOAD: 29.5 10⁶BTU/YR

AUXILIARY ENERGY: 4.04 BTU/DD/FT²

YEARLY SOLAR FRACTION: 56%

COLLECTOR: Southwest-facing windows, storage mass wall, skylight Area: 314.8 FT²

STORAGE: Water-filled cylinders, storage mass wall Capacity: 8,716 BTU/°F

CONTROLS: Operable shades, windows and vent; automatic vent flap at storage wall

BACKUP: 30,000 BTUH electric heat pump

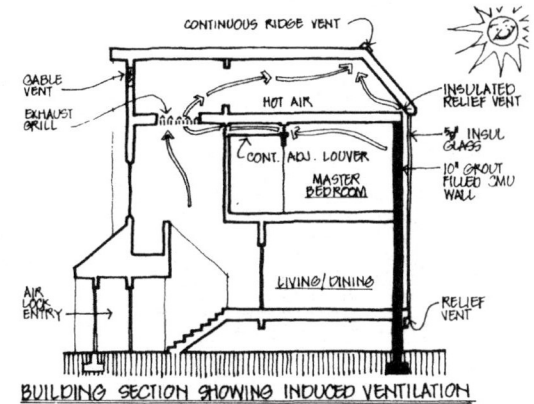

BUILDING SECTION SHOWING INDUCED VENTILATION

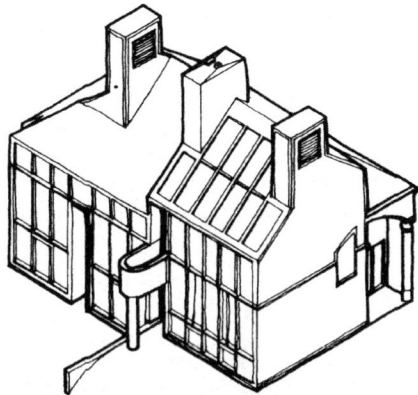

This contemporary 2-story, 3-bedroom house is in the $110,000 price range. Evergreen trees to the north of this detached building provide winter wind protection.

Open market sale.

Reston, VA

Walter F. Roberts, Jr., Architect
Reston, VA

HEATED AREA: 2,523 FT²

NUMBER OF DEGREE DAYS: 4,962

NET THERMAL LOAD: 47.4 10⁶BTU/YR

AUXILIARY ENERGY: 1.12 BTU/DD/FT²

YEARLY SOLAR FRACTION: 84%

COLLECTOR: South-facing windows, solarium, and solar mass wall Area: 730 FT²

STORAGE: Concrete solar mass wall and chimney wall, stone floor and rock bed Capacity: 86,895 BTU/°F

CONTROLS: Automatic roof vents, motor-operated solar curtain, operable night curtains

BACKUP: 19,857 BTUH electric lamps

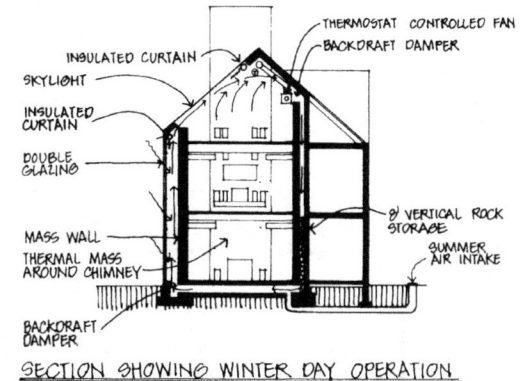

SECTION SHOWING WINTER DAY OPERATION

This is a 1-story, 2-bedroom wood frame house in the $20,000 range. Evergreens to the north provide a natural windbreak; trees to east and west for shading.

Private client.

Middlebury, VT

Harris Hyman, R. Finkle/H. Hyman
Rochester, VT

HEATED AREA: 916 FT²

NUMBER OF DEGREE DAYS: 7,988

NET THERMAL LOAD: 52.3 10⁶BTU/YR

AUXILIARY ENERGY: 4.6 BTU/DD/FT²

YEARLY SOLAR FRACTION: 56%

COLLECTOR: Solar mass wall, south-facing windows Area: NA FT²

STORAGE: Concrete wall and floor Capacity: 20,352 BTU/°F

CONTROLS: Thermal shades, vents

BACKUP: 35,000 BTUH wood stove; 34,130 BTUH electric baseboard

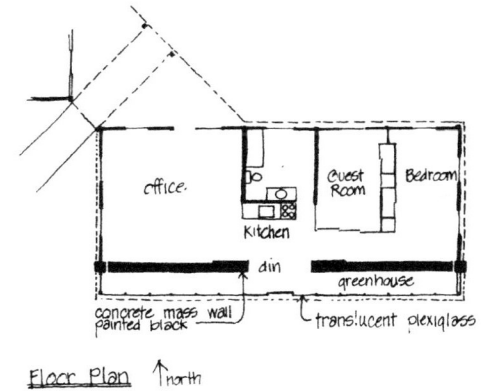

Floor Plan ↑north

concrete mass wall painted black

translucent plexiglass

This 2-story, 3-bedroom house is built into a south-facing slope. The exterior is stucco and cedar siding. It is priced in the $50,000 range. Evergreens and shrubs to the north will shield the home from winter winds. Abundant trees on south, east, and west provide summer shade.

Private client.

Madison, WI

Don Schramm, Prado
Mount Horeb, WI

HEATED AREA: 1,632 FT²

NUMBER OF DEGREE DAYS: 7,596

NET THERMAL LOAD: 78 10⁶BTU/YR

AUXILIARY ENERGY: 4.3 BTU/DD/FT²

YEARLY SOLAR FRACTION: 50%

COLLECTOR: South-facing windows Area: 328 FT²

STORAGE: Terrazzo floor and concrete walls with concrete cylinders at windows Capacity: 7,952 BTU/°F

CONTROLS: Operable insulating shades and shutters

BACKUP: 40,940 BTUH electric baseboard; 5120 BTUH heatlamps; 15,000 BTUH wood stove

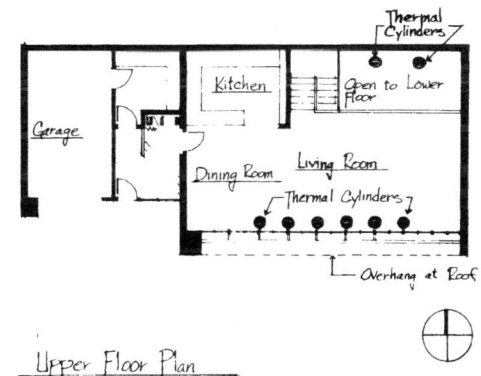

Upper Floor Plan

This is a new 2-story, 3-bedroom detached home with cedar butt siding. There will be an evergreen windbreak on the north side of this level lot.

Private client.

Harrisville, WV

David Reynolds
Manchester, NH

HEATED AREA: 2,300 FT²

NUMBER OF DEGREE DAYS: 4,688

NET THERMAL LOAD: 81.1 10⁶BTU/YR

AUXILIARY ENERGY: 6.86 BTU/DD/FT²

YEARLY SOLAR FRACTION: 35%

COLLECTOR: Solar mass wall, solarium
Area: 380 FT²

STORAGE: Solar mass wall
Capacity: 8,341 BTU/°F

CONTROLS: Insulating curtains

BACKUP: Wood burning stove

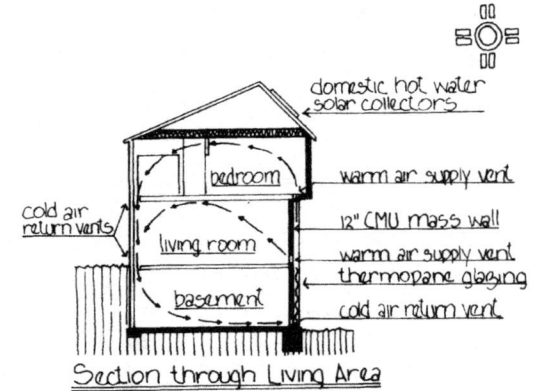

domestic hot water solar collectors
warm air supply vent
cold air return vents
bedroom
12" CMU mass wall
living room
warm air supply vent
thermopane glazing
basement
cold air return vent

Section through Living Area

This contemporary 2-story, 3-bedroom wood frame house has a rustic flavor which is compatible with the surrounding open sagebrush country at the edge of the Rocky Mountains. The exterior has cedar siding. Price is in the $80,000 range.

Open market sale.

Riverton, WY

Michael Framson, Framson General Construction
Slater, CO

HEATED AREA: 1,668 FT²

NUMBER OF DEGREE DAYS: 8,433

NET THERMAL LOAD: 83.8 10⁶BTU/YR

AUXILIARY ENERGY: 1.64 BTU/DD/FT²

YEARLY SOLAR FRACTION: 72%

COLLECTOR: Solar mass wall, solarium
Area: 432 FT²

STORAGE: Concrete wall, rock bed
Capacity: 22,460 BTU/°F

CONTROLS: Curtains, fans, operable windows

BACKUP: 24,000 BTUH electric heat pump

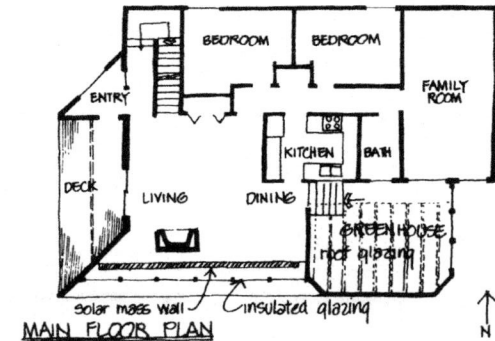

BEDROOM   BEDROOM
FAMILY ROOM
ENTRY
KITCHEN   BATH
DECK   LIVING   DINING
GREENHOUSE hot glazing
solar mass wall   insulated glazing
MAIN FLOOR PLAN
N

CHAPTER 5
SOLARIUM

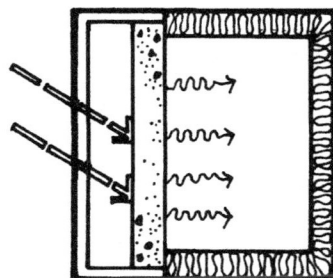

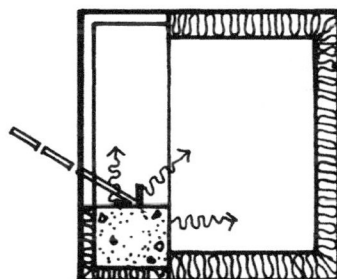

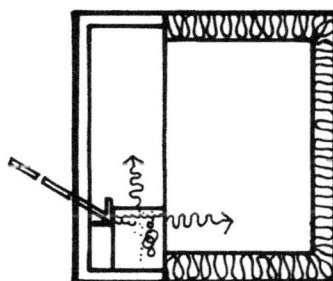

DEFINITION
The SOLARIUM passive building type collects solar radiation in a secondary space which is separate from the living space, and also stores heat for later distribution. This Solarium offers both the potential separation of the collector-storage system from the living space, or the direct gain "live-in" situation which maximizes the use of low temperature solar gain. Thus, in concept, a Solarium passive solar home is midway between a Direct Gain home, in which the living space is the collector of heat, and a Mass or Water storage building which collects heat indirectly for the living space. An atrium, a sun porch, a greenhouse, and a sunroom all represent potential examples of Solarium.

REQUIREMENTS AND VARIATIONS
The requirements of a Solarium passive solar home center on the glazed 'collector' space which must be both attached yet distinct from the living space. Provided with a strong southern exposure, the collector space must be thermally linked to a solar storage mass for heat retention and later distribution.

The Solarium can be variable in its spatial and functional relationships to the primary living spaces of the building. It may vary from a minimum addition to a building with one small contact surface, to extending the entire south side of the building, to being contained within the building with an interface on several sides. The specific location of the Solarium will depend on the building design, spatial organization, and sun orientation. A storage mass is necessary in the Solarium to retain heat for non-sunshine hours. Massive floors, walls, benches, rock beds, and covered pools of water can all provide effective solar heat storage. Temperature levels within the Solarium will depend upon the other uses to which it may be put. If it is generally unoccupied, temperatures can be permitted to rise as high as the capacity of the storage materials will permit. If, on the other hand, it is to be used as a greenhouse or other living space, the temperature restrictions of a Direct Gain home would apply. Finally, if the space includes an indoor swimming pool for storage, pool temperatures must be controlled for swimmer comfort.

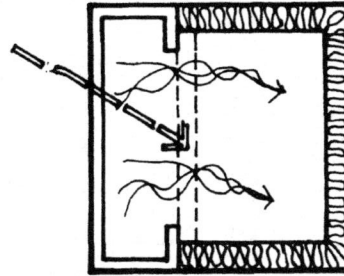

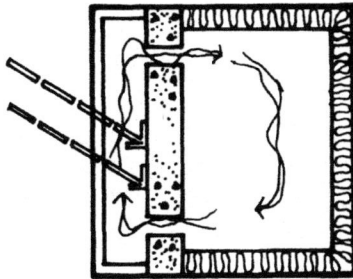

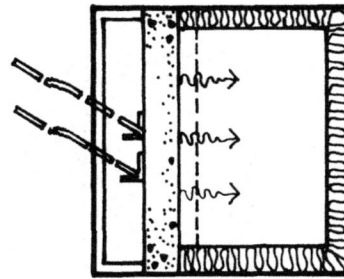

CONTROLS

The most mandatory control consideration for this passive solar building type is the design of the link between the Solarium and the living space. The walls which interface a Solarium and living space require built-in flexibility in order that these spaces can be thermally connected and separated as desired. The kind of distribution—radiation, convection, or conduction—will be determined by these interfaces and differentiate the Solarium passive solar home from the Direct Gain type. In addition, as in other passive solar building types, shading should be provided to prevent overheating of glazed spaces during the summer; and some form of moveable insulation would prevent unnecessary heat losses on winter nights or cloudy days. Humidity control is also an important consideration to prevent mold development in the storage mass in the plant-or water-occupied Solariums.

The important issue to understand with the SOLARIUM passive solar homes is that the occupant needs only to be in direct contact with the distribution component of the solar heating (thus ISOLATED GAIN), unlike the DIRECT GAIN home where the occupant is in contact with the collection, storage, and distribution of solar heat; or the INDIRECT GAIN home where the occupant is in contact with both storage and distribution.

The Solarium passive solar home suggests yet another Isolated Gain solar heating method; the Thermosiphon.

THERMOSIPHON

DEFINITION

The use of the Thermosiphon passive solar home also includes a collector space which intercedes between the direct sun and the living space and is distinct from the building structure. A thermosiphoning heat flow occurs when a cool air or liquid naturally falls to the lowest point (in this case below the collectors) and, once heated by the sun, rises up into an appropriately placed living space or storage mass, causing somewhat cooler air or liquid to fall again, so a continuous heat-gathering circulation is begun. Since the collector space is completely separate from the building space, the Thermosiphon system begins to resemble the active systems frequently seen on today's market. However, no external power from fans or blowers are necessary to move the heat transfer medium. The Thermosiphon principle has been applied in numerous solar domestic hot water systems, and it offers equally great potential for space heating application.

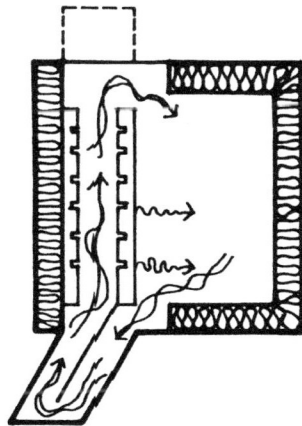

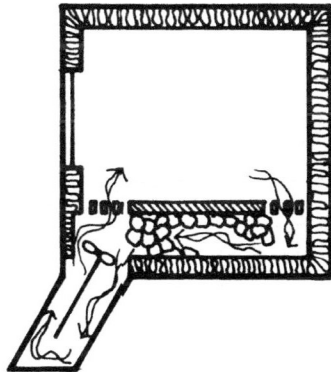

REQUIREMENTS AND VARIATIONS

The basic elements of the Thermosiphon home include a collector space, usually a storage mass, and a method of distribution. Solar heat is collected on a dark metal or wood absorber surface, heating up the adjacent fluid (usually air), which then rises naturally into a storage mass for convective or radiant distribution. In the Thermosiphon solar home the collector location is not fixed by the building and thus can take maximum advantage of sun exposure. Since the collector area is separate from the building facade, the house is also flexible in its wall and opening design. The solar storage mass can be located under the house floor, below windows, or in prefabricated wall elements. The storage location and material is the element of most variation and offers building and system design flexibility. Distribution is provided by radiation from the storage mass and/or by convection (naturally rising air movements) from storage or directly from the collector. The spatial arrangement of the building is critical in providing effective heat distribution.

CONTROLS

In the Thermosiphon building type, the link or contact area between the collector space and solar storage is not great and can be easily blocked or disconnected to prevent air flow in adverse collector conditions (such as unwanted heat loss or overheating). The area of interface between the storage mass and the building, however, will determine the speed with which the living space can be heated through radiation and convection. The greater the contact area between the storage and the living space, the easier the radiant distribution will be, but the control against untimely or overabundant space heating is more crucial. For convective distribution from the storage mass of a Thermosiphon passive solar home, controls similar to those used in the Trombe building types are required, including operable dampers and insulation panels.

Oxford, OH

Fuller Moore
Oxford, OH

HEATED AREA: 1,600 FT²

NUMBER OF DEGREE DAYS: 4,806

NET THERMAL LOAD: 36.4 10⁶BTU/YR

AUXILIARY ENERGY: 1.63 BTU/DD/FT²

YEARLY SOLAR FRACTION: 74%

Open market sale.

CONTEXT
The Solargreen Project is conceived by its designer, Fuller Moore, as a passive solar dwelling for all seasons. Located in the southwestern portion of Ohio, the design responds to both the characteristically cold winters and hot summers and also considers the transitional spring and fall seasons. The design is flexible in size and layout to appeal to a broad market base, and it is intended specifically for the speculative home builder. This three-bedroom, two-floor, 1,600 square foot home with attached garage includes a solarium which serves both as a solar collector and a greenhouse It makes use of primarily conventional wood frame construction, with insulated concrete block below grade and a poured-in-place concrete wall for the primary solar heat storage mass.

CONSERVATION
The design exhibits a strong relationship between energy conservation and solar utilization. The concept, called "thermal layering", established an inter-related series of living zones which are ordered in response to the activities of people and the climatic demands of different seasons. Those rooms in which people spend the most time are adjacent to the solarium, which is the principal

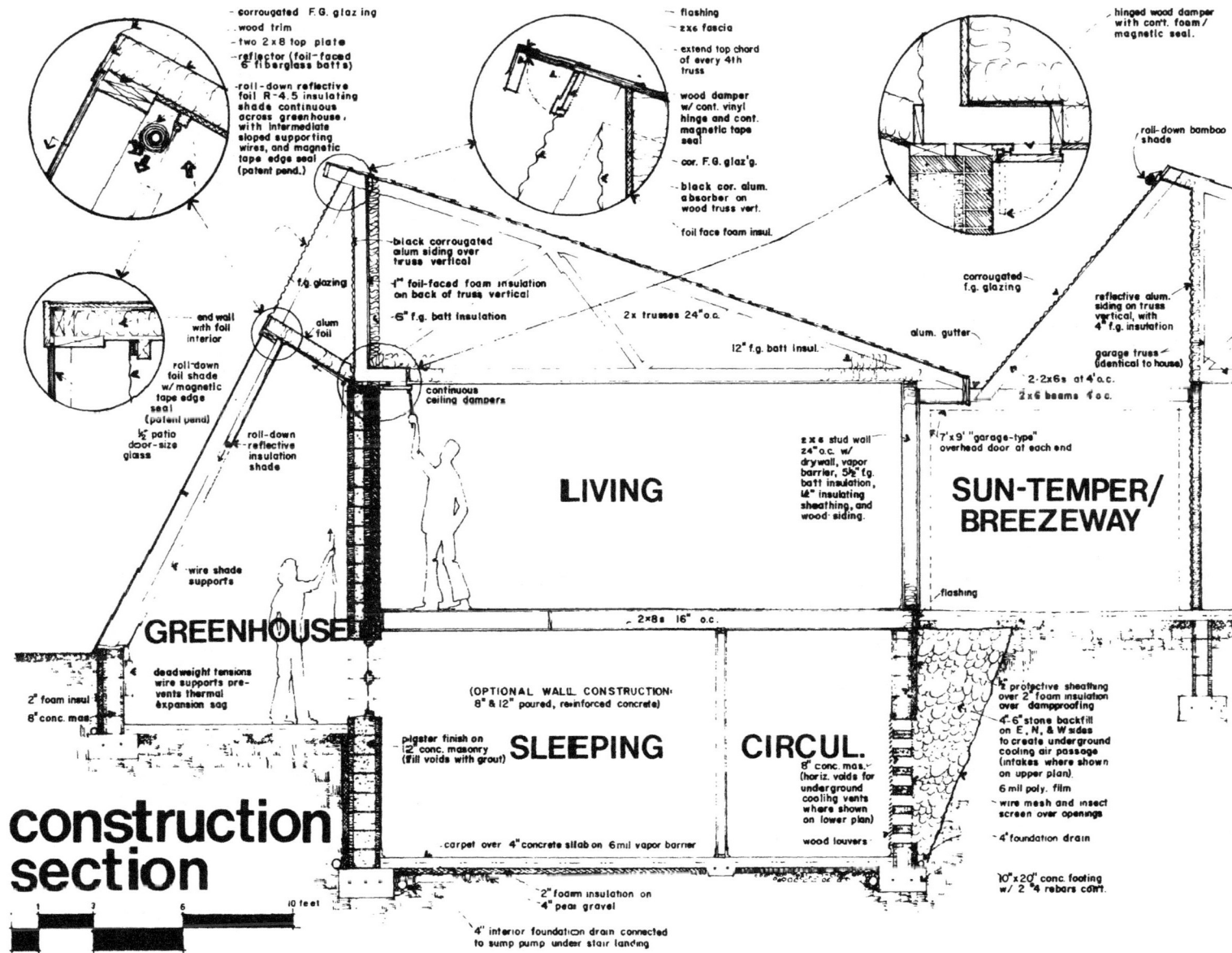

construction section

GREENHOUSE

LIVING

SUN-TEMPER/
BREEZEWAY

SLEEPING

CIRCUL.

- corrougated F.G. glazing
- wood trim
- two 2 x 8 top plate
- reflector (foil-faced 6 fiberglass batts)
- roll-down reflective foil R-4.5 insulating shade continuous across greenhouse, with intermediate sloped supporting wires, and magnetic tape edges seal (patent pend.)

- flashing
- 2x6 fascia
- extend top chord of every 4th truss
- wood damper w/ cont. vinyl hinge and cont. magnetic tape seal
- cor. F.G. glaz'g.
- black cor. alum. absorber on wood truss vert.
- foil face foam insul.

- hinged wood damper with cont. foam/ magnetic seal.

- roll-down bamboo shade

end wall with foil interior

f.g. glazing

alum foil

roll-down foil shade w/ magnetic tape edge seal (patent pend)

½ patio door-size glass

roll-down reflective insulation shade

wire shade supports

deadweight tensions wire supports prevents thermal expansion sag

2" foam insul

8" conc. mas.

- black corrougated alum siding over truss vertical
- 1" foil-faced foam insulation on back of truss vertical
- 6" f.g. batt insulation

continuous ceiling dampers

2 x 6 stud wall 24" o.c. w/ drywall, vapor barrier, 5½" f.g. batt insulation, ½" insulating sheathing, and wood siding.

2x trusses 24" o.c.

12" f.g. batt insul.

alum. gutter

corrougated f.g. glazing

reflective alum. siding on truss vertical, with 4" f.g. insulation

garage truss (identical to house)

2-2x6s at 4' o.c.
2x6 beams 4' o.c.

7'x9' "garage-type" overhead door at each end

flashing

2 x 8s 16" o.c.

(OPTIONAL WALL CONSTRUCTION: 8" & 12" poured, reinforced concrete)

plaster finish on 12" conc. masonry (fill voids with grout)

8" conc. mas. (horiz. voids for underground cooling vents where shown on lower plan)

wood louvers

carpet over 4" concrete slab on 6 mil vapor barrier

protective sheathing over 2" foam insulation over dampproofing

4"-6" stone backfill on E, N, & W sides to create underground cooling air passage (intakes where shown on upper plan)

6 mil poly. film

wire mesh and insect screen over openings

4" foundation drain

10"x20" conc. footing w/ 2 #4 rebars cont.

2" foam insulation on 4" pea gravel

4" interior foundation drain connected to sump pump under stair landing

10 feet

thermal layering concept

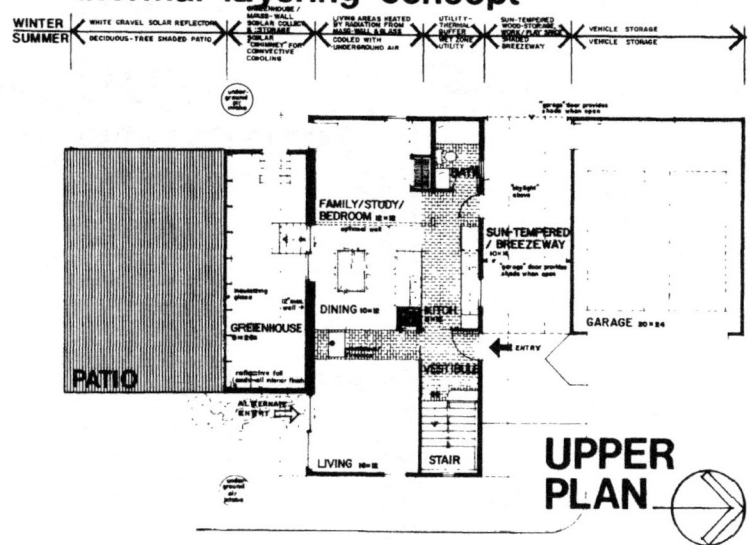

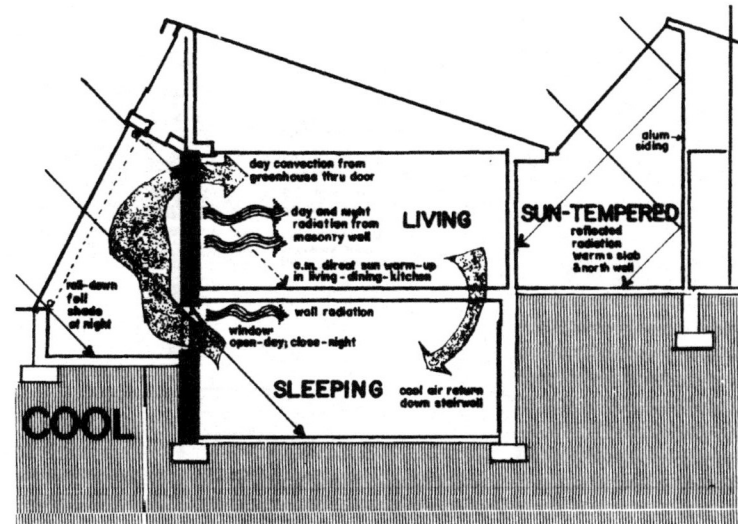

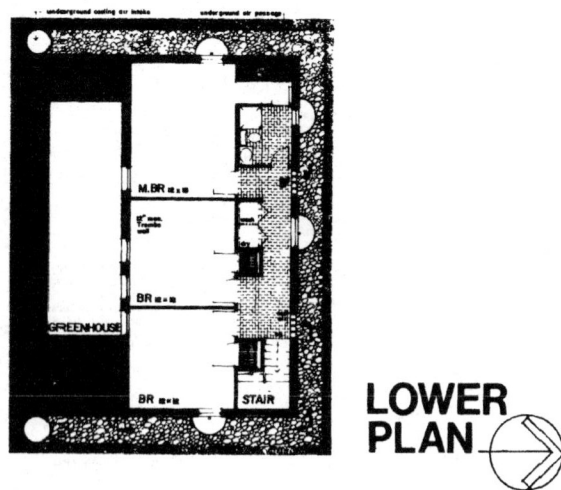

UPPER PLAN

LOWER PLAN

collector of solar heat during the heating season and which established a convective flow of outside air during the cooling season. Moving north from the primary living areas, a utility and circulation zone buffers the main area from the north. Wind sheltering is provided by the garage on the north side.

A particularly interesting feature is a sun-tempered breezeway which connects the garage with the house. During the heating season, this space is enclosed by closing the garage doors on the east and west ends. The skylight permits the sun to warm the space which is used for firewood storage and also serves as an entry vestibule from the garage to the house. During the summer when the garage doors of the breezeway are opened, they block the skylight, keeping the space cool.

The lower level of the plan, which contains the sleeping area, is below grade to reduce heat losses. It is constructed of concrete block with insulation on the exterior. The upper level of the house uses 2 x 6 frame construction with full thickness fiber glass insulation and insulated sheathing on the exterior. The roof construction is wood trusses with 12" of fiberglass. Windows are double glazed and make use of insulating shades.

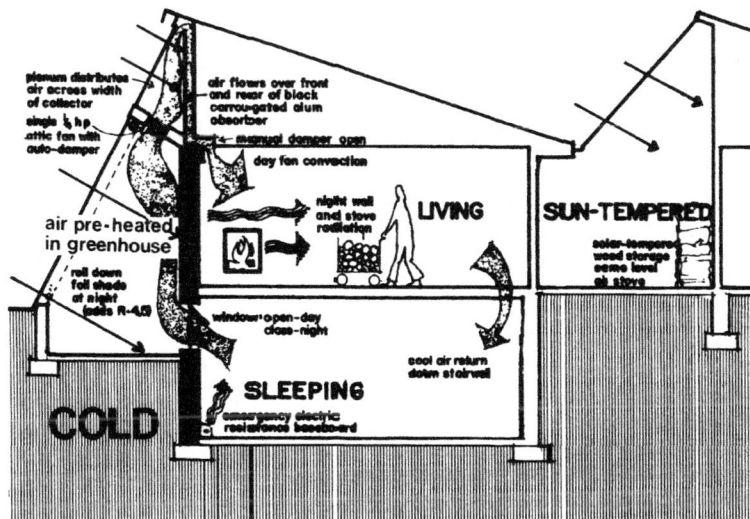

During the colder portions of the heating season, some additional heat may be gained on sunny days by the 142 square foot attic collector, which is located above the solarium. When this heat is needed, an automatic damper opens and a small fan circulates air from the solarium over the front and rear sides of a black corrugated aluminum absorber, and into the living spaces. Since this portion of the system is not connected with a storage mass, any heat collected must be used immediately or overheating of the living spaces will occur. This attic collector space will become quite hot during those sunny days when air is not being circulated through it. This portion of the solar collection system is secondary in importance to the solarium and south-facing glass, and its contribution to the heating requirements is less significant.

Additional heat is provided from a wood burning stove which is centrally located on the upper level of the home, and the stove is easily accessible from the sun-tempered wood storage area connecting the house and the garage. Auxiliary heat can also be provided by the electric resistance baseboard units which are located in the lower level.

HEATING

During cool periods, when some heating is required, the solarium collects solar energy as available. The occupant raises the rolling foil shade, permitting the sun to strike the heavy masonry wall and floor of the solarium. By opening the windows which connect the lower sleeping area with the solarium, and the door which connects the upper living area with the solarium, a convective air flow is established. The air which is heated in the greenhouse flows through the upper living area; as it cools, it flows down the stairwell, through the bedrooms, and returns to the solarium. To improve this flow of air, additional vents between floors can be added. Still, humidity control in the house may be a potential problem if substantial watering is done in the greenhouse. At night, the occupant closes the doors and windows which connect the house to the solarium and rolls down the foil shade. The solar energy absorbed by the masonry wall between the solarium and living area is conducted through the wall and radiated into the house. There is also some direct gain of heat through the south-facing windows in the upper level, but with no mass available to store this heat, it serves primarily to meet heating loads which occur during available sunshine.

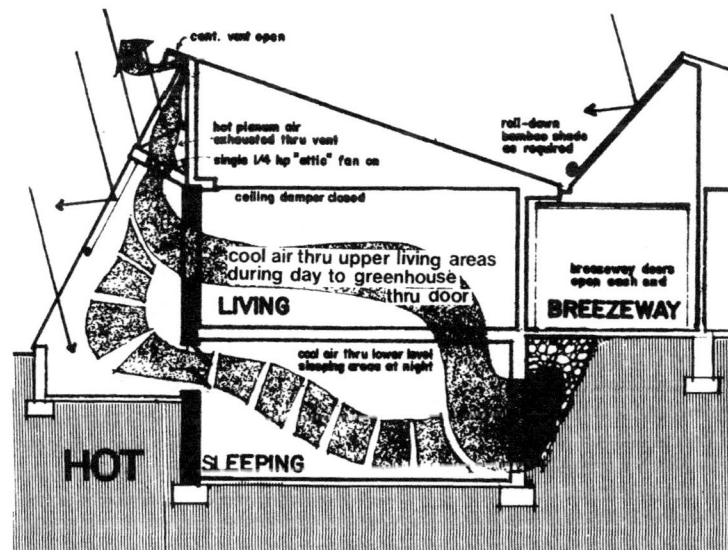

COOLING

During warm days in the transitional seasons, cooling is required in the house during the day. To facilitate this passive cooling, the occupant lowers the foil shade in the solarium so that solar energy cannot be absorbed by the masonry wall. Since the attic collector is not shaded, the air in this space becomes quite warm and established a convective flow of air through a continuous, manually operated vent at the ridge line. This thermal chimney is designed to draw outside air in through the underground ducts which cool this air by contact with the lower earth temperature. It is unlikely, however, that dehumidification of this outdoor air will occur, since the contact with the cooler earth surface will not lower the air temperature to the point at which moisture is extracted. During the warm period of the year, overhead doors are opened in the connecting link between the garage and house. This blocks the passage of the solar energy and establishes a breezeway for cooling. Since the sleeping level is below grade, it tends to remain cool.

During hot periods of the year, the natural ventilation system is augmented by the attic fan which draws air from the solarium to the attic collector. During this season, windows connecting the upper and lower living areas with the solarium are opened, permitting the passage of air from the interior of the house to the solarium and out through the attic collector area. Air is once again drawn in through the underground air passage, which provides cooling. During this portion of the year, a roll-down bamboo shade is lowered over the glazing which covers the connecting link between the garage and the living areas, further blocking solar penetration into the breezeway. Deciduous trees to be planted on the south side of the house will provide a patio area for use during the warm seasons.

CONCLUSION

It is expected that the passive solar features of this design will contribute more than 50 percent of the building's annual heating requirements. In addition, the design is expected to provide the complete cooling requirements for the building, although during periods of extreme heat interior temperatures probably will be somewhat higher than normal.

This design exhibits an excellent relationship between the solar elements and the living spaces. Through manual adjustments of the passive solar components, the occupant can control the home's response to climatic forces and can maintain comfort levels in the living spaces. The simplicity and flexibility of this design permits it to be expanded to a variety of configurations, with different size and siting requirements. Access can be made from a street located in any direction from the design, as shown on the opposite page. The floor plan can be modified to include only two bedrooms or as many as six. This flexibility contributes to the design's attractiveness to the speculative market and identifies it as an outstanding example of passive solar architecture.

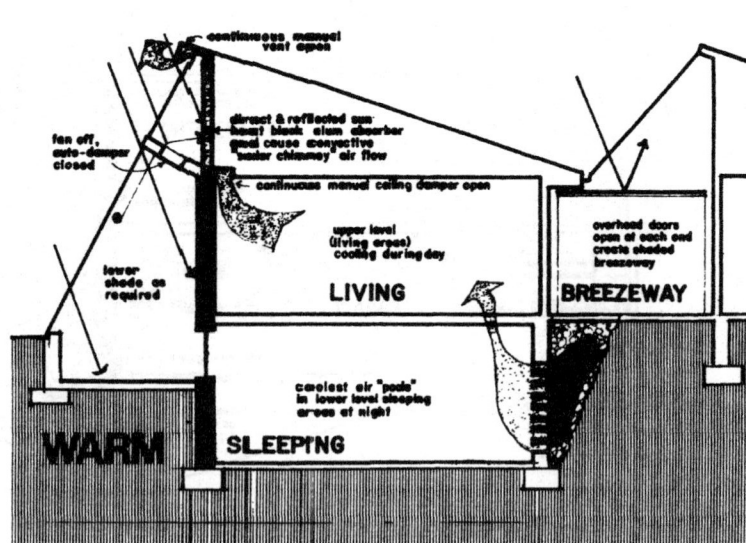

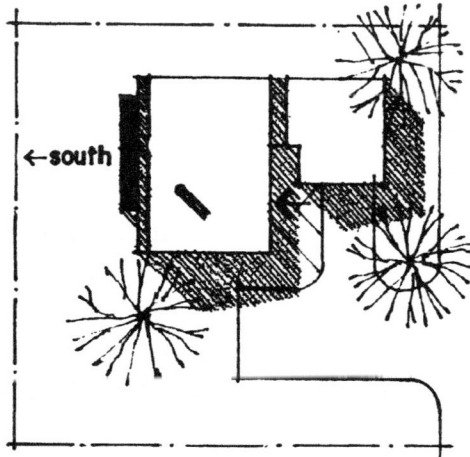
NORTH ACCESS

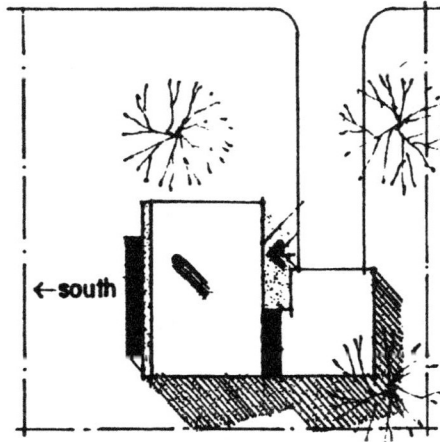
WEST ACCESS

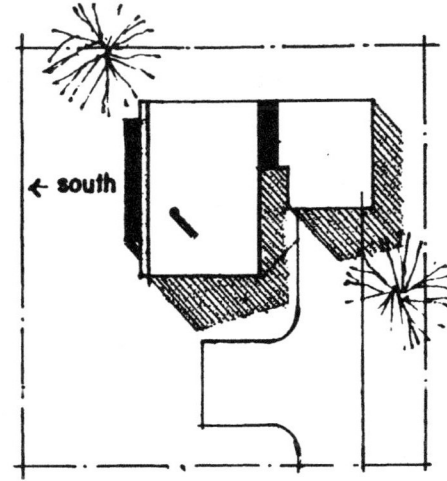
EAST ACCESS

SITE VARIATION

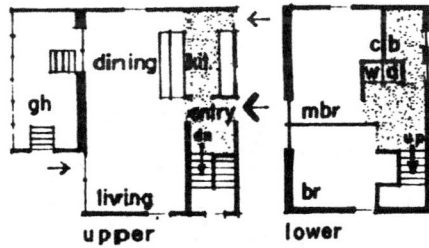

**2 BEDROOMS -
1120 sf**

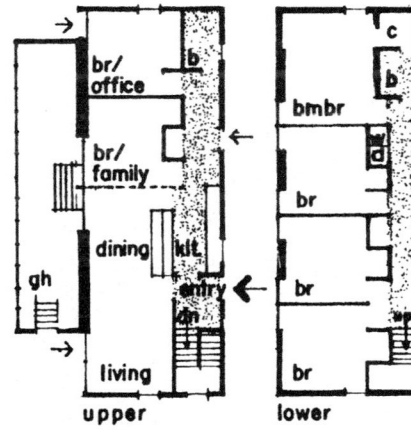

**5 or 6 BEDROOMS -
2080 sf**

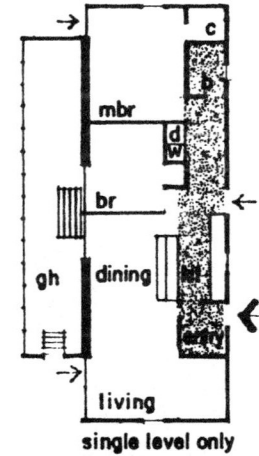

**2 BEDROOMS -
1120 sf**

PLAN VARIATION

Lima Township, MI

Gary A. Cook
Ann Arbor, MI

HEATED AREA: 2,000 FT² plus solarium

NUMBER OF DEGREE DAYS: 6,267

NET THERMAL LOAD: 61.5 10⁶BTU/YR

AUXILIARY ENERGY: 1.8 BTU/DD/FT²

YEARLY SOLAR FRACTION: 73%

Private client.

CONTEXT

This 2000 sq. ft. home is designed for a site outside Ann Arbor, Michigan. The climate for the region has 6,267 heating degree days with average wind speeds of 10 mph. The site slopes south toward a peat bog and is lightly wooded with deciduous trees. The house contains a large living room, 3 large bedrooms, 3 full baths, study, kitchen, pantry and dining room as well as a very large (9' x 48') attached solarium. The construction is wood frame above grade and concrete block below grade.

CONSERVATION

The house is oriented 7° east of south because Michigan skies tend to be clearer in the morning than afternoon. Although coniferous trees would be better windbreaks, large deciduous trees to the west help to break the winter winds. The massing of the house fits nicely into the topography with the collector area and thermal storage placed down the hill. The simple rectangular shape helps to keep both heating loss and building costs low.

The airlock vestibule is positioned at the northeast corner of the building and has a door on the east for winter entry and a door on the west for summer entry. The entrance is at the second level with a stair that runs directly down to the living areas.

The walls above grade are extremely well insulated with 5-1/2" of glass fiber and 1" of styrofoam for an R-25 wall. Below grade concrete walls are covered with 2" to 3" of styrofoam. The R-38 roof is achieved with 11" of glass.fiber (double 5-1/2" batts). The double glazed windows have a built-in Venetian blind which can be closed at night to achieve an R-factor of 3.

HEATING

The solar system consists of a solarium and a thermosyphoning rock bed. The house is directly modeled on the Paul Davis House in New Mexico.

The steeply-sloping site allows a thermosyphoning air loop collector and rock bed to be positioned below the living space. There are two modes: one from the collector to thermal storage and another from storage to the living space. To control both modes a series of dampers are operated manually. In the collector to storage mode, dampers to and from the collector are opened by a cable plunger in the solarium above. Heat is allowed to enter the living space from the collector via top and bottom dampers at the back of the rock storage which are opened. At night, heat can be thermosyphoned from the rock bed into the living space by closing off the set of dampers to the collector while the set of dampers at the rear of the rock bed remain open. Warm air supplied from the collector or rock bed is allowed to circulate freely through the living spaces. There is a cool-air return duct which takes air from eight floor registers at the north side of the living room, underground to the plenum at the bottom of the rock bed. The rock bed serves only the first floor which is somewhat thermally isolated from the second floor.

The 1500 cubic foot rock bed is sized to store roughly 400,000 BTU's on a clear winter day. The 432 sq. ft. collector has corrugated aluminum panel reflectors at top and bottom. They are opened manually during the day to act as reflectors and closed at night. The double glazing collector, tilted at 60° above horizontal, is constructed with an absorber plate of expanded black metal lath placed diagonally across the air channel to maximize heat transfer from the solar heated metal to the passing air stream.

The solarium has vertical double glazing with four operable skylights on its roof. Direct sunlight is stored by a 9" thick concrete and brick floor and a 4" brick north wall. The solarium supplies warm air through a high register on its north wall into bedroom above. Typically, air is returned to the greenhouse via the stairwell at the north and back through the French doors that open into

SECOND FLOOR

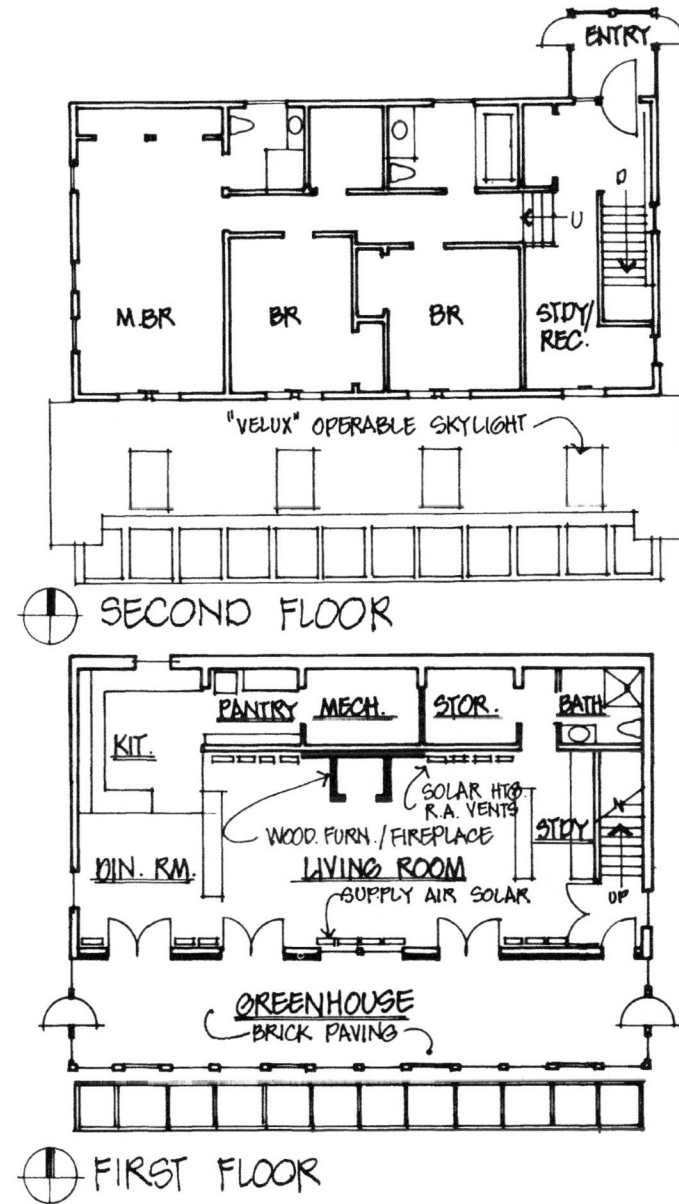

FIRST FLOOR

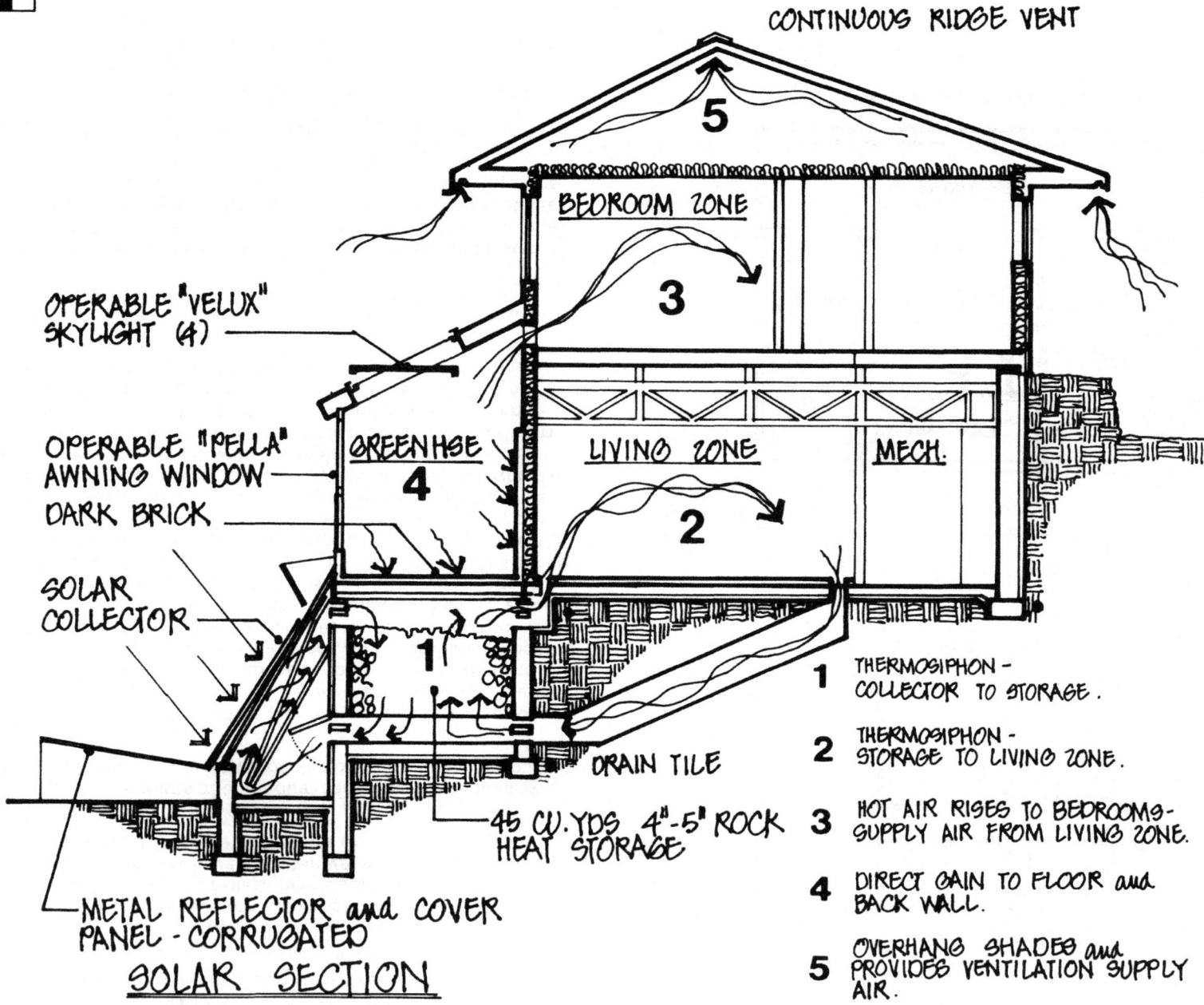

CONTINUOUS RIDGE VENT

BEDROOM ZONE

5

3

OPERABLE "VELUX"
SKYLIGHT (4)

GREENHSE

4

LIVING ZONE

MECH.

OPERABLE "PELLA"
AWNING WINDOW

DARK BRICK

2

SOLAR
COLLECTOR

1

DRAIN TILE

45 CU. YDS 4"-5" ROCK
HEAT STORAGE

1 THERMOSIPHON -
COLLECTOR TO STORAGE.

2 THERMOSIPHON -
STORAGE TO LIVING ZONE.

3 HOT AIR RISES TO BEDROOMS-
SUPPLY AIR FROM LIVING ZONE.

4 DIRECT GAIN TO FLOOR and
BACK WALL.

5 OVERHANG SHADES and
PROVIDES VENTILATION SUPPLY
AIR.

METAL REFLECTOR and COVER
PANEL - CORRUGATED

SOLAR SECTION

the solarium. If the stairwell door is left open and allowed to complete this loop, warm air supplied from the collector-rock bed will be allowed to rise to the second floor which will increase heat stratification in the house and possibly overheat the second floor. However, if a return air duct from the bedrooms directly to the greenhouse were added, excessive stratification would be avoided and warm air from the collector would remain on the first floor where it is needed. When functioning properly, the solarium will provide heat to the second floor, while the thermosyphon collector-rock bed system will heat the first floor. Together, the two systems are calculated to provide about 75% of the winter heating needs. This savings represents approximately 45 million BTU's. Auxiliary heat is supplied by both a forced hot-air furnace rated at 45,000 BTUH and an airtight wood/coal stove rated at 50,000 BTUH.

COOLING

The cooling load in this climate is not high. There are eight operable windows on the south wall and another ten on the east, west and north walls to provide good cross ventilation, particularly on the second floor. The massive thermal walls and floor of the first level should help to stabilize summer temperatures and prevent uncomfortably hot afternoons.

CONCLUSION

The solarium works well in conjunction with the rock bed. Depending on the regime required by the plants, the combination could be improved further if the roof of the rock bed was insulated so that it did not radiate heat to the solarium at night. It appears that the 18″ thermosyphon collector air channel is too wide and the performance could be increased further by reducing it. In an air system of this type, care must be taken by the builder to insure that the rock bed vents seal tightly and that the rock bed is very well insulated. Raising and lowering the reflectors at the head and sill of the collectors may be an unnecessary task as the rock bed loses very little heat back through the collector at night.

This design displays excellent energy conservation features and is well sited. The isolated air loop collector has the advantage of negligible heat loss at night because it is shut off from the house. Summer heat gain from the collectors to the house is also avoided. Additionally, the system allows a good view to the south because it is below the living levels.

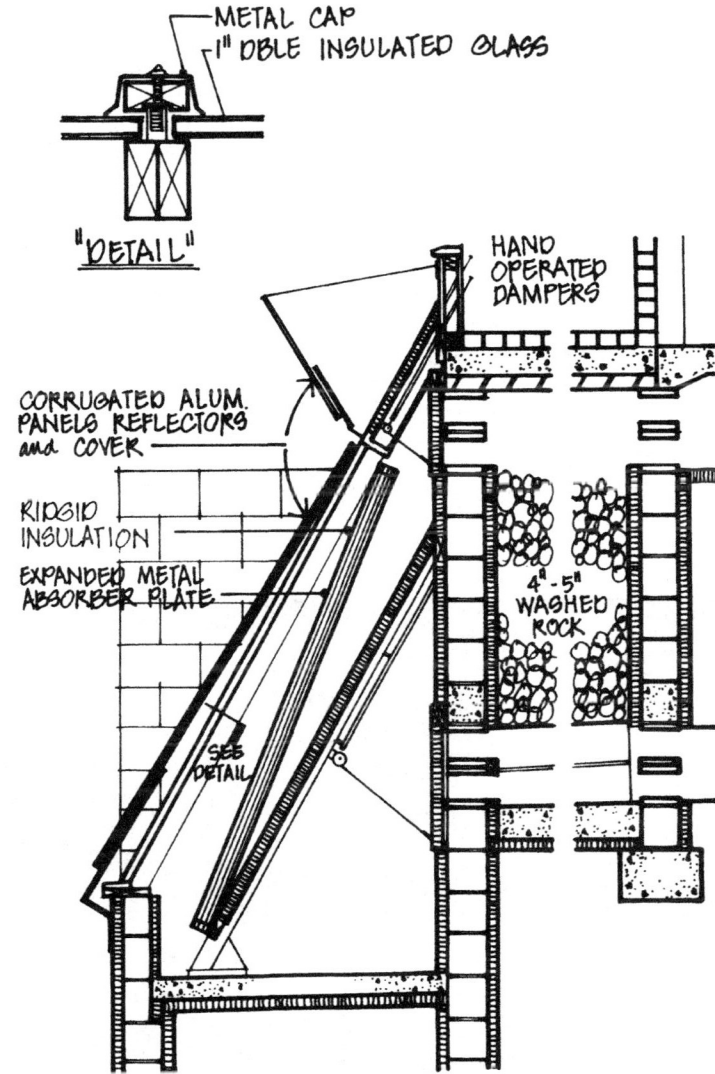

SOLAR COLLECTOR DETAIL

Hopewell, NJ

Vinton Lawrence, Harrison Fraker Architect
Princeton, NJ

HEATED AREA: 2,680 FT^2

NUMBER OF DEGREE DAYS: 4,911

NET THERMAL LOAD: 62.3 10^{6}BTU/YR

AUXILIARY ENERGY: 1.69 BTU/DD/FT^2

YEARLY SOLAR FRACTION: 79%

Private client.

CONTEXT

This solarium house is located in a rural area of Central New Jersey which experiences about 5,000 heating degree days. The building is sited near the top of a ridge at the edge of a gently sloping meadow which is surrounded by dense woods. It is a 3-story wood frame house with 3 bedrooms, study, living room, dining room, kitchen, and large work room. There is a garage that is connected by a breezeway to the house and an entry courtyard.

CONSERVATION

The building is oriented so that the solarium faces due south. The sloping site allows part of the basement to be below grade on the north side. The main entrance is sheltered from north winter winds by the garage and breezeway and incorporates an air-lock vestibule. There is a tall dense stand of woods to the north, east and west that provides further buffering from winter winds. By locating the building near but not on top of the ridge, the house can also take advantage of summer winds for cooling. The shape of the house is a cube partially buried in the ground to cut heat loss and to allow entry at the second level. Summer heat gain can be reduced by elongating a building in an east-west direction

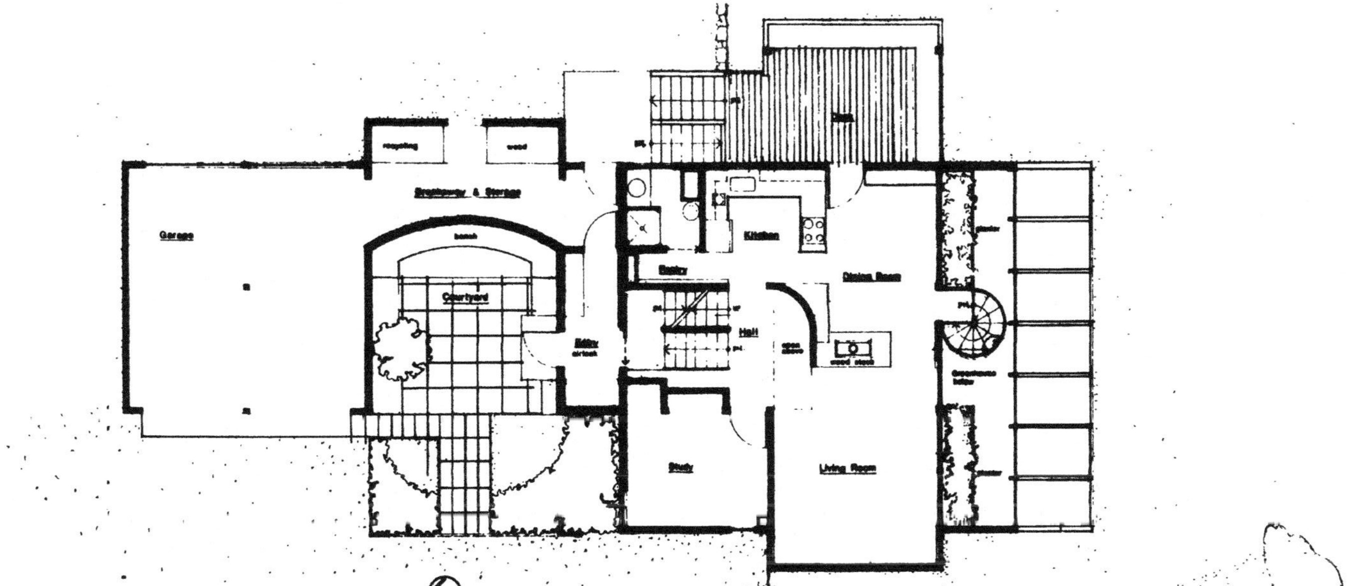

GROUND FLOOR PLAN

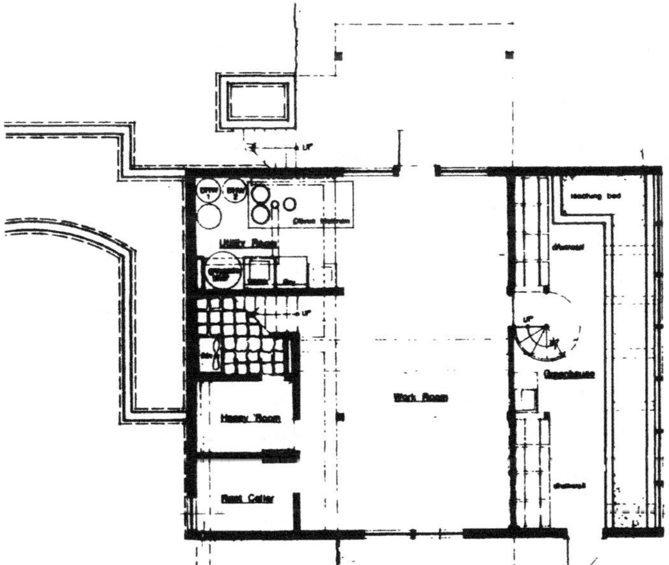

BASEMENT PLAN

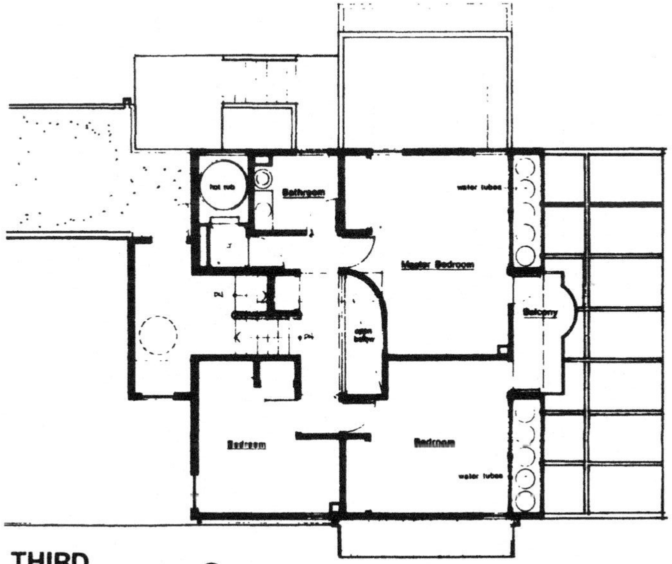

THIRD FLOOR PLAN

151

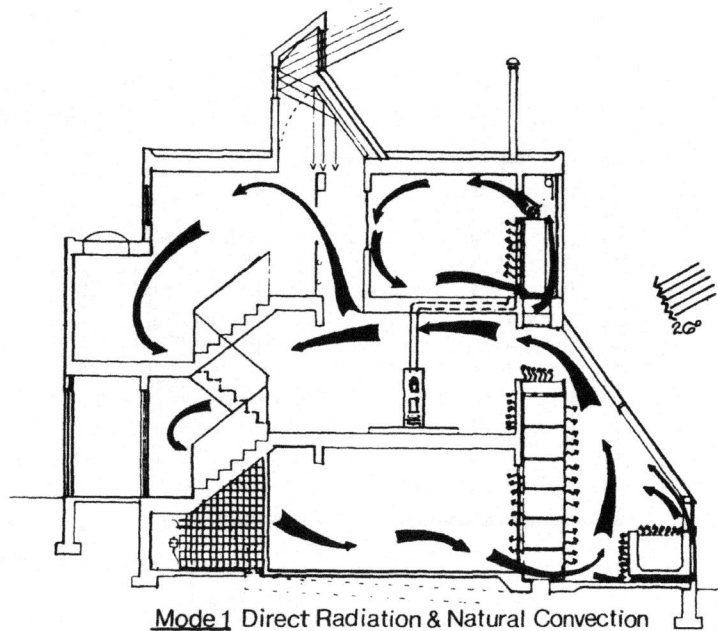

Mode 1 Direct Radiation & Natural Convection

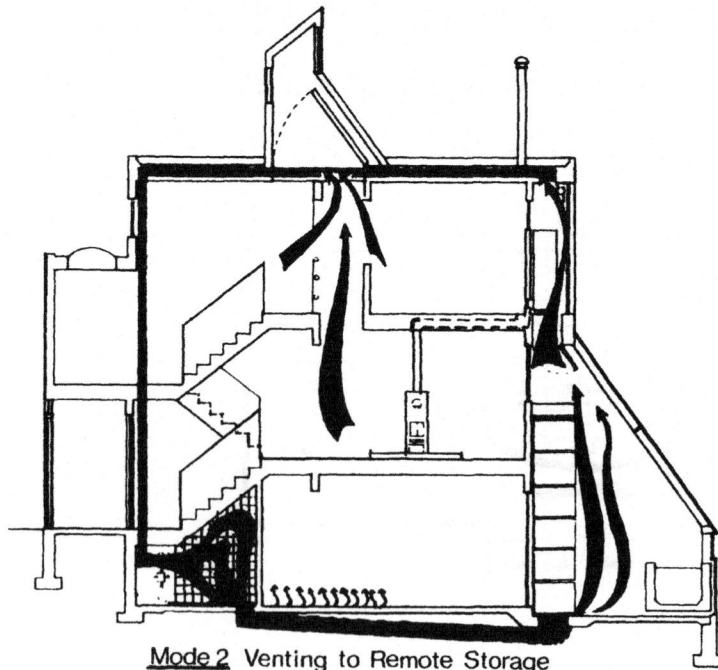

Mode 2 Venting to Remote Storage

which keeps the size of the east and west walls at a minimum. However, with sufficient shading of windows on the east and west walls, the cube is an energy conserving shape because of its low exterior surface to floor ratio. A heat loss for the home of 6.6 BTU/DD per square foot of building is achieved without excessive insulation. Ureaformaldehyde foam is blown into the 2 x 6 wood frame wall to give an R-26 wall. The roof, constructed of 2 x 10 wood joists with 9-1/2″ glass fiber batt insulation has an R factor of 40. All windows are double-glazed and have curtains that cover them at night.

The solarium glazing uses a moveable insulating shade to create an R of 4.6 at night. Four inches of urethane insulation is applied to the exterior of the foundation walls and is covered with an epoxy stucco. Weatherstripping all of the doors and windows cuts the air infiltration rate to an estimated 1/2 air change per hour. Night set-back of the thermostat is credited with cutting the heat loss by 15%. This winter loss translates into 17,635 BTU/DD or a peak loss of about 48,000 BTU/hr., when the outside temperature is 0°F.

HEATING
The dominant solar component is the attached two-story greenhouse which leans against the southern side of the building. The greenhouse or solarium has 333 square feet of tilted glazing and 100 square feet of vertical glazing above. There is an additional 120 square feet of water storage wall which consists of fiberglass water tubes. Both of these areas are glazed with a double-wall clear acrylic plastic. There is also 43 square feet of south-facing, double-glazed windows. Together these glazed areas collect slightly over 100 million BTU during an average winter of which about 50 million is used. The average solar heating fraction for the months of October through May is about 80%. Auxiliary energy consumed by the woodstove back-up system is approximately 13 million BTU. The 50 million BTU supplied by the solar system is the equivalent of about 660 gallons of oil (@19 gal. = 75,000 BTU) or about 14,500 Kwh.

The primary modes of operation of the solar system are for heating. Although only 4 modes are described below and in the accompanying diagrams, as the occupants gain familiarity with the system, they will be able to use combinations of these modes to "fine-tune" the system to suit their needs more closely.

1. Direct Radiation and Natural Convection from Collection Areas to Living Space
This mode is the primary operating mode in the heating season, providing all the heating needed for an average sunny day. When operating in this mode, the following controls and conditions are

152

necessary: a) A convective loop from the solarium and water wall to the house is started manually by opening windows, doors and vents between the house and solarium as needed; b) The solarium drumwall and the water wall radiate heat to the house; c) The return air duct from the solarium to remote storage is automatically shut.

2. Venting of Excess Heat Gain from Collection Area to Remote Storage

When the solarium and/or the house air temperature rises above a set limit of 80°F., the excess hot air is circulated to a remote storage—800 gallons of water jugs below the stairs. In this mode the following controls are necessary: a) First, doors, windows and vents between the solarium and interior rooms are closed in this mode; b) The return air damper in the solarium duct to remote storage automatically opens when the temperature exceeds set limit of 80°F.; c) The return air damper from the water wall to remote storage opens automatically and a fan is turned on by a differential thermostat; d) The solarium supply air damper opens while the house main supply damper from auxiliary or remote storage remains closed. Automatic control operates these two dampers through a mechanical linkage so that when one is closed the other is open and vice-versa; e) The return air damper at the top of the light well opens to draw excess space heat.

3. Fan Assisted Circulation from Remote Storage to Living Space

On those occasions when the primary passive mode cannot completely satisfy the heating demand of the house, a fan has been provided to assist in the distribution of heat from the remote storage and recirculation of hot air from the top of the central stair well. In this mode the following controls and conditions are necessary: a) The solarium and house are thermally isolated zones in this mode; b) A fan operating at 1700 cfm maximum (speed selected by user) is controlled automatically by a thermostat in the living space; c) The damper to the main house supply duct from remote storage is opened; d) The damper in the return air duct from living space to remote storage is opened; e) The solarium return air damper to remote storage is closed by demand from the house.

4. Night Time Mode

On winter nights, moveable insulating curtains are closed and the cupola insulating shutter is lowered to minimize the building loss. A fan circulates heated air to all rooms as needed: the insulating curtain is power-operated by manual switch or time clock; the insulating shutter at the top of light well is closed manually; and the insulating shade outside the water wall is power-operated by manual switch or time clock.

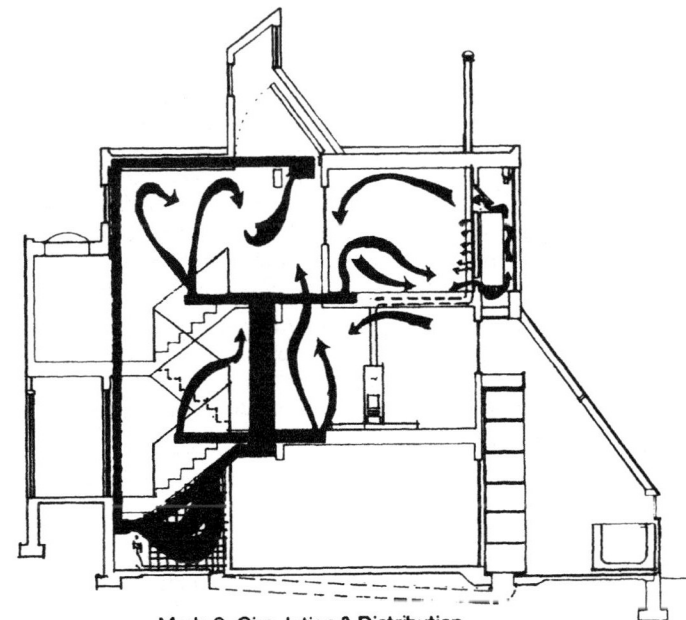

Mode 3: Circulation & Distribution

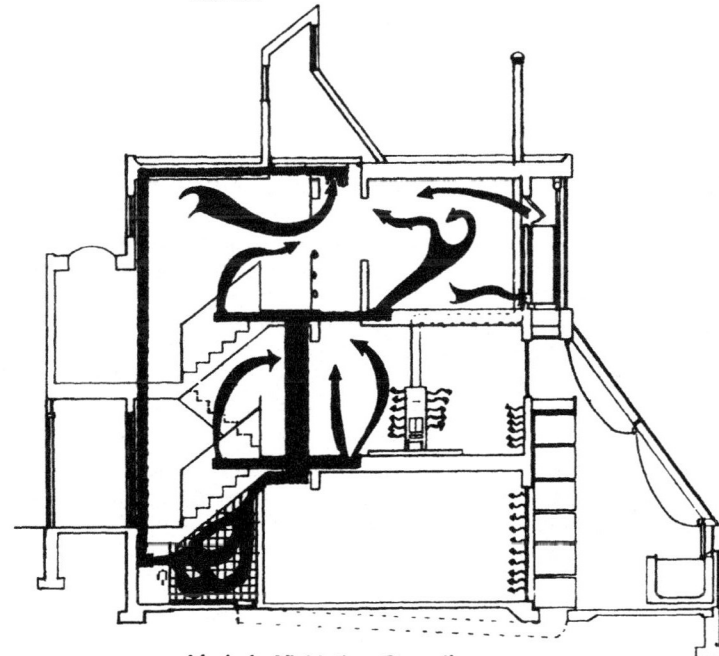

Mode 4: Night-time Operation

For back up, a wood stove (with thermostat) is manually loaded and lighted as required with a maximum capacity of 55,000 BTUH. An alternate back-up heating source such as an oil-fired furnace, wood furnace, etc. could easily be integrated into the duct system as required by different users and codes.

COOLING

Due to the topography of the site, prevailing breeze will usually provide sufficient natural ventilation. When this breeze is insufficient rising solar heated air in the water wall, solarium and the cupola creates sufficient pressure to "induce" or draw cool air from the well. The fan provided in the ducting system helps recirculate both cool air in summer and redistribute heat in winter. The remote heat storage capacity of 800 gallons of water provided in the basement draws house heat for both heating and cooling modes. In addition, the water pipe from the house well is housed in a 12" diameter culvert which can double as an air intake duct for the cooling system. The air in the culvert is cooled by heat loss to the earth as well as to the cold well water in the pipe. Based on a water usage of 120 gallons per day and a temperature of 50°F., the available cooling from this source is estimated at 115,000

BTU/day. Combined with the remote storage the overall capacity of the cooling system is calculated at 12,000 BTUH which is a 1-ton air conditioner.

CONCLUSION

This complex solar home should probably not be attempted until the builder or designer has had some experience in passive solar applications. Although it promises to provide both a high level of comfort and performance, its numerous modes of operation make it a difficult design for the beginner.

Although the house has been designed for a specific client, it can be adapted to different sites, climates and lifestyles. A conventional back-up furnace can be substituted for the wood stove. The entire system of remote thermal storage can be eliminated as it does not add that much capacity (15%) and adds considerably to the cost. As is, the system will work better if the remote storage is charged and discharged from the same plenum (reverse flow fans). (It is not known yet whether the temperatures in storage will be high enough for blowing into the rooms without creating discomfort.) Without the remote heat storage, the water wall will continue to heat the bedrooms through the night because the night curtain is on the exterior side of the tubes and will keep heat within the living space. In addition, with a thermal curtain at the perimeter of the solarium the steel water drums and plant bed in the solarium can heat the living space at night by opening the solarium to the house. (It is questionable whether enough heat can be radiated throuh the closed glazing and gypsum dry wall from the stacked water drums.) In less severe winter climates, where not much heat is needed to maintain the solarium at an adequate temperature for plants, the water drums can be positioned within the living space, on the other side of the back wall of the solarium to provide better heat storage and distribution.

The passive cooling system is a large marketing asset. The "cool pipe" provides considerable cooling, although after several hours the earth around the culvert will warm up and reduce the cooling effect. Without the cool-pipe, the cupola and numerous other vents and windows provide excellent natural ventilation.

The architectural style of the residence is contemporary with natural wood siding. Marketability to homebuyers who are looking for self-sufficiency is enhanced by the integration of a composting toilet, gray-water recycling and greenhouse. This represents a promising solution for sites without sewer or difficult septic conditions. The house responds to an emerging market of homebuyers who want simplicity, independence, and frugality as well as contemporary architectural style and grace.

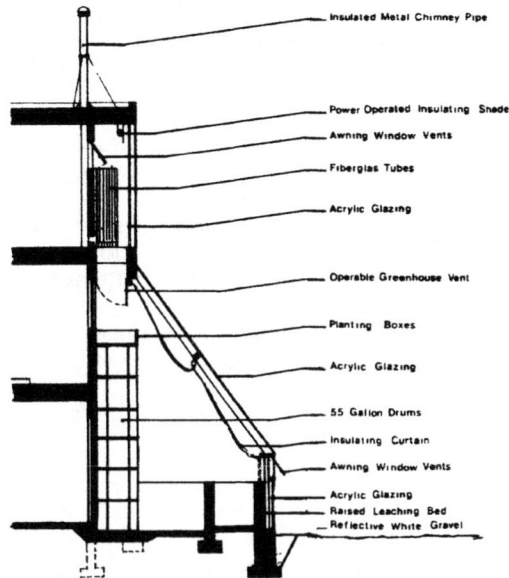

Insulated Metal Chimney Pipe
Power Operated Insulating Shade
Awning Window Vents
Fiberglas Tubes
Acrylic Glazing
Operable Greenhouse Vent
Planting Boxes
Acrylic Glazing
55 Gallon Drums
Insulating Curtain
Awning Window Vents
Acrylic Glazing
Raised Leaching Bed
Reflective White Gravel

SECTION

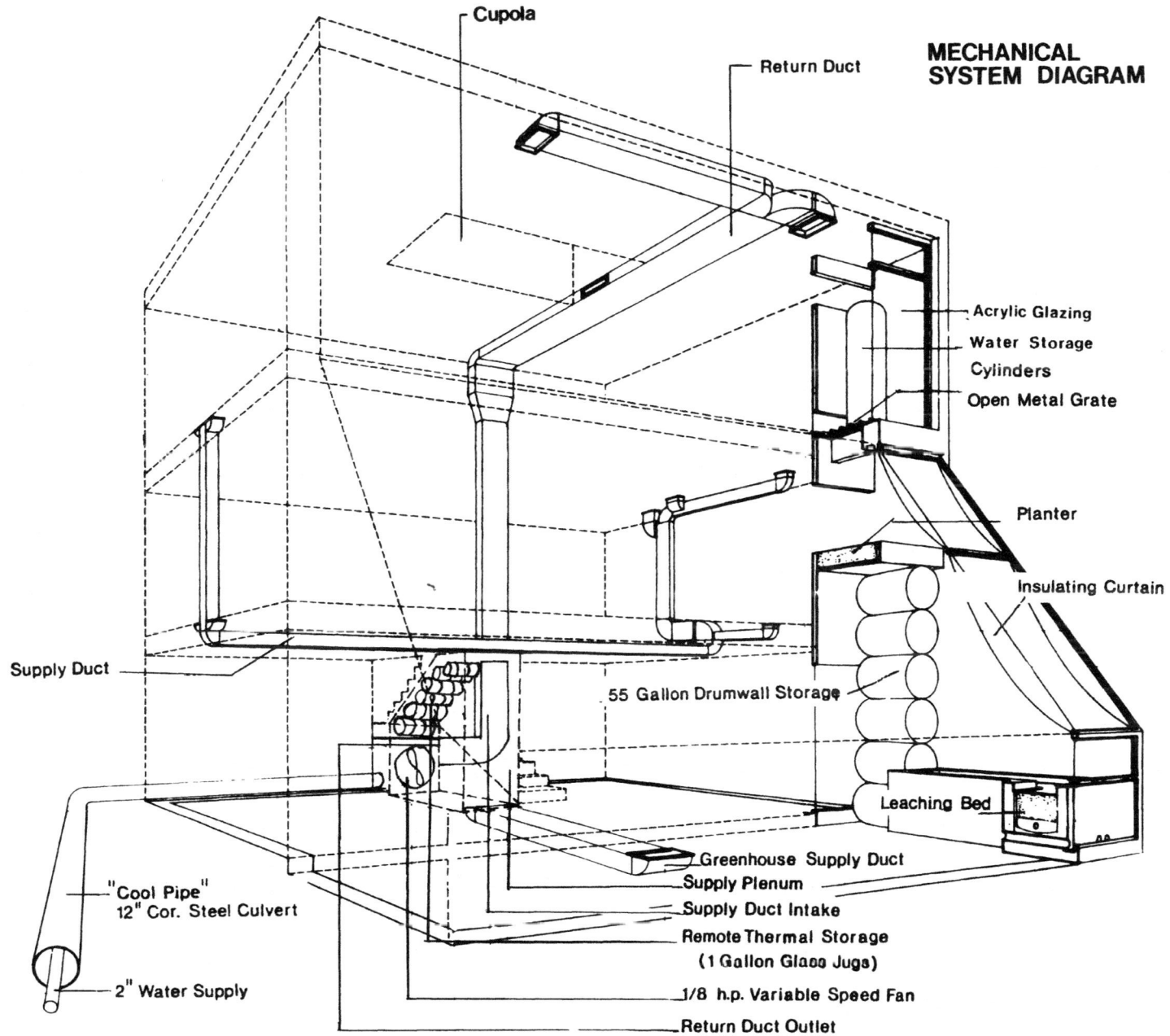

Cupola

Return Duct

MECHANICAL SYSTEM DIAGRAM

Acrylic Glazing

Water Storage Cylinders

Open Metal Grate

Planter

Insulating Curtain

Supply Duct

55 Gallon Drumwall Storage

Leaching Bed

"Cool Pipe" 12" Cor. Steel Culvert

Greenhouse Supply Duct

Supply Plenum

Supply Duct Intake

Remote Thermal Storage (1 Gallon Glass Jugs)

2" Water Supply

1/8 h.p. Variable Speed Fan

Return Duct Outlet

155

Raleigh, NC

Mike Funderburk, Sunshelter Design
Raleigh, NC

HEATED AREA: 1,450 FT²

NUMBER OF DEGREE DAYS: 3,352

NET THERMAL LOAD: 34.9 10⁶BTU/YR

AUXILIARY ENERGY: 3.44 BTU/DD/FT²

YEARLY SOLAR FRACTION: 81%

Private client.

CONTEXT
This design was conceived in response to the 3,350 heating degree days, 2,200 cooling degree day climate typical of North Carolina's Central Piedmont region. The need for both heating and cooling in the Raleigh area dictated a scheme that is compact enough to enable effective heating, yet sufficiently open to allow natural ventilation and summer cooling.

The building envelope of the 1,450 sq. ft., two-bedroom, two-bath home is of conventional wood frame construction. The design is derived from the two-story I-shaped homes built in the Southeast Piedmont during the turn of the century. Although the traditional room arrangement has been retained, the design has been modified to incorporate a greenhouse/solarium passive solar heating system.

CONSERVATION
The house is situated in an open area near the northern edge of the wooded site to take maximum advantage of the available sunshine. It is elongated in the east-west direction, providing a shape and orientation which minimizes yearly heating and cooling requirements. Frequently occupied spaces are separated from the greenhouse atrium by a massive interior masonry wall. This wall absorbs incoming solar radiation through the solarium and acts as a radiant heat source for the adjacent living areas. This arrangement simplifies the distribution of stored solar energy and provides the occupied spaces with a natural light.

Double glazing is used throughout the home. The main south-facing glazing areas are protected against heat loss by thermal curtains which can also act as sun shades during periods of overheating.

Operable windows strategically placed throughout the home work in conjunction with a ridge vent which can be opened as needed to provide ventilation.

Other heat-producing appliances such as the woodstove, cookstove, gas range and gas hot water heater are located within or adjacent to central mass walls. This arrangement takes advantage of the thermal dampening effect of the mass, lowering instantaneous heat gain from these items and storing their heat for distribution over longer periods.

Two thermostatically operated, 7,000 cfm ceiling fans with reversible blades are used to recirculate stratified warm air during the heating season and aid in ventilating during the summer months. The fans use very little energy and add to the character of the home.

Wind sheltering is provided by mature trees and other vegetation. An airlock vestibule is also included for further protection against infiltration.

HEATING
The direct gain passive heating system in this design consists of the greenhouse/solarium collector area coupled to an interior mass wall and concrete and brick floor mass storage. The thermal curtains are raised manually to allow sunlight to penetrate this area each day and are lowered in the evening to retain the stored heat. Convection currents circulating through the solarium to the living areas assist the natural radiant distribution of heat from the storage mass. The ceiling fans cycle on at a predetermined temperature to further aid in moving warm air throughout the home. A wood stove and electric hot water heating are provided for auxiliary heat.

COOLING

Cooling in the Raleigh area is largely dependent on interior air movement and control of solar heat gain. The use of operable windows, continuous ridge vents and ceiling fans ensure adequate air movement throughout this home. The two-level, open solarium space allows the stratification of warm air during the summer, further isolating the occupants from possible discomfort. The use of a ridge vent in combination with the natural stack effect will provide adequate heat removal regardless of wind direction. The thermostatically operated fans can be employed to exhaust hot air from the home during periods of objectionable overheating.

CONCLUSION

This design combines a number of passive space conditioning techniques which are logically selected and implemented to cope with the changing seasons of the North Carolina Piedmont. The use of a solarium as the collection space provides useable living space in addition to passive heating. The central position of the interior mass wall affords radiant heating to the majority of the occupied spaces while providing a convenient structural element. The two-story spaces which allow heat to stratify serve the dual purpose of reinforcing the stack effect for summer ventilation while providing a common gathering space for redistribution of stratified heat in winter. The variation in spaces which has resulted from the designer's attempt to reinforce the natural energy flows in this home also adds visual interest and marketing appeal.

The home as presented would benefit from the addition of mass in the form of water-filled containers or increased mass wall area, in order to avoid overheating the solarium on clear winter days. The mass wall which separates the solarium from the remainder of the living spaces should be provided with doors and clear glazing to allow the option of completely isolating the living areas from this space. In so doing, the problem of overheating could be reduced and heat supplied to the living areas could better be protected during extended cold periods.

The thermal curtains included in the design will not be as effective as rigid shutters unless a tight seal can be achieved. Care must be taken to select fabrics which are not subject to ultraviolet degradation. This is also true of furnishings and other fabrics which may be used in the solarium space.

Generally, this design is a compact, well-conceived scheme which blends passive solar space conditioning with a respect for the traditional architecture of the region. Future expansion has been well considered, making this home a wise choice for growing families.

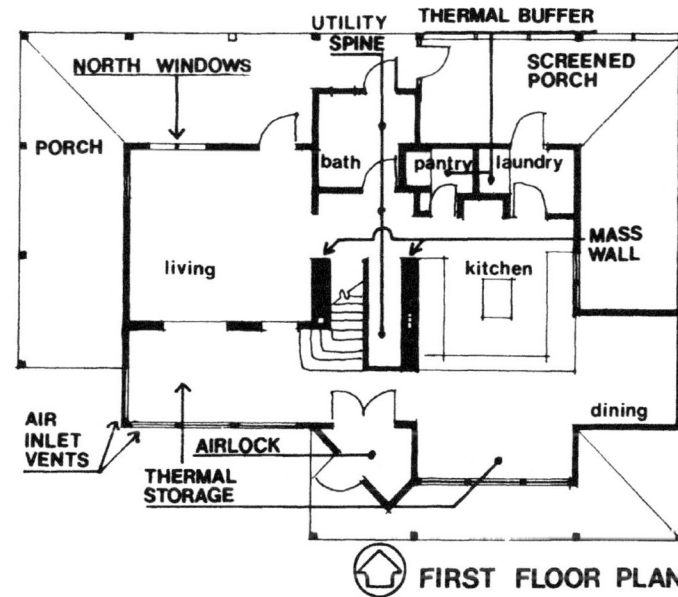

FIRST FLOOR PLAN

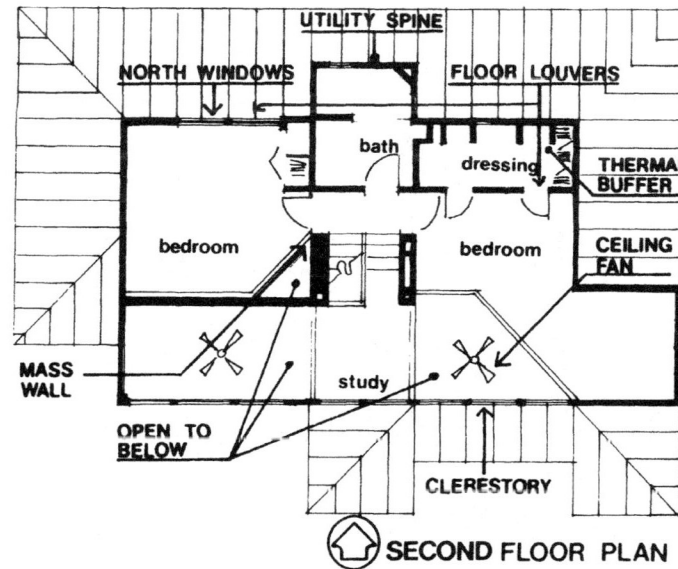

SECOND FLOOR PLAN

summer day

In the summertime, most of the incoming solar radiation is shaded by building overhangs and deciduous trees directly to the south of the house. To reduce thermal stress due to high temperatures and humidity, ventilation is highly emphasized. Cross ventilation takes place when air enters the south and west inlet vents (fig. a), and exits through exhaust vents (fig. b) on the north side of the house. Ventilation by the stack effect takes place when low lying air is heated and rises to be exhausted by ridge vent (fig. c) at highest part of house. Ceiling fan (fig. d) provides mechanical assist for this ventilation by reversible blades which blow upward in summer, downward in winter.

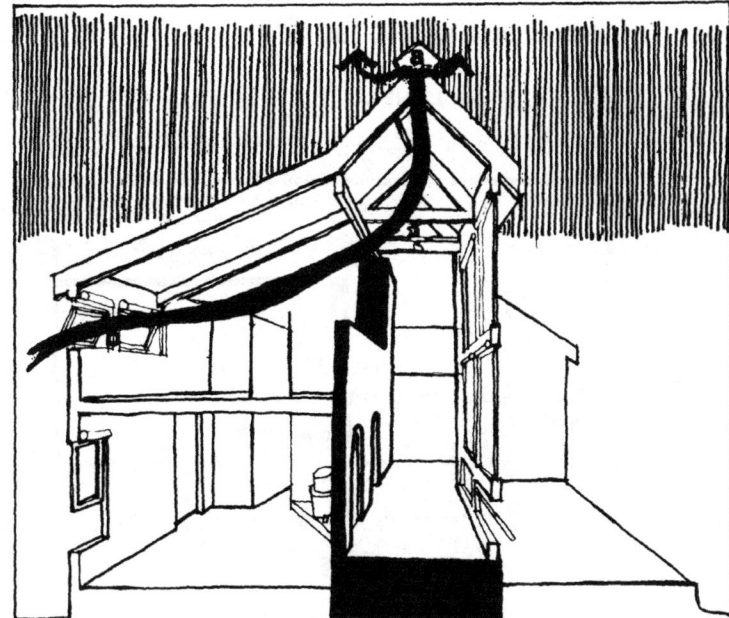

summer night

Any heat gain on the mass surfaces is radiated to a certain extent to the night sky. Heat also is convected to the upper ridge area where it is exhausted through the ridge vents (fig. a). Vents (b) adjacent to sleeping areas are opened to allow air to flow from these spaces to the ridge area. This dissipation of interior heat allows the house to cool down to only 5°-10° above the minimum night temperature. The mass starts the following day at a relatively low temperature, reducing the need for mechanical assist till late afternoon, if at all.

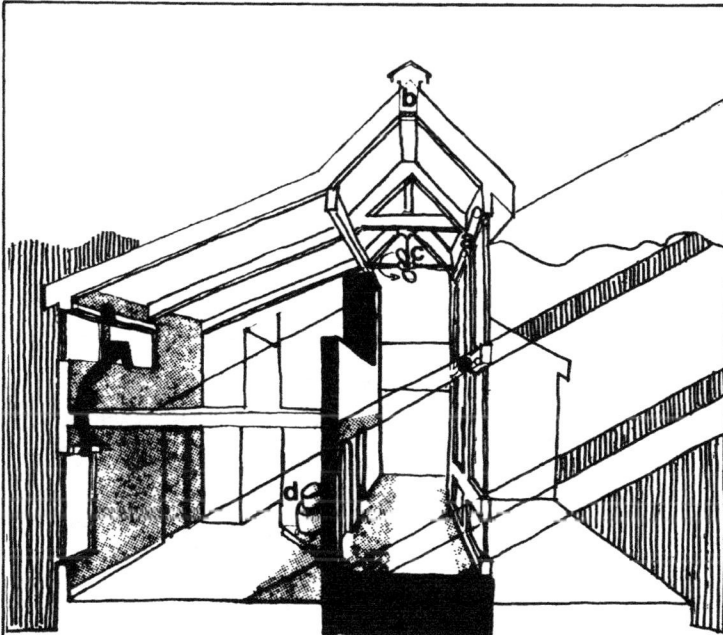

winter day

Sunlight enters south glass and strikes mass areas on floor and mass walls. Areas shown not shaded are in sun at noon on Dec. 21. Roll up "window quilt" shades are raised at dawn and lowered at dusk (fig. a). Insulating door at ridge vent is left closed throughout the heating season (fig. b). Ceiling fan (fig. c) recirculates stagnant warm air from the high ridge area to lower level living areas. When sun is not shining, backup heat is provided by an airtight woodstove (fig. d). Further backup heat is provided by closed-loop hot water electric heaters. Louvers located in second floor allow heated air to rise to bedroom spaces (fig. e).

winter night

Solar radiation stored in mass walls and floor (fig. a) radiates to living spaces to provide heat when the sun is not shining. An airtight woodstove provides backup heat when needed. Moveable insulation (fig. b) is rolled down on all major windows to prevent nighttime radiative heat losses. The ceiling fan (fig. c) transfers stagnant warm air to the living areas of the home. Floor louvers (fig. d) allow air to convect from first floor to second floor living spaces.

Santa Rosa, CA

R. Fernau & Berkeley Solar Group
Berkeley, CA

HEATED AREA: 1,150 FT²

NUMBER OF DEGREE DAYS: 2,912

NET THERMAL LOAD: 26.4 10⁶BTU/YR

AUXILIARY ENERGY: 4.55 BTU/DD/FT²

YEARLY SOLAR FRACTION: 58%

CONTEXT
This 1,100 sq. ft. residence is designed for a slightly sloping site in northern California. The area has a mild climate of approximately 3,000 degree days heating loads with partly cloudy winters and hot summer periods. The building is two stories of light-frame construction with a stucco exterior. Sited for southern exposure and shading from trees on the west, the home employs a greenhouse and water wall for winter heating. The building uses a belvedere for induced ventilation and thermal mass to store the typically cool night temperatures for summer cooling.

CONSERVATION
Because of the mild climate, rigorous conservation techniques were not required. This allowed north glass to be placed to complement the views and several bays to project from the basically cubical form. However, many conservation techniques are used such as: an entry vestibule, insulating curtains, under floor insulation, a greenhouse buffer zone, double glazed windows throughout, and south window shades. The basic shape of the building, a two-story square plan, is extremely efficient at reducing overall surface area and therefore heating loads.

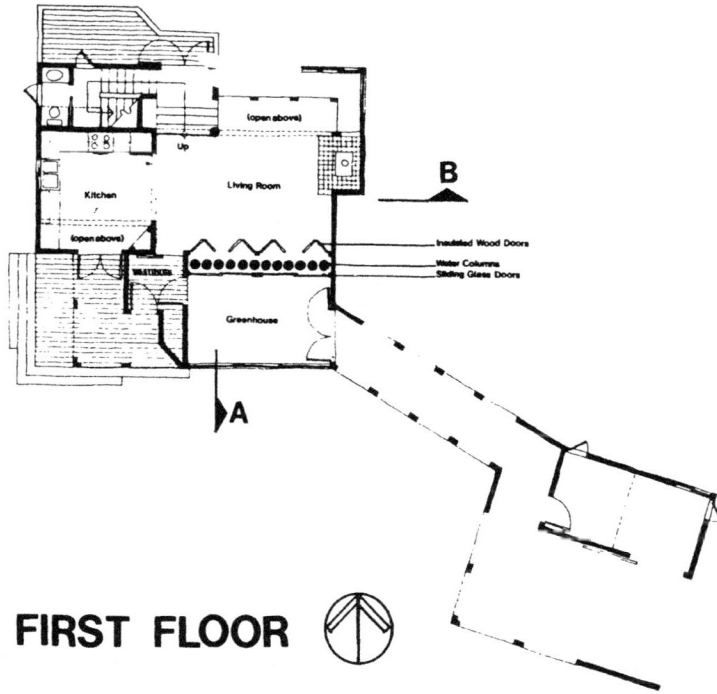

FIRST FLOOR

(open above)

Up

Kitchen

Living Room

B

(open above)

WATERBOX

Insulated Wood Doors

Water Columns

Sliding Glass Doors

Greenhouse

A

possible sunlight striking the storage mass. In addition to the greenhouse water wall of 125 sq. ft., there are 89 sq. ft. of south windows allowing sunlight to directly heat the second floor bedrooms. Unfortunately, these south windows have fixed projections which will partly shade them in winter. Additionally, there is no significant quantity of mass in the rooms they face, meaning the entering sunlight will not be stored for night use. These windows are therefore only effective for daytime heating, which constitutes only 30% of the total heating load.

The size of the water wall plus effective window area equals approximately 150 sq. ft. This area should produce about a 45% solar participation in the heating load. Given the mild climate, a larger contribution could be achieved without reducing cost effectiveness by enlarging the system. The quantity of mass in the 12" diameter water-filled steel culverts is calculated at 5,841 BTU/°F. This is 47 BTU/°F per sq. ft. of south glass, an adequate quantity of storage for the amount of transmitted heat.

HEATING

The passive solar heating system uses a greenhouse collector with water wall storage mass. Some of the sunlight which enters the greenhouse is absorbed by the brick floor and some passes through the sliding glass doors to be absorbed by water tubes in the living room. The sunlight stored in the brick heats the greenhouse directly and in turn reduces the heat loss from the house and water tubes. The heat stored in the water tubes is used to heat the house by direct radiation and natural convection loops. Many solar greenhouses in more severe climates place more of the heat storage mass in the greenhouse in order to keep that space at a reasonable temperature through the night. In this climate, however, where less energy is needed to keep the greenhouse warm, the storage mass shoud be placed in the living space where it will experience smaller losses to the exterior. It should be noted that the east and west walls of the greenhouse will shade the water wall in the morning and afternoon, reducing the total

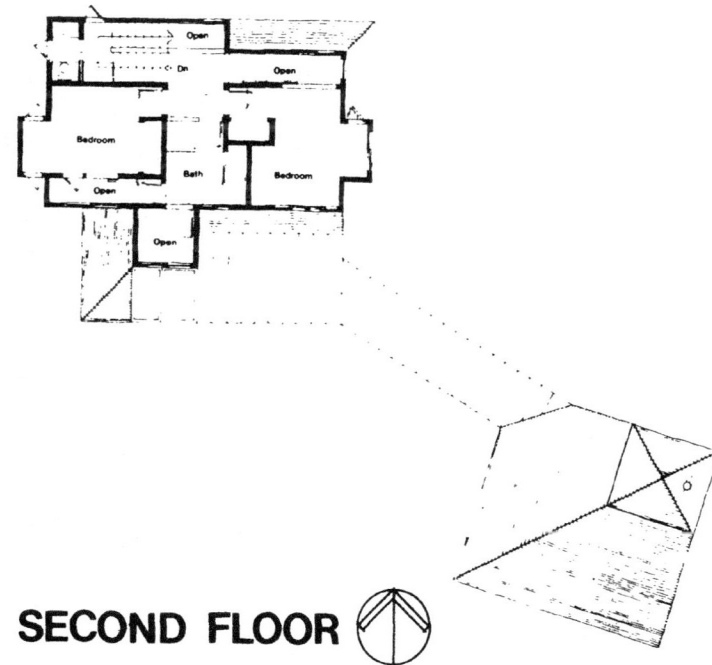

Open

Dn

Open

Bedroom

Bath

Bedroom

Open

Open

SECOND FLOOR

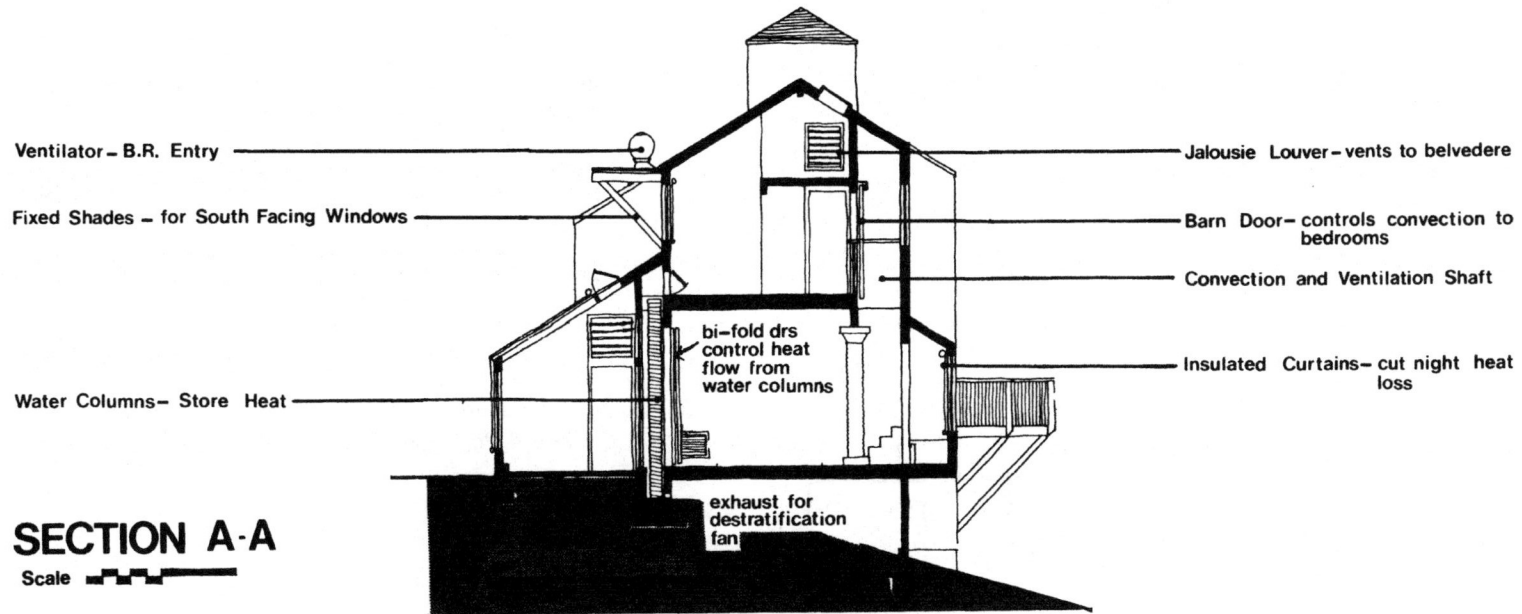

Ventilator – B.R. Entry

Fixed Shades – for South Facing Windows

Water Columns – Store Heat

bi-fold drs control heat flow from water columns

Jalousie Louver – vents to belvedere

Barn Door – controls convection to bedrooms

Convection and Ventilation Shaft

Insulated Curtains – cut night heat loss

exhaust for destratification fan

SECTION A·A

Scale

The heat distribution system of this design is a fine example of controlled natural heat flows. Direct radiation and convection from the warmed tubes to the living room is controlled by a bank of insulated by-fold doors. This allows the owner to close off the heat source during the day when the mass is receiving sunlight and the building doesn't require heat and open the doors in the evening to provide the necessary radiant warmth. One of the bedrooms above has an opening with an insulated shutter to provide direct convective heat flows from the water tubes. Both bedrooms have vertical air shafts and internal windows to allow heat to rise from the first floor, convectively heating those spaces. Doors on sliding "Barn Door" hardware allows those interior windows to be closed when heat is not required or when acoustic privacy is desired. At night these internal windows can be opened, allowing the natural temperature stratification to heat these rooms. Overheating can be vented off by the belvedere or by opening exterior windows. Excess heat in the second floor can also be recycled to the main floor by a small thermostatically operated fan. Heat from the backup system, a simple wood burner, can be distributed by the same mechanisms. Operation of the by-fold doors, shutters,

and internal windows will require a greater degree of participation by the owners than a standard forced-air furnace. However, the energy savings will be great, and the temperature control afforded by those features will be better than many other passive systems.

COOLING
In this climate area, most of the hot summer days are accompanied by cool nights. Storing this "coolth" at night for use the next day employs the same thermal mass used for winter heating. This is accompanied by opening the northeast windows, by-fold doors, and sliding glass doors around the water tubes to allow cross ventilation to cool the storage mass at night. During the day, those cool tubes will absorb excess heat, keeping the temperature down and creating a cool atmosphere.

This stored "coolth" is aided by the natural ventilation afforded by the windows, internal air shafts, and belvedere. The same features used for winter heat distribution allow the summer air to rise through the spaces and be exhausted by the belvedere. This traditional architectural device helps induce ventilation during breeze-

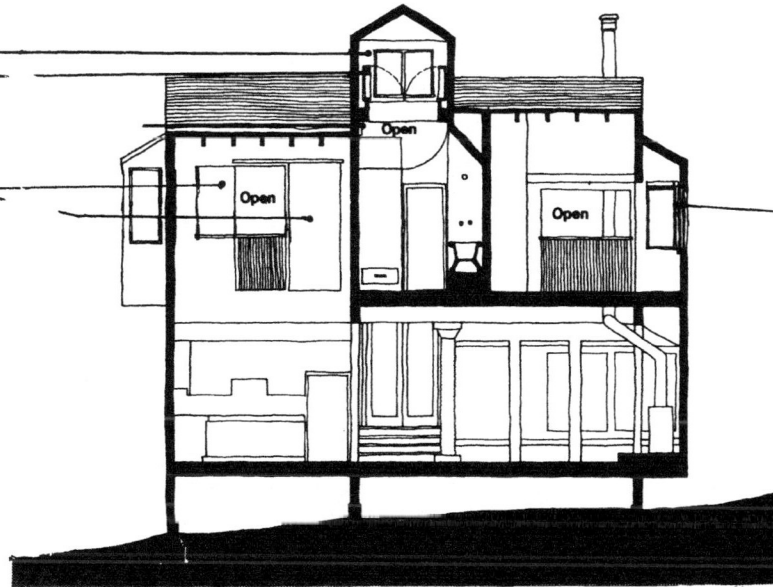

Belvedere — summer venting
Insulated Skylights

Shutter & Barn Door — control heat
from below

Insulated Curtains — cut night
heat loss

Open
Open
Open

SECTION B-B
Scale

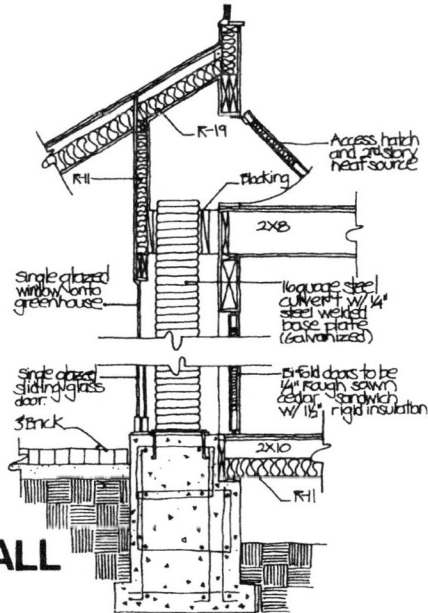

R-19
R-11
Access hatch
and auxiliary
heat source
Blocking
2×8
Single glazed
window onto
greenhouse
16 gauge steel
culvert w/ 1/4"
steel welded
base plate
(galvanized)
Single glazed
sliding glass
door.
Bi-fold doors to be
1/4" rough sawn
cedar, sandwich
w/ 1 1/2" rigid insulation
3 Brick
2×10
R-11

DETAIL
WATER WALL

less periods by its height and resulting stack effect. Finally the greenhouse is converted into a cool, screened porch by removing its front glass, inserting screens, and rolling down a canvas shade over its transparent roof. In this form, the greenhouse provides a cool buffer for the living room and shades the water tubes. In addition, the brick floor will now store night "coolth", making the porch an extremely pleasant room on summer days.

CONCLUSION
By economy of size this house provides a good example of a low-cost, low-energy-use residence for childless couples or older owners. It demonstrates a well-integrated design which combines a highly controllable passive solar system, an effective summer cooling system, and an excellent natural ventilation scheme. The double uses of many of the features such as internal mass, air shafts, and greenhouse make them extremely cost effective. Although the owners will have to operate various insulating doors, shutters, and vents, the degree of temperature control will be higher than most passive systems. The spacial amenities of the greenhouse/screened porch will be a strong marketing point as well as providing for the energy systems.

163

The following project pages are devoted to a brief description of the other SOLARIUM homes which were selected for awards. Each project is shown in perspective and accompanied by either a plan or a section. The project information extracted from the grant application is as follows:

Location of the home

Designer's name and firm
City and state of the designer*

HEATED AREA in square feet

NUMBER OF HEATING DEGREE DAYS

NET THERMAL LOAD in millions of British Thermal Units per year

AUXILIARY heating load in British Thermal Units per degree day per square foot

YEARLY SOLAR FRACTION: the percentage of heating energy provided by solar

COLLECTOR: Description and number of square feet

STORAGE: Description and Capacity in British Thermal Units per degree Fahrenheit

CONTROLS: Description

BACKUP: Type and capacity in British Thermal Units per hour

*Space limitations did not permit printing complete address. If you would like to contact the designer or builder concerning any of these projects, simply contact the National Solar Heating and Cooling Information Center (PROFESSIONALS FILE) by calling 800-523-2929 or 800-462-4983 (if calling from Pennsylvania), or by writing P.O. Box 1607, Rockville, MD 20850.

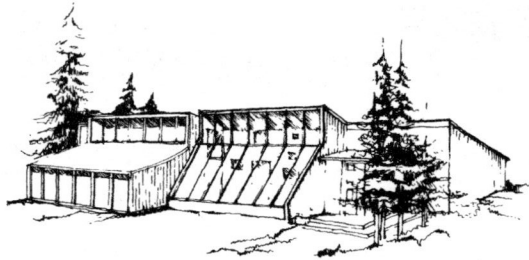

This 1-story, 3-bedroom wood frame house has rough-sawn Douglas fir siding and an asphalt strip shingle roof. Shading is provided by evergreens on the northwest corner of this level lot. Price is in the $70,000 range.

Private client.

Flagstaff, AZ

Michael Frerking, Environmental Architecture
Chino Valley, AZ

HEATED AREA: 2,100 FT²

NUMBER OF DEGREE DAYS: 7,177

NET THERMAL LOAD: 132 10^6 BTU/YR

AUXILIARY ENERGY: 3.37 BTU/DD/FT²

YEARLY SOLAR FRACTION: 84%

COLLECTOR: Solarium clerestory windows, solar mass wall Area: 750 FT²

STORAGE: Solar mass wall, rock storage wall, concrete and brick walls and floors Capacity: 31,200 BTU/°F

CONTROLS: Curtains, shutter, fan, vents

BACKUP: 44,000 BTUH electric baseboard; 75,000 BTUH wood stove

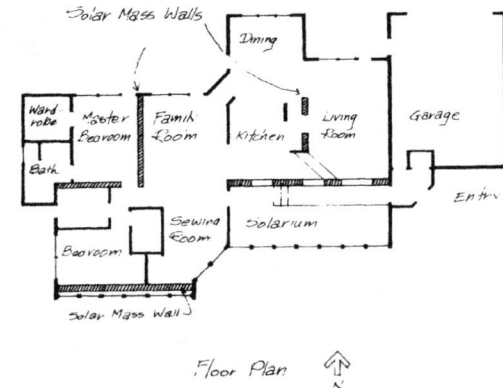

Floor Plan

This 2-story, 2-bedroom home has masonry walls and is priced in the $70,000 range. Earth berming will cover 68% of the exterior wall surface. The sod roof planted with native grasses will protect the interior from summer heat.

Open market sale.

Tempe, AZ

James J. Hoffman, James Hoffman Design Group
Tempe, AZ

HEATED AREA: 1,600 FT²

NUMBER OF DEGREE DAYS: 4,929

NET THERMAL LOAD: 64.5 10⁶BTU/YR

AUXILIARY ENERGY: 3.26 BTU/DD/FT²

YEARLY SOLAR FRACTION: 90%

COLLECTOR: Solarium, south-facing windows
Area: 244.9 FT²

STORAGE: Sand-filled masonry walls, concrete slab, water tanks, rock bed
Capacity: 27,981.9 BTU/°F

CONTROLS: Operable windows and doors, dampers, awnings

BACKUP: Fireplace, electric evaporative cooler

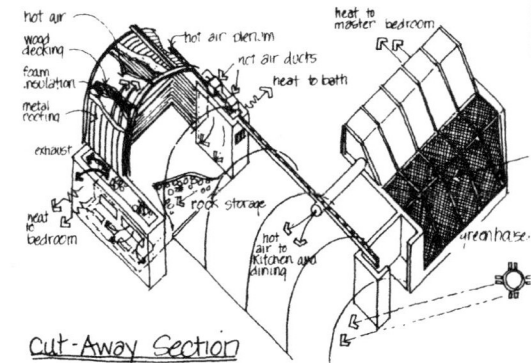

Cut-Away Section

Anderson, CA

Jonathan Allan Stoumen, Architect
Miranda, CA

HEATED AREA: 1,520 FT²

NUMBER OF DEGREE DAYS: 2,415

NET THERMAL LOAD: 11.6 10⁶BTU/YR

AUXILIARY ENERGY: 2.76 BTU/DD/FT²

YEARLY SOLAR FRACTION: 96%

COLLECTOR: Solarium glazing Area: 240 FT²

STORAGE: Concrete floors and walls; masonry fireplace; rock storage
Capacity: 26,041 BTU/°F

CONTROLS: Thermal curtains, manual vents

BACKUP: 20,000 BTUH fireplace

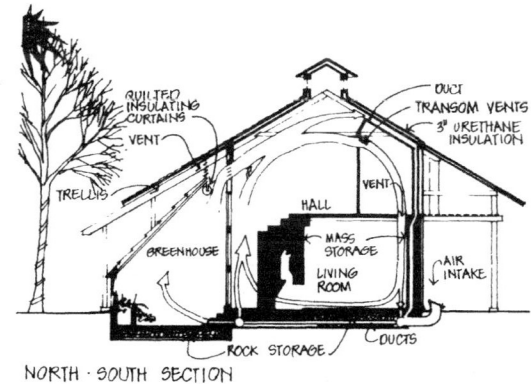

This new, 2-story, 3-bedroom detached home with solarium is priced in the $40,000 range. The low-profile structure with clay tile roof is shielded from winter winds by heavily wooded areas around the site.

Private client.

NORTH · SOUTH SECTION

A solarium has been retrofitted on the south side of this 100-year-old wood frame house. The 2-story, 4-bedroom home is on a site with a variety of mature trees and dense foliage. It is valued in the $85,000 range.

Retrofit.

Pleasanton, CA

Rob Anglin
Livermore, CA

HEATED AREA: 900 FT²

NUMBER OF DEGREE DAYS: 2,582

NET THERMAL LOAD: 40.1 10⁶BTU/YR

AUXILIARY ENERGY: 7.82 BTU/DD/FT²

YEARLY SOLAR FRACTION: 63%

COLLECTOR: Solarium, solar mass wall, south-facing windows, glass roof Area: 270 FT²

STORAGE: Brick walls and floors; rock bin; sand and concrete floors Capacity: 10,531 BTU/°F

CONTROLS: Fan, vents, evaporative cooler vertical sails

BACKUP: 24,500 BTUH gas heater; fireplace

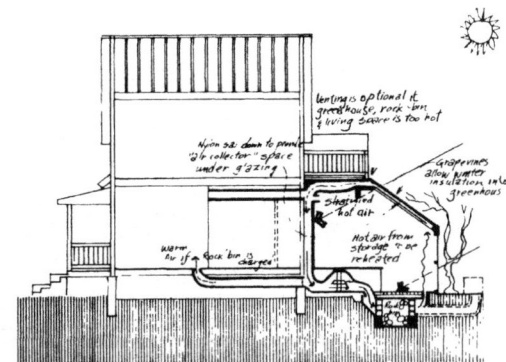

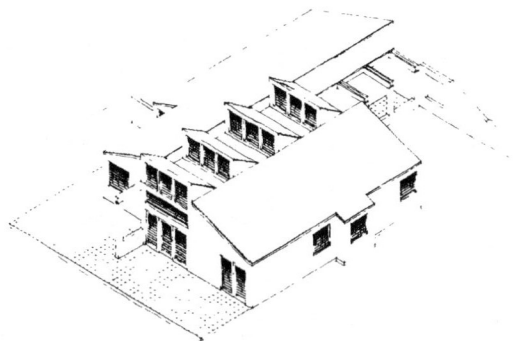

This 1-story, 3-bedroom wood frame house is situated on a flat suburban lot. Trees shade the east and west sides of the house, and a garage buffers the north side. The house is priced in the $45,000 range.

Open market sale.

Santa Cruz, CA

Peter Calthorpe, Calthorpe/Fernau/Wilcox
San Francisco, CA

HEATED AREA: 1,108 FT²

NUMBER OF DEGREE DAYS: 2,863

NET THERMAL LOAD: 20.8 10⁶BTU/YR

AUXILIARY ENERGY: 4.9 BTU/DD/FT²

YEARLY SOLAR FRACTION: 81%

COLLECTOR: Solarium and south-facing windows Area: 322 FT²

STORAGE: Concrete solar mass wall, concrete slab Capacity: 22,500 BTU/°F

CONTROLS: Operable doors and vents

BACKUP: 11,600 BTUH electric baseboard

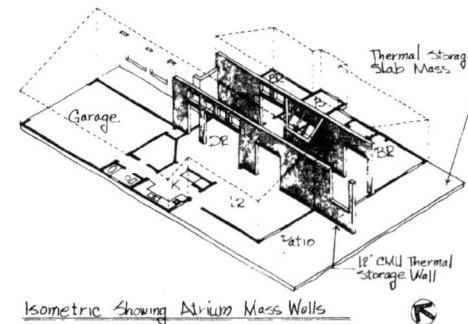

Isometric Showing Atrium Mass Walls

Woodland, CA

Jim Plumb
Sacramento, CA

HEATED AREA: 1,600 FT²

NUMBER OF DEGREE DAYS: 2,600

NET THERMAL LOAD: 27 10⁶BTU/YR

AUXILIARY ENERGY: 3.37 BTU/DD/FT²

YEARLY SOLAR FRACTION: 76%

COLLECTOR: South-facing window
Area: 261 FT²

STORAGE: Concrete floor with solar mass
water wall Capacity: 19,344 BTU/°F

CONTROLS: Operable windows and draperies

BACKUP: 35,000 BTUH gas furnace

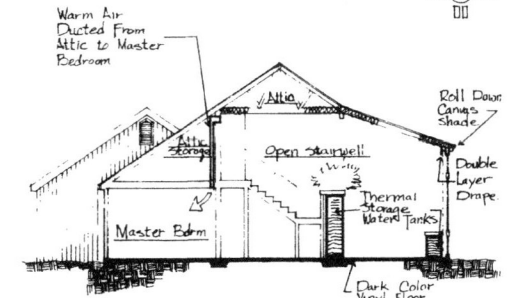

Section at Stairwell

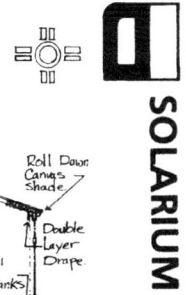

This wood frame, 2-story house with gable roof has 3 bedrooms and a solarium. Trees will be planted on this flat site for winter windbreak. Vines will shade south windows and patio in the summer. This home is in the $75,000 range.

Open market sale.

Boulder, CO

Bruce Downing, Downing/Leach Associates
Boulder, CO

HEATED AREA: 1,250 FT²

NUMBER OF DEGREE DAYS: 6,283

NET THERMAL LOAD: 16.6 10⁶BTU/YR

AUXILIARY ENERGY: 1.75 BTU/DD/FT²

YEARLY SOLAR FRACTION: 81%

COLLECTOR: Solarium, skylight, sunscoop
Area: 219 FT²

STORAGE: Concrete and gravel base and
solarium floor, slump block walls,
slump block fireplace core
Capacity: 3,477 BTU/°F

CONTROLS: Operable windows and doors,
insulated lids, fans

BACKUP: 12,000 BTUH gas forced air; 40,000
BTUH fireplace; 500 BTUH electric
bathroom heatlamp

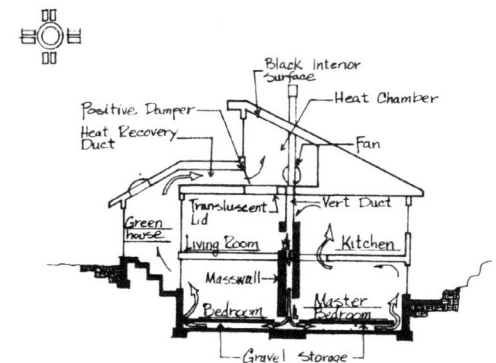

North South Section

Located in a large subdivision, this 2-story, 3-bedroom attached townhouse has its lower level below ground. Features include slump block walls, a landscaped courtyard, and evergreens to buffer northerly winter winds. It is priced in the $70,000 range.

Open market sale.

This 2-story, 3-bedroom wood frame house is located in a hilly area partially wooded with Ponderosa pine. A garage on the north protects the home from winter winds. The price is in the $65,000 range.

Open market sale.

Buena Vista, CO

James L. Moore
Buena Vista, CO

HEATED AREA: 1,660 FT²

NUMBER OF DEGREE DAYS: 7,812

NET THERMAL LOAD: 72.2 10⁶BTU/YR

AUXILIARY ENERGY: 2.33 BTU/DD/FT²

YEARLY SOLAR FRACTION: 73%

COLLECTOR: Solar mass wall, solarium
Area: 504 FT²

STORAGE: Solar mass wall Capacity: N/A

CONTROLS: Solarium roof panels, insulated drapes, fan, operable doors and windows

BACKUP: 7,000 BTUH electric baseboard; fireplace; wood stove; wood thermo-release drain

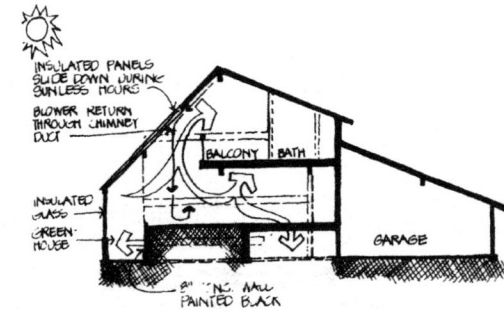

SECTION

This new 3-story, 3-bedroom house is designed to meet the needs of a mountain-dwelling family. The price is in the $82,000 range. Tall stands of conifer trees and a steep south-sloping site provide effective windbreaks.

Private client.

Jefferson County, CO

Lawrence Atkinson, AIA
Denver, CO

HEATED AREA: 2,445 FT²

NUMBER OF DEGREE DAYS: 7,432

NET THERMAL LOAD: 56.2 10⁶BTU/YR

AUXILIARY ENERGY: 1.60 BTU/DD/FT²

YEARLY SOLAR FRACTION: 96%

COLLECTOR: South-facing solariums, windows and water storage tanks
Area: 945 FT²

STORAGE: 18-inch diameter water tubes
Capacity: 11,297 BTU/°F

CONTROLS: Operable insulation, vents, greenhouse covers; automatic fans

BACKUP: 71,000 BTUH electric heater; two 55,000 BTUH wood stoves

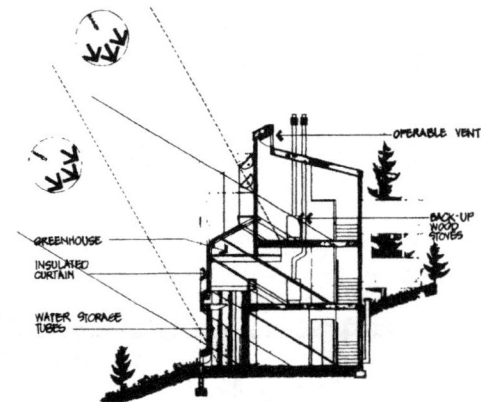

This is a retrofit solarium on a 2-story, 3-bedroom detached home in the $75,000 price range. The addition will have cedar siding to match the existing structure. The home is located on a flat, sparsely vegetated site.

Retrofit.

Golden, CO

Peter O'Connor
Golden, CO

HEATED AREA: 1,530 FT²

NUMBER OF DEGREE DAYS: 6,016

NET THERMAL LOAD: 74.3 10⁶BTU/YR

AUXILIARY ENERGY: 6.3 BTU/DD/FT²

YEARLY SOLAR FRACTION: 55%

COLLECTOR: Solarium, windows Area: N/A

STORAGE: Slump block/concrete wall, concrete floor Capacity: 3,783 BTU/°F

CONTROLS: Moveable insulation, window and thermal covers, solarium vents, operable windows

BACKUP: 140,000 BTUH and 125,000 BTUH gas furnace

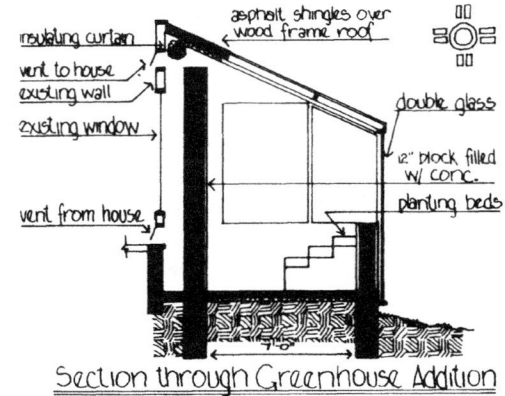

Section through Greenhouse Addition

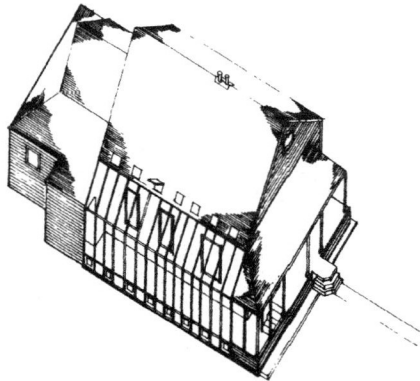

This detached, 2-bedroom retrofit is in the $45,000 price range. The building is elongated on an east-west orientation, with a solarium facing due south. Earth berming on the north side of the building and entry via southwest or east airlocks prevent winter wind infiltration.

Retrofit.

Longmont, CO

Barry Sulam, B. Sulam/L. Deutsch
Longmont, CO

HEATED AREA: 1,376 FT²

NUMBER OF DEGREE DAYS: 6,360

NET THERMAL LOAD: 83.3 10⁶BTU/YR

AUXILIARY ENERGY: 5.67 BTU/DD/FT²

YEARLY SOLAR FRACTION: 59%

COLLECTOR: South-facing solarium Area: 440 FT²

STORAGE: Plastic storage tubes—water filled (45 gallons each) Capacity: 9,486 BTU/°F

CONTROLS: Automatic vents; moveable urethane boards

BACKUP: 80,000 BTUH gas forced air furnace

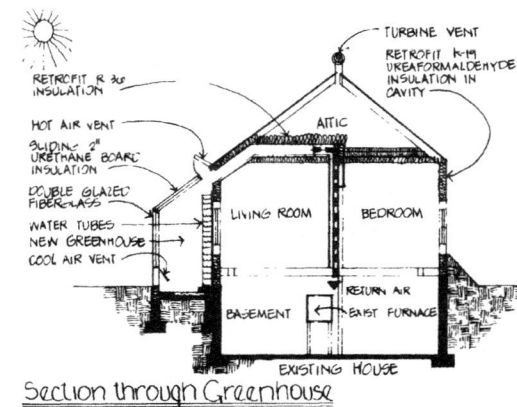

Section through Greenhouse

This new 3-story, 3-bedroom (with solarium) detached house is located in a heavily wooded area and is in the $110,000 price range. The trees on the gently sloping site provide shading during the summer along with winter wind protection.

Private client.

Fairfield, CT

Carl Mezoff, Sunborne Designs
Stamford, CT

HEATED AREA: 2,051 FT²

NUMBER OF DEGREE DAYS: 5,102

NET THERMAL LOAD: 45.6 10^6BTU/YR

AUXILIARY ENERGY: 2.15 BTU/DD/FT²

YEARLY SOLAR FRACTION: 80%

COLLECTOR: Solarium, skylights, south-facing windows, solar mass wall
Area: 465 FT²

STORAGE: Solar mass wall, water drums, brick floors Capacity: 6,387 BTU/°F

CONTROLS: Skylight insulation, solarium door, ducts, operable windows, fan

BACKUP: 26,000 BTUH electric resistor; wood stove; fireplace

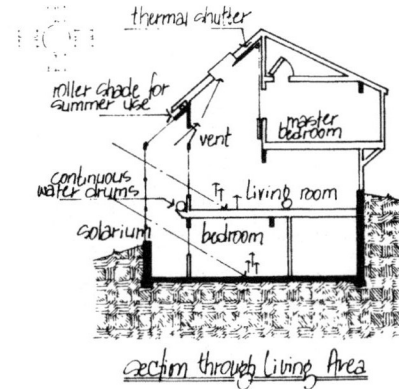

section through Living Area

This contemporary 2-story, 3-bedroom house will be protected from winter cold by extensive earth berming. The exposed walls will be made of stucco and concrete. It is in the $135,000 price range.

Private client.

Ames, IA

David Block
Ames, IA

HEATED AREA: 2,200 FT²

NUMBER OF DEGREE DAYS: 6,774

NET THERMAL LOAD: 43 10^6BTU/YR

AUXILIARY ENERGY: 2.92 BTU/DD/FT²

YEARLY SOLAR FRACTION: 86%

COLLECTOR: Solarium, south-facing windows
Area: 500 FT²

STORAGE: Concrete walls and floors, cored slab floor Capacity: 16,271 BTU/°F

CONTROLS: Insulating panels, dampers, louvers, sliding door

BACKUP: 44,000 BTUH gas furnace; 50,000 BTUH fireplace

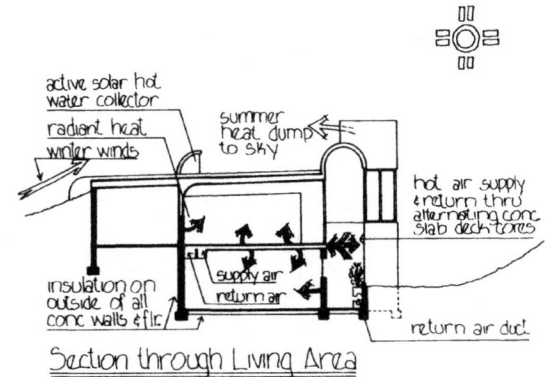

Section through Living Area

This contemporary, 2-story wood frame house has 3 bedrooms and is priced in the $70,000 range. Trees will be planted on this slightly sloping site to provide summer shade and winter wind protection.

Open market sale.

New Albany, IN
James Rosenbarger/Terry White
New Albany, IN

HEATED AREA: 1,600 FT²

NUMBER OF DEGREE DAYS: 4,605

NET THERMAL LOAD: 48 10⁶BTU/YR

AUXILIARY ENERGY: 4 BTU/DD/FT²

YEARLY SOLAR FRACTION: 48%

COLLECTOR: Solarium Area: 380 FT²

STORAGE: Rockbed Capacity: NA BTU/°F

CONTROLS: Fans, vent door, louvers, vents

BACKUP: 22,000-40,000 BTUH wood stove; 31,000 BTUH electric baseboard heat

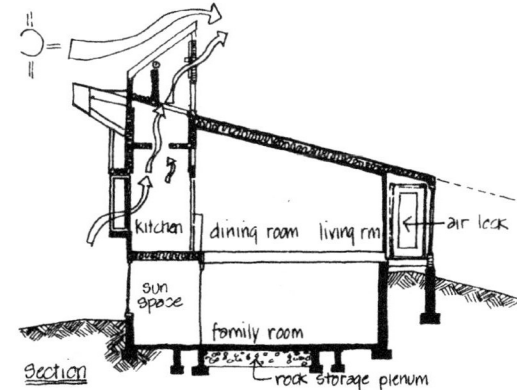

section — rock storage plenum

This renovated 1-story, 2-bedroom building with greenhouse is priced in the $125,000 range. The sloping site is heavily wooded for summer shade and winter wind protection.

Retrofit.

Beverly Farms, MA
L. Bradley Cutler, Associated Architects
Boston, MA

HEATED AREA: 2,000 FT²

NUMBER OF DEGREE DAYS: 5,627

NET THERMAL LOAD: 80.9 10⁶BTU/YR

AUXILIARY ENERGY: 5.21 BTU/DD/FT²

YEARLY SOLAR FRACTION: 44%

COLLECTOR: Solarium Area: 312 FT²

STORAGE: Rock bed Capacity: 48,111 BTU/°F

CONTROLS: Operable skylight thermal shades, lined curtains, and doors

BACKUP: 50,000 BTUH wood stove; 50,000 BTUH water heater; 100,000 BTUH oil heaters (2)

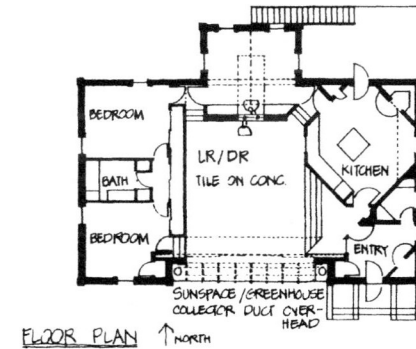

FLOOR PLAN ↑NORTH

Manchester, MA

DeFrancesco & Baker Associates
Boston, MA

HEATED AREA: 3,208 FT²

NUMBER OF DEGREE DAYS: 5,529

NET THERMAL LOAD: 76.8 10⁶BTU/YR

AUXILIARY ENERGY: 2.54 BTU/DD/FT²

YEARLY SOLAR FRACTION: 48.6%

COLLECTOR: South-facing windows, solarium, clerestory, and solar mass wall Area: 556 FT²

STORAGE: Concrete walls and floor Capacity: 14,536 BTU/°F

CONTROLS: Operable dampers, fans, vents, and panels

BACKUP: 100,000 BTUH gas furnace; 30,000 BTUH and 15,000 BTUH wood stoves

SECTION

This new 2-story, 3-bedroom house is priced in the $108,000 range. The house has a modified Saltbox architectural design, with low maintenance cedar shingle and plywood exterior siding. Earth berming insulates the front face of the solarium.

Private client.

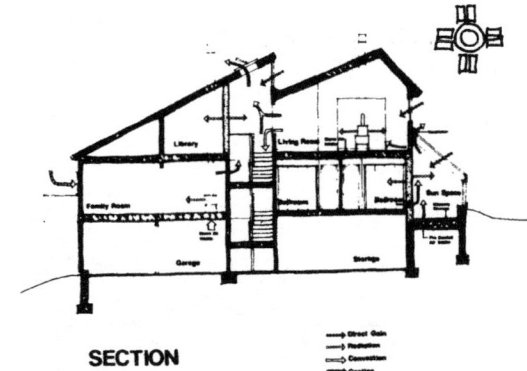

Southampton, MA

Seigfried Porth and Larry O'Connor
Southampton, MA

HEATED AREA: 1,600 FT²

NUMBER OF DEGREE DAYS: 6,851

NET THERMAL LOAD: 66.9 10⁶BTU/YR

AUXILIARY ENERGY: 3.64 BTU/DD/FT²

YEARLY SOLAR FRACTION: 61%

COLLECTOR: South-facing windows and solarium Area: 351 FT²

STORAGE: Concrete slab, rock under slab, concrete chimney Capacity: 12,878 BTU/°F

CONTROLS: Operable shutters, ducts, windows and shades

BACKUP: Wood furnace

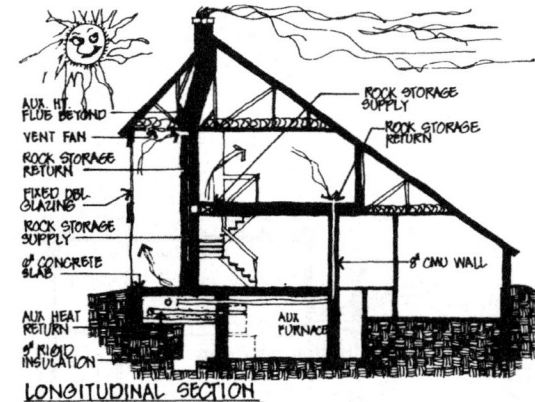

LONGITUDINAL SECTION

This new 2-story, 3-bedroom house has a rustic, contemporary architectural style, using native pine siding, and a saltbox-style roof. The north and west walls of the house are earth-bermed, and conifer trees also provide protection from winter winds. The house is in the $49,000 price range.

Private client.

This new, 1-story, 1-bedroom building with stone veneer and wood siding is in the $62,000 price range. Earth berms and northern site vegetation limit the effect of winter winds, and moveable awnings provide controlled summer shading.

Private client.

Webster Township, MI

Richard McMath, Sunstructures, Inc.
Ann Arbor, MI

HEATED AREA: 1,180 FT²

NUMBER OF DEGREE DAYS: 6,267

NET THERMAL LOAD: 68.3 10⁶BTU/YR

AUXILIARY ENERGY: 5.3 BTU/DD/FT²

YEARLY SOLAR FRACTION: 57%

COLLECTOR: South-facing windows, solar mass wall Area: 364 FT²

STORAGE: Concrete floor slab, stone wall, cavity wall Capacity: 12,598 BTU/°F

CONTROLS: Operable windows, insulating shades, damper, fan, awnings

BACKUP: 100,000 BTUH draft fireplace; 46,000 BTUH gas water heater

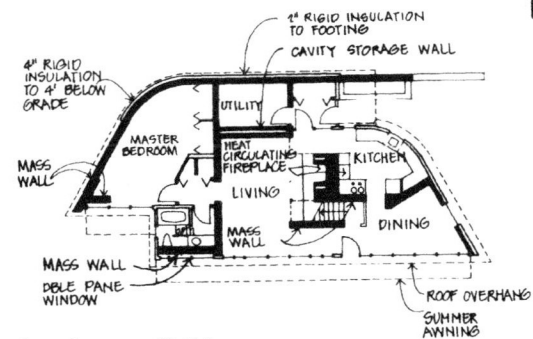

FIRST FLOOR PLAN

This 2-story, 3-bedroom wood frame house is built on a lot heavily wooded with poplar and aspen on three sides. South side of lot is clear of vegetation to provide maximum sun exposure for solarium. The home is priced in the $75,000 range.

Open market sale.

Duluth, MN

Charles Williams
Richmond, KY

HEATED AREA: 1,428 FT²

NUMBER OF DEGREE DAYS: 9,930

NET THERMAL LOAD: 69.6 10⁶BTU/YR

AUXILIARY ENERGY: 4.26 BTU/DD/FT²

YEARLY SOLAR FRACTION: 59%

COLLECTOR: Solarium, south-facing window Area: 400 FT²

STORAGE: Filled block wall Capacity: 11,088 BTU/°F

CONTROLS: Insulating drapes, louvers

BACKUP: 20,000 BTUH fireplace; 45,000 BTUH wood stove; 40,000 BTUH electric baseboard

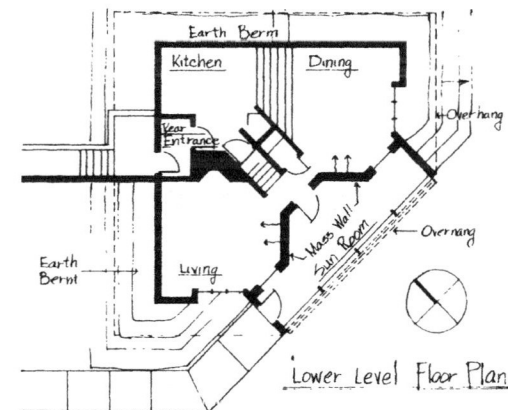

Lower Level Floor Plan

A solarium is added to this renovated inner-city, 3-story brick building. Formerly abandoned, it is converted into an attached, 2-family dwelling with 2-bedrooms each. The front facade maintains its historical appearance in accordance with the Historical Society's guidelines. It is priced in the $40,000 range.

Retrofit.

St. Louis, MO

Warren L. Cargal, Interface Design Group
Clayton, MO

HEATED AREA: 3,150 FT²

NUMBER OF DEGREE DAYS: 4,900

NET THERMAL LOAD: 108 10⁶BTU/YR

AUXILIARY ENERGY: 5.86 BTU/DD/FT²

YEARLY SOLAR FRACTION: 35%

COLLECTOR: Atrium and solarium
　　　　　　Area: 400 FT²

STORAGE: Brick walls Capacity: 18,358 BTU/°F

CONTROLS: Atrium panel, insulative drapes

BACKUP: 30,000 BTUH wood stove; 65,000
　　　　　BTUH electric heat pump

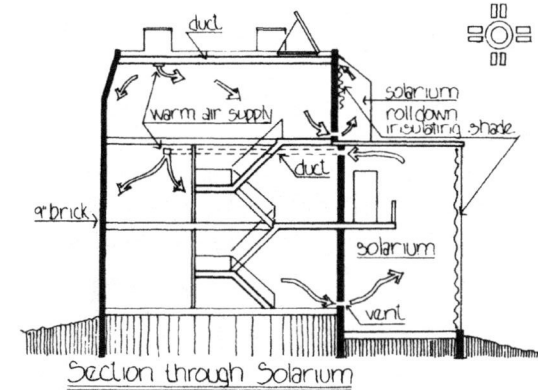

Section through Solarium

Two single—family residences, attached to form a duplex, share a solarium in this retrofitted old mill house. An open site and an east-west axis of the lot permits maximum exposure to the south.

Retrofit.

Carrboro, NC

Steven Fisher, Graphicon
Chapel Hill, NC

HEATED AREA: 2,115 FT²

NUMBER OF DEGREE DAYS: 3,338

NET THERMAL LOAD: 39.0 10⁶BTU/YR

AUXILIARY ENERGY: 6.8 BTU/DD/FT²

YEARLY SOLAR FRACTION: 50%

COLLECTOR: South-facing windows and
　　　　　　solarium Area: 300 FT²

STORAGE: Water container filled brick wall
　　　　　Capacity: 5,011 BTU/°F

CONTROLS: Operable doors, windows, and
　　　　　vents

BACKUP: 18,150 BTUH electric baseboard

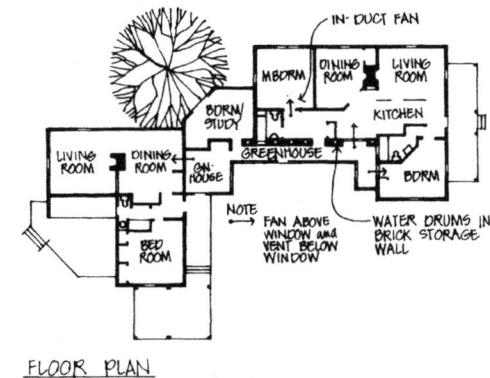

FLOOR PLAN

Morrisville, NC

Donald W. Barnes, Jr., AIA
Raleigh, NC

HEATED AREA: 2,190 FT²

NUMBER OF DEGREE DAYS: 3,338

NET THERMAL LOAD: 13.7 10⁶BTU/YR

AUXILIARY ENERGY: 4.4 BTU/DD/FT²

YEARLY SOLAR FRACTION: 50%

COLLECTOR: South-facing windows, solarium
Area: 272 FT²

STORAGE: Concrete floor and ceiling
Capacity: 8,530 BTU/°F

CONTROLS: Vents, insulated shutters, night
walls, modulators, doors, pump

BACKUP: 10,200 BTUH electric baseboard;
34,100 BTUH electric radiant ceiling

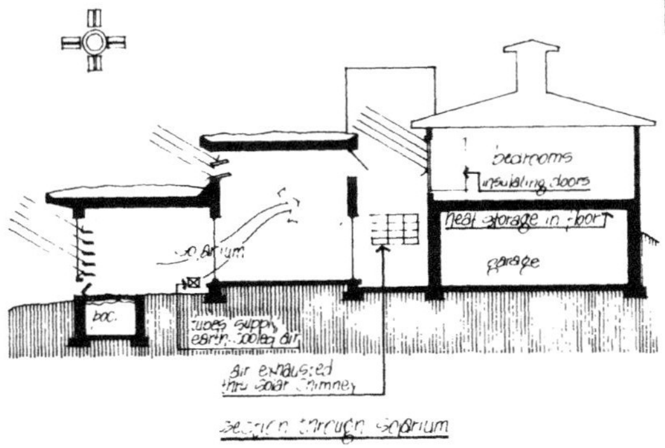

This new 2-story, 2-bedroom detached house
with solarium is in the $60,000 price range.
Trees and shrubs are planted on the crest of a
knoll overlooking the house. The trees on this
rolling farmland shield the house from winter
winds.

Private client.

Raleigh, NC

John Meachem, Sunshelter Design
Raleigh, NC

HEATED AREA: 1,460 FT²

NUMBER OF DEGREE DAYS: 3,352

NET THERMAL LOAD: 40 10⁶BTU/YR

AUXILIARY ENERGY: 1.66 BTU/DD/FT²

YEARLY SOLAR FRACTION: 92%

COLLECTOR: South-facing window
Area: 512 FT²

STORAGE: Solar mass water wall, concrete and
brick floor Capacity: 17,976 BTU/°F

CONTROLS: Moveable window insulation,
ceiling fans

BACKUP: 30,000 BTUH wood stove; 50,000
BTUH gas furnace

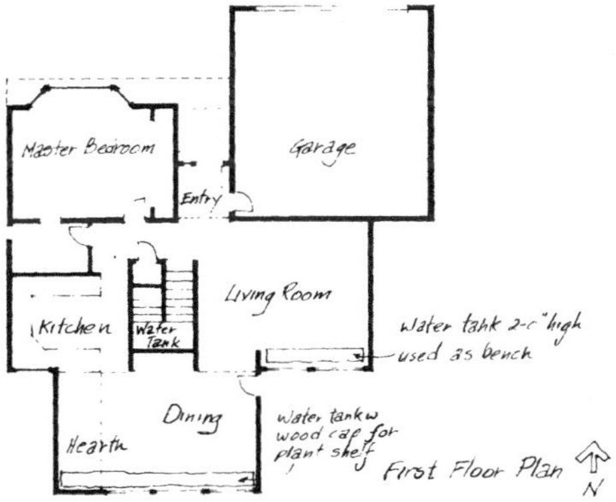

This 2-story, 2-bedroom, ranch-style house
with basement/work area is priced in the
$45,000 range. The house is located in a
heavily wooded area which has been partially
cleared to the south.

Private client.

175

This 2-story house with loft and solarium is priced in the $70,000 range. Room space can be interpreted by owner to have from one to five bedrooms. Mature hardwood forest to north and east shields the home from winter winds.

Open market sale.

Randleman, NC

John Alt
Randleman, NC

HEATED AREA: 1,812 FT²

NUMBER OF DEGREE DAYS: 3,731

NET THERMAL LOAD: 63.3 10⁶BTU/YR

AUXILIARY ENERGY: 3.2 BTU/DD/FT²

YEARLY SOLAR FRACTION: 79%

COLLECTOR: South-facing windows
Area: 500 FT²

STORAGE: 55-gallon water drums, rock crawl space Capacity: 25,224 BTU/°F

CONTROLS: Insulating curtains, shutters, insulating panels, fans, weather-stripping

BACKUP: 55,000 BTUH wood stove

Section, Winter

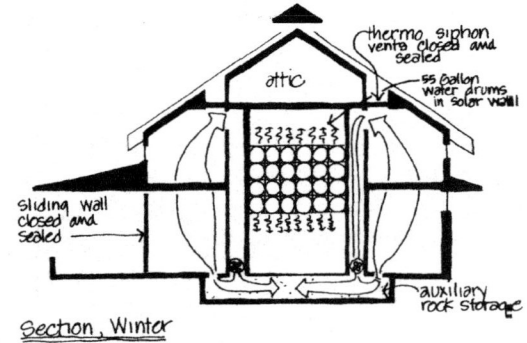

This contemporary 1-bedroom building has natural wood siding and is in the $80,000 price range. The low-profile home is partially buried in the side of a hill for natural insulative protection.

Private client.

Pennington, NJ

Ted Bickford and Charly Lowery,
Harrison Fraker Architect
Princeton, NJ

HEATED AREA: 2,640 FT²

NUMBER OF DEGREE DAYS: 4,911

NET THERMAL LOAD: 60.3 10⁶BTU/YR

AUXILIARY ENERGY: 1.12 BTU/DD/FT²

YEARLY SOLAR FRACTION: 92%

COLLECTOR: South-facing solar mass wall, solarium Area: 480 FT²@60°
372 FT²@90°

STORAGE: Solar mass wall, water barrels, rockbed Capacity: 30,481 BTU/°F

CONTROLS: Operable windows, curtain, and insulation

BACKUP: 46,200 BTUH wood fireplace; 38,500 BTUH electric heating panels

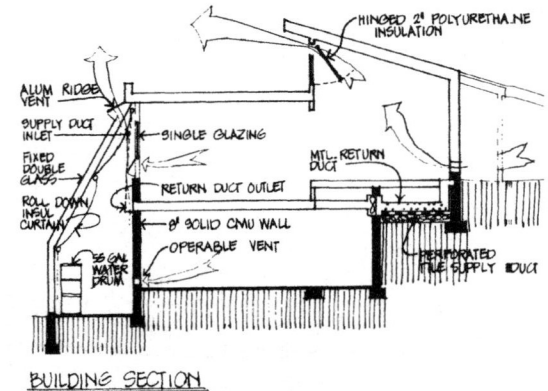

BUILDING SECTION

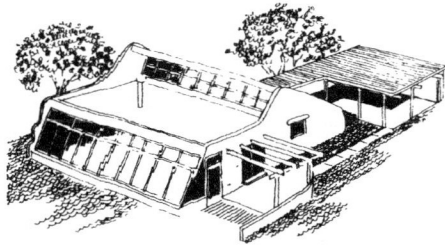

This new 1-story, 2-bedroom with solarium detached house is in the $40,000 price range. Trees and hedges will be planted on the hillside site overlooking a valley to protect the house from winter winds.

Private client.

Los Lunas, NM

Robert W. Richardson
Los Alamos, NM

HEATED AREA: 1,000 FT²

NUMBER OF DEGREE DAYS: 4,292

NET THERMAL LOAD: 58.0 10⁶BTU/YR

AUXILIARY ENERGY: 3.93 BTU/DD/FT²

YEARLY SOLAR FRACTION: 77%

COLLECTOR: Clerestory, solarium Area: 416 FT²

STORAGE: Brick floor, adobe walls
Capacity: 14,460 BTU/°F

CONTROLS: Insulated doors, louvers, operable windows

BACKUP: 40,000 BTUH wood stove; 7,000 BTUH electric radiant heat

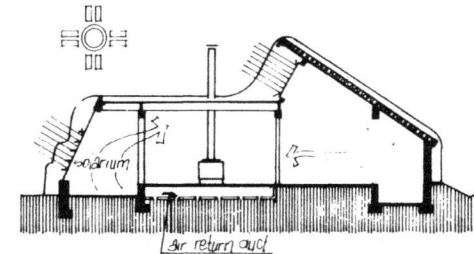

air return duct

cool air drawn through underground duct is heated in the solarium

This new 1-story, 3-bedroom detached house with solarium is priced in the $110,000 range. The flat site is lightly wooded and provides excellent exposure for collecting the sun's heat.

Private client.

Santa Fe, NM

Stephen Merdler, Soltec Associates
Santa Fe, NM

HEATED AREA: 2,080 FT²

NUMBER OF DEGREE DAYS: 5,913

NET THERMAL LOAD: 76.3 10⁶BTU/YR

AUXILIARY ENERGY: 3.66 BTU/DD/FT²

YEARLY SOLAR FRACTION: 72%

COLLECTOR: South-facing windows, solarium, solar mass walls Area: 443.22 FT²

STORAGE: Solar mass wall, adobe pool enclosure, pool/fountain, adobe interior wall
Capacity: 12,677 BTU/°F

CONTROLS: Dampers, shutters, blower, operable windows and doors

BACKUP: 69,966 BTUH electric baseboard; 35,000 and 23,300 BTUH wood stoves

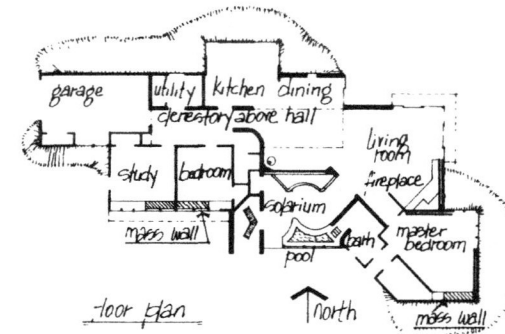

garage utility kitchen dining
clerestory above hall
study bedroom living room
fireplace
mass wall solarium
pool bath master bedroom
floor plan north mass wall

SOLARIUM

This new house overlooking Lake Ontario has a rural architectural style and is designed to be maintenance-free with aluminum-clad windows, stained wood siding, and wood shingles. Designed for a family of four or five, the house is priced in the $65,000 range. Earth berms and windbreaks reduce winter heat losses.

Open market sale.

Henderson, NY

Stephen Yaussi, Moran and Yaussi, Architects
Watertown, NY

HEATED AREA: 1,990 FT²

NUMBER OF DEGREE DAYS: 7,273

NET THERMAL LOAD: 99.4 10⁶BTU/YR

AUXILIARY ENERGY: 5.5 BTU/DD/FT²

YEARLY SOLAR FRACTION: 80%

COLLECTOR: Solarium, skylight, south-facing windows Area: 416.5 FT²

STORAGE: Concrete wall and floor Capacity: 18,530 BTU/°F

CONTROLS: Operable Skylids™, insulating curtain, vent door, and fan

BACKUP: 90,000 BTUH wood furnace; 65,500 BTUH electric duct heater; 23,300 BTUH fireplace

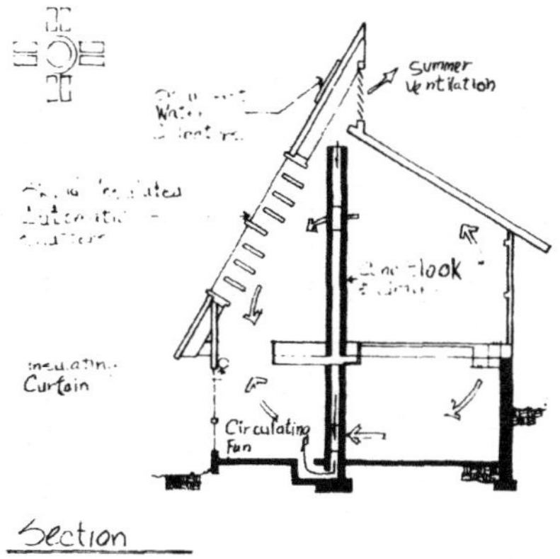

Section

This new, 2-story, 4-bedroom detached wood frame house is priced in the $60,000 range. The home is oriented toward an open southern slope, and it is surrounded on the north, west, and east by mixed species woods.

Private client.

Rogue River, OR

Arden Handshy
Jacksonville, OR

HEATED AREA: 1,724 FT²

NUMBER OF DEGREE DAYS: 5,008

NET THERMAL LOAD: 56.0 10⁶BTU/YR

AUXILIARY ENERGY: 4.6 BTU/DD/FT²

YEARLY SOLAR FRACTION: 47%

COLLECTOR: Solarium Area: 360 FT²

STORAGE: Concrete wall and floor; 1.5-2 inch rock bed Capacity: 9,220 BTU/°F

CONTROLS: Manual dampers, operable insulating shutters

BACKUP: 20,700 BTUH electric baseboard; 10,000 BTUH wood stove

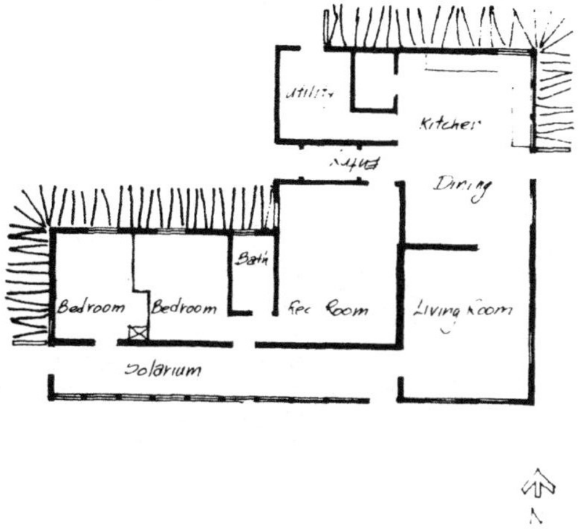

Chicora, PA

Steve Nearhoof, Ecol·lection
Chicora, PA

HEATED AREA: 1,500 FT²

NUMBER OF DEGREE DAYS: 5,905

NET THERMAL LOAD: 73.3 10⁶BTU/YR

AUXILIARY ENERGY: 6.87 BTU/DD/FT²

YEARLY SOLAR FRACTION: 37%

COLLECTOR: Solarium, solar mass walls
Area: 316 FT²

STORAGE: Concrete walls
Capacity: 9,442 BTU/°F

CONTROLS: Shutters, vents, shades, fans

BACKUP: 27,000 BTUH wood stove; 30,000
BTUH gas boiler (water heater)

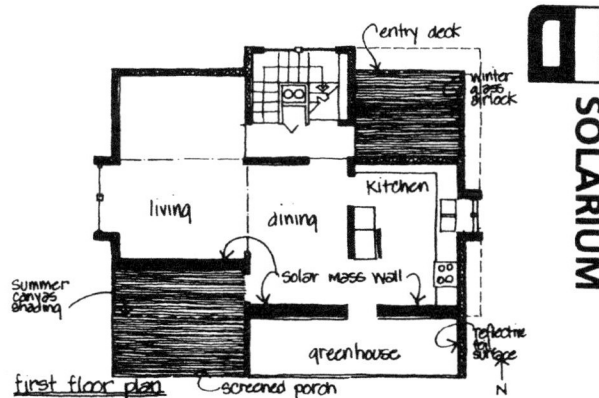

first floor plan

This 2-story, 2-bedroom wood frame house has a natural wood exterior and is a saltbox design. The 18-acre woods to the northwest forms a natural windbreak. The price is in the $25,000 range.

Private client.

Feasterville, PA

Peter-Paul d'Entremont
Feasterville, PA

HEATED AREA: 1,630 FT²

NUMBER OF DEGREE DAYS: 5,364

NET THERMAL LOAD: 48.2 10⁶BTU/YR

AUXILIARY ENERGY: 3.06 BTU/DD/FT²

YEARLY SOLAR FRACTION: 79%

COLLECTOR: South-facing greenhouse,
clerestory, storage mass wall
Area: 910 FT²

STORAGE: Concrete wall and floor; pool of
water Capacity: 37,300 BTU/°F

CONTROLS: Operable shades and ducts;
thermostatically controlled fan
and vent

BACKUP: 50,000 BTUH oil furnace

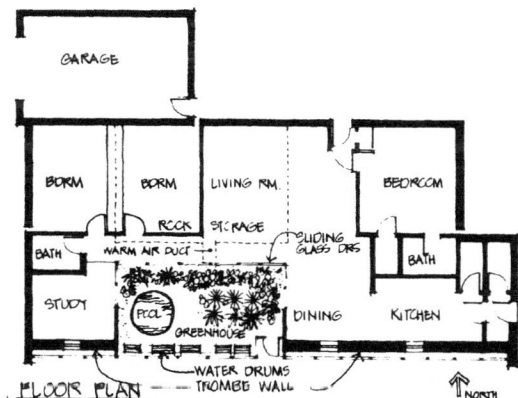

FLOOR PLAN

This new 3-bedroom, 1-story house is in the $75,000 price range. Earth berming on the north and west sides of the house will reduce winter heat losses, while large shade trees on the southwest will cool the house in summer.

Open market sale.

Greenville, SC

Randy Granger, Helio Thermics, Inc.
Greenville, SC

HEATED AREA: 1,568 FT²

NUMBER OF DEGREE DAYS: 2,955

NET THERMAL LOAD: 19.7 10⁶BTU/YR

AUXILIARY ENERGY: 1.28 BTU/DD/FT²

YEARLY SOLAR FRACTION: 100%

COLLECTOR: Solar attic Area: 756 FT²

STORAGE: Rock bed Capacity: 28,947 BTU/°F

CONTROLS: Operable vents

BACKUP: 15,000 BTUH electric furnace

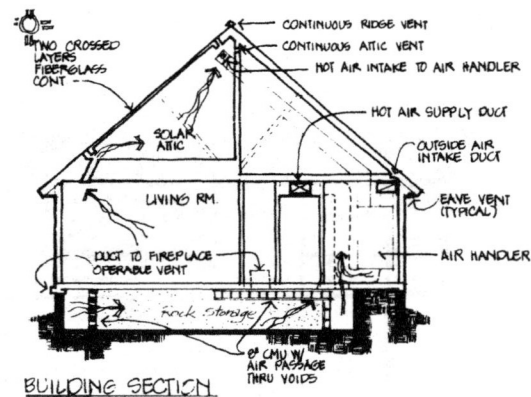

This 1-story, 3-bedroom wood frame house is priced in the $40,000 range. The garage forms a windbreak against northeast winter winds.

Open market sale.

BUILDING SECTION

Oak Ridge, TN

Dan Fenyn, Land Systems, Inc.
Oak Ridge, TN

HEATED AREA: 1,572 FT²

NUMBER OF DEGREE DAYS: 3,507

NET THERMAL LOAD: 25.8 10⁶BTU/YR

AUXILIARY ENERGY: 5.17 BTU/DD/FT²

YEARLY SOLAR FRACTION: 54%

COLLECTOR: South and west side solarium
            Area: 290.1 FT²

STORAGE: Block and plaster wall with concrete
          floor Capacity: 3,589 BTU/°F

CONTROLS: Attic fan switch

BACKUP: 33,000 BTUH electric heat pump

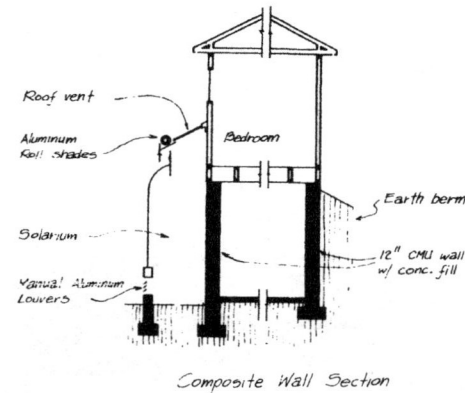

This new 2-story, 3-bedroom detached home includes a solarium and is in the $70,000 price range. The site contains a ridge which crests along the southern boundary, and a creek which runs the full length of the property. Hardwood trees to the south and west shade the summer sun.

Open market sale.

Composite Wall Section

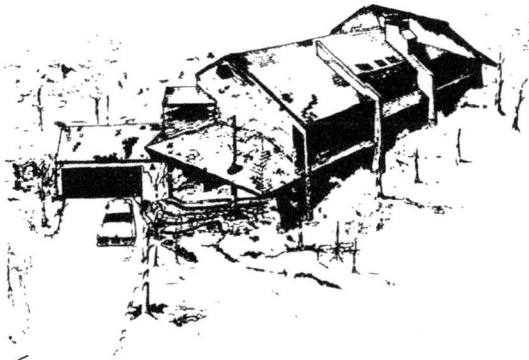

Arlington, TX

Jamie M. Rohe, Concept Consultants, Inc.
Magnus Magnusson, The Ehrenkrantz Group
Dallas, TX

HEATED AREA: 3,212 FT²

NUMBER OF DEGREE DAYS: 2,209

NET THERMAL LOAD: 48.3 10⁶BTU/YR

AUXILIARY ENERGY: 4.15 BTU/DD/FT²

YEARLY SOLAR FRACTION: 43%

COLLECTOR: South-facing solarium and windows Area: 257 FT²

STORAGE: Stone and tile solarium floor, concrete floor, brick walls Capacity: 7,918 BTU/°F

CONTROLS: Operable doors, windows, and shutters

BACKUP: 52,000 BTUH electric heat pump

This new 2-story, 3-bedroom house is priced in the $158,000 range. The garage is located on the west side of the house for reduced solar gain in the afternoon. Large oak trees provide summer shading.

Open market sale.

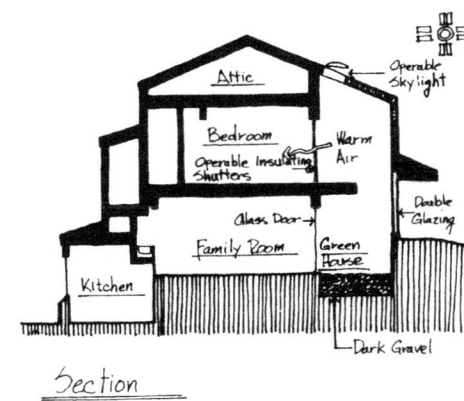

Section

Hockley, TX

L. R. Bachman, Design Technology, Inc.
Houston, TX

HEATED AREA: 778 FT²

NUMBER OF DEGREE DAYS: 1,354

NET THERMAL LOAD: 6.18 10⁶BTU/YR

AUXILIARY ENERGY: 6.12 BTU/DD/FT²

YEARLY SOLAR FRACTION: 75%

COLLECTOR: South-facing skylight, windows, and solarium Area: 288 FT²

STORAGE: Concrete floor Capacity: 6,728 BTU/°F

CONTROLS: Operable windows and doors

BACKUP: 19,000 BTUH electric heat pump

This new house is in the $39,500 price range. The main entry to the house is sheltered from winter winds. Deciduous trees to the south will shade the house in the summer, while evergreen trees to the north will deflect winter winds away from the house.

Private client.

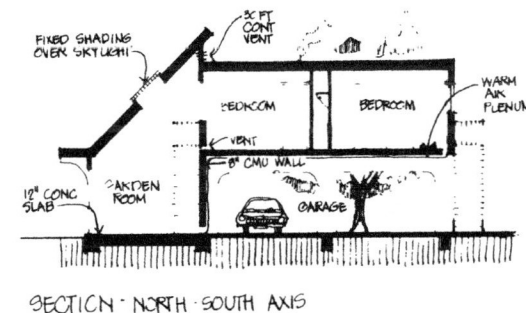

SECTION · NORTH · SOUTH AXIS

Parker County, TX

Ray Boothe, Boothe and Associates, Architects
Forth Worth, TX

HEATED AREA: 1,635 FT²

NUMBER OF DEGREE DAYS: 2,234

NET THERMAL LOAD: 9.55 10⁶BTU/YR

AUXILIARY ENERGY: 3.4 BTU/DD/FT²

YEARLY SOLAR FRACTION: 91%

COLLECTOR: South-facing windows, solarium
Area: 234 FT²

STORAGE: Concrete floor
Capacity: 2,239 BTU/°F

CONTROLS: Dampers, draperies, operable
windows

BACKUP: 32,000 BTUH electric heat pump

Section Thru Kitchen

This 1-story, 2-bedroom house has all exterior
surfaces underground, except those for
southern exposure. It will be priced in the
$65,000 range.

Private client.

Berkeley Springs, WV

R. Ashelman, Natural Sun Homes
Berkeley Springs, WV

HEATED AREA: 2,157 FT²

NUMBER OF DEGREE DAYS: 5,428

NET THERMAL LOAD: 54.3 10⁶BTU/YR

AUXILIARY ENERGY: 2.57 BTU/DD/FT²

YEARLY SOLAR FRACTION: 73%

COLLECTOR: South-facing window, solarium
Area: 396 FT²

STORAGE: Concrete floor, walls, ceiling, rock
bed Capacity: 78,268 BTU/°F

CONTROLS: Shutters, registers, "Skylid" vent,
fan

BACKUP: 75,000 BTUH wood stove; 200 AMP
electric baseboard

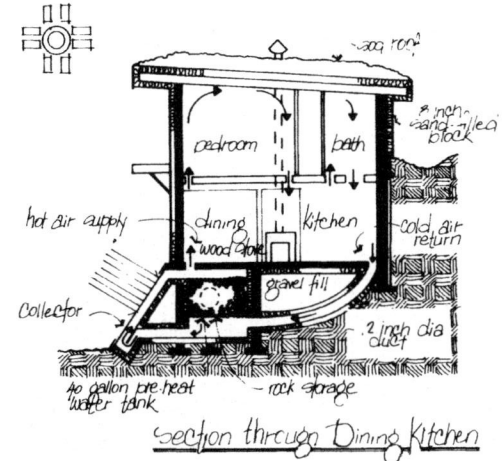

Section through Dining Kitchen

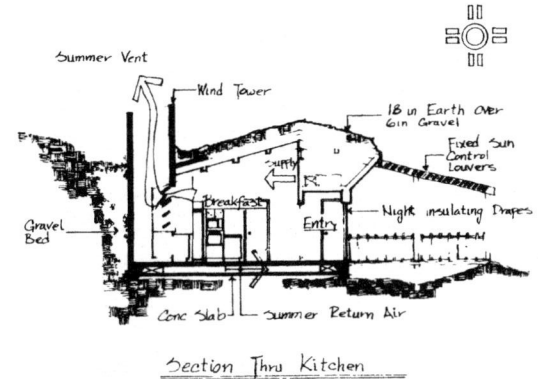

This new 2-story, 3-bedroom detached house
(with solarium) is on a slightly sloping site and
is priced in the $70,000 range. Abundant trees
near the house provide excellent shading to
prevent overheating during the summer.

Open market sale.

Middleton, WI

Bruce K. Kieffer, Northland Country Homes, Inc.
Madison, WI

HEATED AREA: 1,620 FT²

NUMBER OF DEGREE DAYS: 7,605

NET THERMAL LOAD: 64.3 10⁶BTU/YR

AUXILIARY ENERGY: 4.27 BTU/DD/FT²

YEARLY SOLAR FRACTION: 51%

COLLECTOR: South-facing windows, Sunspace, solar mass wall, and skylights
Area: 306 FT²

STORAGE: Concrete and sand floors, water drums, rockbed
Capacity: 15,913 BTU/°F

CONTROLS: Operable dampers, curtains, panels, vents, and shades

BACKUP: 55,000 BTUH gas furnace

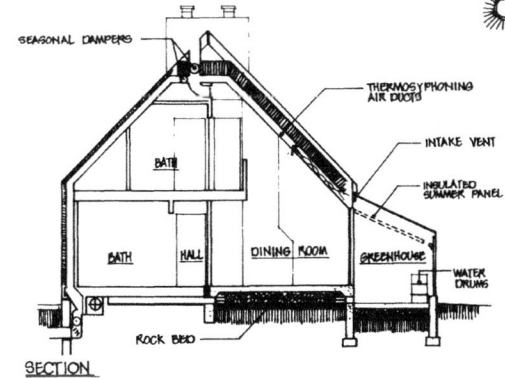

This new 2-story, 3-bedroom detached house is priced in the $46,900 range. Airlock entries reduce winter wind infiltration. Evergreen trees will be planted along the north property line.

Open market sale.

SECTION

CHAPTER 6
HOW TO SELECT THE BEST SOLAR ALTERNATIVE

After having seen 162 winners of the HUD Passive Design Competition, you may ask how one chooses between solar heating by direct solar gain, indirect solar gain, or solarium gain. Although there are many reasons—from overall efficiency to overall appearance—why different designers prefer different systems, we will look at four variables to determine the most appropriate solar heating method for your house designs. Givens, such as the house plan you are presently working with, the site you are building on, and the building codes you must work within, will be first. The climate, which is also a given, will be the second determinant. Then cost and lifestyle will follow as the third and fourth variables to choosing a solar heating concept for speculative homebuilding in your area. For each determinant we will discuss the three basic passive solar heating methods: direct solar gain, indirect solar gain, and solarium gain.

VARIABLE I: THE GIVEN HOUSE PLAN, SITE, AND CODES
The building site may not dictate whether direct gain, indirect gain, or solarium homes are preferable, but it does determine where the collector aperture must be located. If vertical windows are being used as solar collectors, care must be taken to see that those walls are not shaded in the winter by neighboring sites. Through the use of roof apertures (shaded clerestories and roof panels) in addition to south-facing window areas, most sites can be made adaptable to solar design; however, shading is crucial since the roof is highly exposed in the summer when sun is not desired. With all solar apertures, avoid sites or parts of sites which are partly shaded in the winter by neighboring buildings and vegetation.

For all solar-heated home designs, floor plans are best suited to solar heating if the major living spaces—living room, dining room, family room, and even the bedrooms—are to the south, close to the solar collector and storage component. Radiant distribution (and low temperature convective distribution) from solar storage is only effective when the occupant can sit, work, eat, or sleep near the heat source. Clerestory direct solar gain heating or roof-lit indirect solar gain (Trombe wall) heating may be the best choice for one-story house plans which do not have the living spaces at the southern exposure. In two-story houses which are not planned for energy conservation, a non-southern oriented house plan with no potential room openings to the south may require indirect solar gain and/or solarium (isolated) gain solutions. These Trombe wall and solarium solutions will allow the collection of higher solar temperatures (>90°F) which are necessary if solar-heated air is to be blown around for heating a non-solar planned house without causing discomfort from lukewarm air motion.

Code problems with any of these passive solar home designs can be solved logically. Direct gain systems generally pose the fewest problems with local code and planning officials. Indirect solar gain Trombe wall houses may meet some local resistance due to unconventional appearance. This problem will disappear over time as a "solar home" aesthetic becomes popular. As demonstrated in the award winning designs, combining solar systems generally softens the visual impact of passive solar.

VARIABLE 2: THE GIVEN CLIMATE
Direct solar gain homes can pose problems in at least two climates. In very cold climates, the amount of glass exposure required for direct gain heating in the living spaces themselves can lead to tremendous heat losses on cloudy days or at night. Sometimes in a poorly designed house the amount of solar heat collected per house is less than the amount of room heat lost through the win-

A SAMPLE SOLAR SYSTEM CHECKLIST FOR COST ASSESSMENT: DIRECT GAIN

Objective	Function	Component	Specification
Minimum Cost	Collection Roof Skylights Clerestory Wall	{ Single glazing { Fixed windows	0.125 in. Glass; 0.040 in. Fiberglass; 0.040 in. Plastic
	Storage Roof Wall Floor	 Water roof ponds Mass wall Mass floor	60 BTU/sq. ft. collector area min. see Indirect Gain 6 in. Concrete; 4 in. Brick
	Distribution Radiant		
	Controls Heat Gain	 Vegetation	 Deciduous trees planted for summer shading
Maximum Value	Collection Roof Wall	Double glazing Operable windows Diffusion devices Conduction devices Exterior reflectors	0.250 in. Glass; 0.050 in. Fiberglass; 0.050 in. Plastic Diffusion fabric screens; Vertical glazing slats; Reflective venetian blinds; Special irregular glass over plastic; Dark venetian blinds Reflective surface on insulation panel; Light colored gravel or patio surface; Snow
	Storage Roof Wall Floor Freestanding	 Water roof pond Mass wall Mass floor Additional Remote Storage	 see Indirect Gain 8 in. Concrete; 6 in. Brick 55 gallon water drums; 8 in. CMU
	Distribution Radiant Convective	 Fan to storage Exterior vents	 Roof exhaust (hot); Floor supply (cool)
	Controls Heat loss Heat gain	 Moveable exterior insulation Moveable interior insulation Vegetation Roof eave Shading devices	 Solid hinged/Sliding panels; Blown in Beadwall™ Roll down curtains; Bi-fold/sliding doors Deciduous trees planted for summer shading Projection designed for summer shading Fixed/operable louvers

A SAMPLE SOLAR SYSTEM CHECKLIST FOR COST ASSESSMENT: INDIRECT GAIN

Objective	Function	Component	Specification
Minimum Cost	Collection Wall	Single glazing	0.125 in. Glass; 0.040 in. Fiberglass; 0.040 in Plastic
	Storage Wall	Mass wall	12 in. Concrete; 10 in. Brick; 8 in. Adobe; 6 in. Water
	Distribution Radiant		
	Controls Heat gain	{ Roof eave { Vegetation	Projection designed for summer shading Deciduous trees planted for summer shading
Maximum Value	Collection Wall	Double glazing	0.025 (2 at 0.125) in. Glass; 0.050 (2 at 0.025) in. Fiberglass; 0.050 (2 at 0.025) in. Plastic
		Exterior reflectors	Reflective surface on insulation panel; Light colored gravel or patio surface
	Storage Wall Floor	Mass wall Additional remote storage	18 in. Concrete; 14 in. Brick; 12 in. Adobe; 12 in. Water
	Distribution Radiant Convective	Interior vents Exterior vents Fan to interior or exterior	Ceiling supply (warm); Floor return (cool) Roof exhaust (hot)
	Controls Heat loss Heat gain	Backdraft dampers Operable vents Moveable exterior Insulation Moveable interior Insulation Roof eave Shading devices Vegetation	0.006 in. Polyethylene Hinged panels Solid hinged/sliding panels; Blown in Beadwall™ Roll down curtains; Bi-fold/sliding doors Projection designed for summer shading Operable/fixed louvers Deciduous trees planted for summer shading
	Natural light	Windows in mass wall	

dow area. This loss can be reduced by double and triple glazing or by moveable insulation over the glass, which must be included for the successful use of direct gain in very cold climates.

In very hot climates, direct gain heating can cause comparable problems. In addition to glare and too much solar collection in spring and fall, the summer heat transfer through the large areas of glass (even shaded) can make cooling loads excessively large. To prevent this heat transfer, double and triple glazing or moveable insulation again may be necessary to keep out the hot ambient air temperatures. On the other hand, homes in consistently cloudy climates may benefit the most from direct gain heating; direct gain allows the use of low amounts of solar radiation to offset daytime building heat loss.

In conclusion, although direct solar gain systems are adaptable to almost every climate, site, and building style, caution must be taken in extremely cold and extremely hot climates.

Indirect solar gain heating by masonry and water storage walls or roof panels can also be designed to suit almost any climate in the continental United States. However, simple radiant masonry or water storage walls work best in climates with large day-night temperature swings while more complex convective and radiant walls may be required in most other climates. In climates and homes which need daytime as well as nighttime heating, a secondary direct solar gain collector may be necessary to provide immediate, early morning heat; or carefully detailed inside vents and dampers will be required for immediate convective heat transfer from the storage wall. In milder climates which may have a tendency to overheat on spring, fall, and even winter days, houses will require vents to the outside and inside to distribute excessive solar heat gains. As with direct gain systems, indirect solar gain systems in cold climates often function better when the solar storage area is protected from nighttime and cloudy day heat loss. This requires additional construction labor and cost but does not reduce the thermal value of the indirect gain solar mass wall or roof.

Solariums or isolated solar gain homes can be pleasant heat sources in many parts of the country. Hot and humid climates, however, might suffer from the addition of "greenhouse" spaces for passive solar heating. The addition of humidity from a greenhouse in a humid climate can make the solar home uncomfortable. In general, a solarium space which is not intended for full-time occupancy will be able to provide effective heating for a house in almost any climate, since it can be either: (1) completely isolated from the house; (2) opened to the house and additional

heat storage as needed; or (3) occupied when the direct gain temperature is right. Solariums which are intended for continuous use, day and night, are direct gain spaces. In cold climates, they will require adequate internal storage mass and moveable insulation over the glass.

Since most climates are complex rather than simply cold or hot, most passive solar homes will combine the best of direct, indirect, and solarium solutions. However, for simplicity, ease of construction, and cost, it is often prudent to try one of the simpler solutions on your first passive solar home.

VARIABLE 3: COST
To make a decision between direct gain, indirect gain, and solarium heating on the basis of cost, three issues must be evaluated on the local level: the cost of design or redesign, the cost and availability of materials, and local building construction techniques. Generally, those systems which are less complex and those which allow incremental introduction (heating one of two rooms in the first model house) will be the most cost effective for the speculative housing market. To aid in the cost estimation of each system, however, a small chart illustrates the range of components inherent in the direct solar, indirect solar, and solarium home.

VARIABLE 4: LIFESTYLES
Lifestyles and taste of the residential buying market are going to be the final criteria for selecting a passive solar system for your house designs. The flexibility and interest of a potential buyer is important in the effective operation of many of these passive solar homes. In many cases these houses are designed to appeal to the individuals who buy a sailboat over a speedboat, for whom the sport of moving a man-made object by natural forces is indeed a thrill.

In direct solar gain homes, several lifestyle changes might be considered. In order to collect enough heat directly during the day to last through the night (or even the next cloudy day), room temperatures must be allowed to get pretty high to "charge" the storage areas. As soon as the sun goes down, room temperatures will drop and the radiant storage will offer comfort even though ambient room temperatures may be low. In the direct gain home, therefore, temperatures inside may fluctuate from 90°F to 60°F—a 30°F swing. This is not a serious problem for owners who work all day and only arrive home in the evening. They can allow room temperatures to get very high throughout the day in order to ensure steady solar heat throughout the evening.

The operation of a direct gain home is relatively simple. Each sunny morning, the moveable insulation over the glass area is

A SAMPLE SOLAR SYSTEM CHECKLIST FOR COST ASSESSMENT: SOLARIUM

Objective	Function	Component	Specification
Minimum Cost	Collection Roof Wall	Single glazing	0.125 in. Glass; 0.040 in. Fiberglass; 0.040 in. Plastic
	Storage Roof Wall Floor Freestanding	Mass wall Mass floor	see Indirect Gain 4 in. Concrete; 4 in. Brick
	Distribution Radiant		
	Controls Heat gain	Vegetation	Deciduous trees planted for summer shading
Maximum Value	Collection Roof Wall	Double glazing (Glazing sloped at Lat. +15°) Exterior reflectors	0.250 (2 at 0.125) in. Glass; 0.050 (2 at 0.025) in. Fiberglass; 0.050 (2 at 0.025) in. Plastic Light colored gravel or patio surface; Snow
	Storage Roof Wall Floor Freestanding	Mass wall Mass floor	see Indirect Gain 8 in. Concrete; 6 in. Brick
	Distribution Radiant Convective	Interior vents Exterior vents Fan to remote storage	Ceiling supply (warm); Floor return (cool) Roof exhaust (hot); Floor supply (cool)

	Controls
	Heat loss
	
	Heat gain
		Moveable insulation
		  at greenhouse skin
		Moveable insulation
		  at building skin
		Vegetation
		Shading devices
			Roll down curtains; solid hinged panels; Blown in Beadwall™
			
			Roll down curtains; Bi-fold and Sliding doors
			
			Deciduous trees planted for summer shading
			Roll down canvas; Operable/fixed louvers

opened to let in the sunshine. Each evening, the insulation is closed to protect the living spaces from heat loss. On a seasonal basis, shading devices, awnings, trellises, and overhangs may be put in place to protect the large glass areas from unnecessary sunshine in summer and sometimes spring and fall.

An indirect solar gain system—a masonry or water storage wall—implies other involvements for the occupant. Although room temperatures are more controlled and exhibit less temperature swing, room air temperatures will still be cooler since radiant heat satisfies the major comfort needs of the occupants.

In some climates, especially those with large day-night temperature swings, no daily operation is required by the homeowner. In other climates, though, the operation of the indirect solar gain home can be complicated. Each morning moveable insulation outside the storage wall is opened, and dampers for convective air distribution are shut so that the solar storage can charge. If immediate heating is required, dampers and fans may be opened and closed throughout the day. Each night, the moveable insulation is closed to cover the storage area, and dampers and fans shut off so lower temperature radiant heating can work at its best. Seasonally, a summer mode could be added in which dampers to the outside are opened, allowing the solar-heated air in the collector-storage wall to vent and draw cooler air through the house. Alternatively, moveable insulation would be put in place daily over the glass area and opened nightly to expose the mass to the cooler night sky.

Isolated gain solarium spaces attached to a house can become a simple solar-heated room, occupied when the temperature is right and ignored when the temperature is wrong. Or the solarium may be a very complex solar collector-storage system for effectively heating the house. In a solarium home, doors, windows, and vents may be opened each day to transfer heat to the living space and additional solar storage mass. As temperatures in the living spaces get too high, these openings would be closed and the solarium either vented or "supercharged" with sunshine. At night, the vented solarium could be cut off from the living space and its storage mass. On the other hand, the supercharged solarium, protected by extensive exterior insulation, could be reopened to the living spaces. A solarium combines the possibilities of a direct gain "live in the sunshine" space with an indirect gain "collect the sunshine" space, but it also combines several modes of operation, providing for the most intriguing but complex owner involvement.

Living in a passive solar home can be very exciting, stimulating, and in the end, very economical. It may involve daily operations, opening and closing areas of the house, so as to use and store heat and sunshine most effectively. It may also involve covering and uncovering window areas to prevent winter heat loss and summer heat gain as well as enhance summer air movement throughout the house. It is not less of an art of raising the right sail for the amount of "wind" to power your well-designed sloop, or in this case, your well-designed solar home.

CHAPTER 7
MOST COMMONLY-MADE THERMAL MISTAKES

During the review of over 550 entries to the HUD Passive Solar Design Competition, *three general mistakes* appeared consistently:

I. In many submissions there was *insufficient contribution* of useful heat from the solar "system" to the living spaces. This could be a result of a small, undersized collector area, *or* inadequate storage mass coupled with a larger collector area.

In some cases, the houses had windows placed evenly on north, south, east, and west facades, ignoring the heating benefits and solar control of south-facing windows. Solar collection area must be carefully sized and placed in the building with adequate control to prevent excessive overheat in summer or excessive heat loss in winter.

Solar storage must also be carefully sized and placed in the building, and storage materials must be chosen for heat capacity and potential distribution capability. In many of the light wood frame houses submitted, no water or masonry elements were added to provide solar storage for nighttime or extended periods of solar heating. When storage was added it was often undersized, not adjacent or directly coupled to the collector area, and almost always left off the "less structurally supporting" second floors. Rock storage beds were often included with little understanding of how one efficiently charges and discharges this kind of storage, or the temperatures needed for effective distribution. All storage materials (especially materials such as sand) must be evaluated for their ability to absorb, conduct, hold, and emit solar heat gain. Granular materials such as rock and sand exhibit very little grain-to-grain heat conduction. Storage must be charged by air flow.

If a conventional residential home in most parts of the U.S. can acquire 10%-15% daytime solar heating without storage, reorientation, and reorganization, a passive solar residential home can surely achieve more than 30%-40% solar contribution towards the total heating needs of the home. A basic understanding of the *collector* aperture to *storage* mass relationship is necessary to ensure adequately sized and located collector and storage—balanced for the best solar contribution. (See Chapter 9 for more on calculations.)

II. The second most commonly made mistake was *inadequate or inefficient distribution* systems based on an understanding of the *storage mass*-to-*living space* relationship. The distribution of solar heat needs to be logically conceived and properly executed.

When radiant distribution to the living space is used, the storage mass must be adjacent to the people and rooms needing heat, not the less-used spaces such as closets and stairwells.

When convective distribution is used, the logical flow of hot and cold air must be understood. Stratification must be anticipated and handled appropriately. Drawing arrows to indicate heat flow does not guarantee heated air flow throughout the house. A key to success is placing storage in the right position for distribution. Remember that hot air rises and can best be used if it doesn't have to travel long distances. It should be understood that convective distribution relies on natural temperature differences or on appropriate fan sizing to take heat from a storage mass and distribute it to the room air.

The best passive solar heat distribution in your homes will provide as simple and direct a link as possible between collection and storage, as well as between storage and living space.

III. The third most commonly made mistake was *poor detailing* of critical controls which allow the user to regulate heat flow. In many cases the final success of passive solar systems (after adequate collection and storage and logical distribution) depends upon the controls which speed, slow, or stop the flow of heat from coming in or going out of the house as needed. These controls—registers, backdraft dampers, moveable insulation, exhaust vents, etc.—are often not stock items and need to be carefully placed, sized, and detailed to ensure the proper operation of your passive solar home.

To help you in the design or construction of future passive solar homes, here is a preliminary checklist of common mistakes to avoid by system type: direct solar gain, indirect solar gain, and solarium.

A. DIRECT SOLAR GAIN
1. COLLECTION: (KEY WORDS: SIZING, PLACEMENT, CONTROL)
• Do not oversize collector areas if you are in a predominantly warm or hot climate. Remember direct gain solar collection for nighttime heating must also be "lived in" during the day.

• Do not pour direct solar gain on the occupant's head. Make every attempt to diffuse and redirect sunlight to the storage mass around the room; or consider direct gain systems for rooms without daytime occupancy so that overheating and glare accompanying efficient solar storage will not be a problem to the occupant.

• Include summer shading devices to prevent overheating from the solar collection/storage systems. In some climates, spring and fall shading will also be necessary so that peak summer sun position (June 21st) will not be adequate. Proper orientation (avoid due west glazing) and proper tilt (vertical glass is the easiest to shade) are the first two steps to both effective winter collection and summer shading. Operable rather than fixed louver shading devices (set for December 21st) will allow solar collection throughout the heating season. Adequately sized overhangs, exhaust vents, operable windows, and deciduous tree planting can all prevent overheating.

• Clerestories can be used effectively to bring natural light and heat to north rooms.

2. STORAGE: (KEY WORDS: SIZING, PLACEMENT, MATERIALS)
• Do not undersize solar storage. If storage is inadequate, the solar gain which potentially could be stored for nighttime heating and next day heat contribution will instead cause uncomfortable overheating and be wasted through opened windows.

• Try to provide storage mass in relation to all collector areas. Often second floors lack storage mass although they have significant potential for collecting energy.

• As much of the storage mass as possible should be located where it will be exposed to *direct* sunlight.

• Do not use loose rocks or sand as direct gain storage for incoming sunshine. A limited amount of heat absorption of the top layer can be relied on, but the conductive distribution of heat down through these materials (charging) is very poor. Rock beds charged with hot air from the top of a direct gain living space should only be used for secondary storage (for overheat prevention) and preferably with radiant heat discharge. Water and solid masonry storage should be used in the living spaces themselves for direct radiant and convective distribution. Direct gain storage should not be covered with materials such as carpet, linoleum, or fabrics which prevent solar absorption and heat radiation.

3. DISTRIBUTION (KEY WORDS: RADIANT DISTRIBUTION, CONVECTIVE DISTRIBUTION, CONTROL)
• Do not place direct gain solar storage away from occupied spaces. For direct radiant distribution, solar storage must be in contact with the occupant. Convective distribution requires a temperature differential to move the heated air. Provide adequate solar heat distribution throughout the house, especially in areas which do not receive direct sunlight. Place rooms in the path of natural heat flow, higher than collection. Small fans can be used to circulate solar-heated air to remote spaces, or existing mechanical distribution systems can be integrated for successful distribution of passively gained solar heat.

• Do not expose storage mass such as floor slabs and vertical walls to the outside without insulation to prevent heat loss. The solar storage mass will radiate most easily to the coldest side unless prevented by a thermal break.

• Do not leave large glass areas designed for direct solar gain exposed at night. Much of the stored heat in the house will flow back out through the glazed area to the cool night air. Although double and triple glazing will limit this flow, moveable insulation over the glass is a preferred barrier to heat loss through large glass areas.

B. INDIRECT SOLAR GAIN
1. COLLECTION (KEY WORD: CONTROL)
• Include summer shading and/or summer exhaust vents to prevent excessive heat storage or degradation of collector glazings in overheated periods.

2. STORAGE (KEY WORD: SIZING)
• Do not make the Trombe wall too thick for effective radiant distribution to the house at night. While some Trombe walls can be 12" to 18" thick, remember that walls thicker than 12" may

radiate stored heat many hours later than desired.

3. DISTRIBUTION (KEY WORDS: RADIANT DISTRIBUTION, CONVECTIVE DISTRIBUTION, CONTROL)

• Avoid insulating a radiant storage wall from the space it is heating. Radiant storage walls are invaluable to passive solar heating since they function well at low temperatures. Caution must be taken not to block radiant heat transfer with closets, bookshelves, and finished wall materials. In many climates, heat loss to the outside should be prevented with moveable insulation over the glass.

• Do not try to heat by convective distribution alone without adequate heat transfer surface and thermal isolation of the Trombe wall at night. Rough masonry surfaces and smaller water storage containers will provide better convective heat transfer. Moveable insulation over the glass to prevent excessive heat loss to the outside is necessary if nighttime heating is to be provided by Trombe wall convection alone.

• Do not forget the details and controls (vents and backdraft dampers) required for effective convective Trombe wall operation. Their size, location, and functions must be clearly understood.

• In both convective and radiant Trombe wall heating systems, adequate distribution should be provided throughout the house.

C. SOLARIUM—ISOLATED SOLAR GAIN

1. COLLECTION (KEY WORD: CONTROL)

• Prevent excessive summer overheating of the solarium space. Consider vertical glazed areas for most effective shading in summer or provide vents and moveable shading devices.

2. STORAGE (TEMPERATURE LIMITATIONS OF THE SOLARIUM; KEY WORDS: STORAGE, SIZING, PLACEMENT, MATERIALS)

• Decide whether the solarium must be thermally regulated for the plants or materials within, or if it can be allowed to fluctuate for maximum solar collection and storage:

To keep "occupied" solariums comfortable, solar storage must be provided within to prevent overheating, and moveable insulation must be added to all glazed areas to prevent excessive heat loss. Balance the storage mass in the solarium with the temperature needed and place all other solar storage within or adjacent to the living space.

In "unoccupied" solariums, temperatures may be allowed to fluctuate considerably, with care taken only to prevent heat loss from the solar storage to ensure that the solar heat collected is distributed to the house and not to the colder outdoors.

• Understand that remote rock storage beds located in basements or crawl spaces present a difficult challenge. When the design calls for "charging" the remote rock storage bed with heat from a solarium/greenhouse space, the builder must consider the fact that most solarium spaces (particularly those which are meant to be occupied) will not reach temperatures above 100°-120°F. The problem is further compounded as soon as the hottest air is drawn off the top of the solarium and stratification disappears. A rock storage bed can be used effectively if it is placed in direct thermal contact with the living spaces (for radiant floor distribution or large venting plenum distribution); and/or if it is charged by a higher temperature "collector" than the typical solarium. Fan-driven 80°F air from the remote rock storage may feel cool and uncomfortable to occupants. If a rock storage bed is discharged by air flow, the air flow direction should be reversed.

• Again, evaluate storage materials for their heat absorption, conduction, storage capacity, and emission capabilities. Do not overestimate the capabilities of loose earth, loose gravel, or sand as heat storage materials.

3. DISTRIBUTION (KEY WORD: PLACEMENT)

• Care must be taken in locating the solarium space. To simplify distribution, the collector/storage arrangement must be adjacent to the living spaces which need heat. The greater the interface with the major living spaces—living room, bedrooms, family room, dining room—the better. Entry halls, stairwells, bathrooms, and utility rooms are secondary and do not need to touch the solarium directly.

CONCLUSION

In conclusion, examine carefully each component in the solar system—collection, storage, and distribution—to determine sizing, location, materials and controls appropriate to your climate's heating and cooling needs.

In all passive solar homes (with direct, indirect, or solarium gain), the interface between the passive solar heating system and the auxiliary (mechanical or woodstove/fireplace) must also be well thought-out. The auxiliary or backup can provide a good means of heat distribution—reinforcing the passive solar system, not duplicating or fighting solar collection, storage, and distribution.

Some accommodation should be made for designs which balance percentage of solar contribution for winter heating against cost-effectiveness, complexity, and summer cooling requirements. At present, in almost all climates, much can be said for designing and building a passive solar home for less than 100% contribution.

CHAPTER 8
CONSTRUCTION DETAILS

While passive solar design is appealing to many people precisely because it can be achieved with conventional construction techniques and components, it is essential that the role of several new construction details be clearly understood. Studies have conclusively shown that the thermal performance of conventional, energy conscious, and passive solar homes is dependent on careful attention to thermal details such as insulation, shading, wall sections, etc. This chapter begins to identify areas and details within the anatomy of the single-family home which may be new to builders and which deviate from standard practice. In some cases these construction details are simply good energy-conscious construction. In other cases, the details may be useful to explain the construction of many of the passive system components discussed throughout the text.

Several of the details use unfamiliar materials, but more typically standard construction materials are combined in non-conventional ways. Field experience has shown that while the structural and weatherproofing characteristics of most assemblies are understood by the majority of builders, few appreciate the thermal criteria for the selection of materials. Several areas of detailing which are critical to a home's thermal performance were not treated adequately in the designs submitted to the competition. These include foundation and wall insulation, Trombe wall details, moveable insulation and shading devices.

FOUNDATION, WALL, AND STORAGE INSULATION
Wall insulation standards for passive solar homes generally exceed the R-value provided by R-11 batts in a 2 x 4 stud wall cavity. Ways of increasing this thermal resistance include filling the cavity with insulating foam (beware of toxicity problems), adding 1" or 2" of foam sheathing to the wall exterior, and/or increasing the effective width of the insulated cavity by adding 2 x 4 horizontal strapping outside a standard wall or using 2 x 6 wall studs. A polyethylene vapor barrier to reduce cold air infiltration, taped at electrical boxes, should be used on the inside surface of the wall studs if a low permeability rated sheathing such as plywood or polystyrene is used. Also let-in diagonal bracing must be specified for racking resistance or some other provision for lateral bracing must be provided if a non-structural sheathing is used on the exterior.

Foundation and basement walls must also be insulated. In most cases this can be most effectively accomplished on the exterior of this wall with foam plastics which are of sufficient density to withstand the lateral loads from soil pressure. Due to warmer temperatures deep in the ground during the winter, it is becoming standard practice to step down exterior foundation insulation to thinner sections once below the frost line; 2" to 1" in mild climates and 4" to 2" in more severe ones.

Frame construction commonly begins at least 8" above finish exterior grade. Exterior insulation should extend, when possible, from the framed wall down over the entire foundation wall area. Where foam insulation is used above grade, it must be protected from objects and the elements. Cement stucco or wire mesh, a fiberglass-reinforced epoxy stucco, some fabric flashing materials, or sheet metal skirting can be applied over the foam to protect it. A common moisture problem area is where the frame wall meets the foundation wall. The wall should be flashed out over the surface treatment of the foam insulation. Sometimes 2 x 6 wall studs are offset in this area to lap over the foundation insulation, minimizing moisture problems and creating a cleaner appearance. However, 2 x 4 studs should not project beyond the foundation wall.

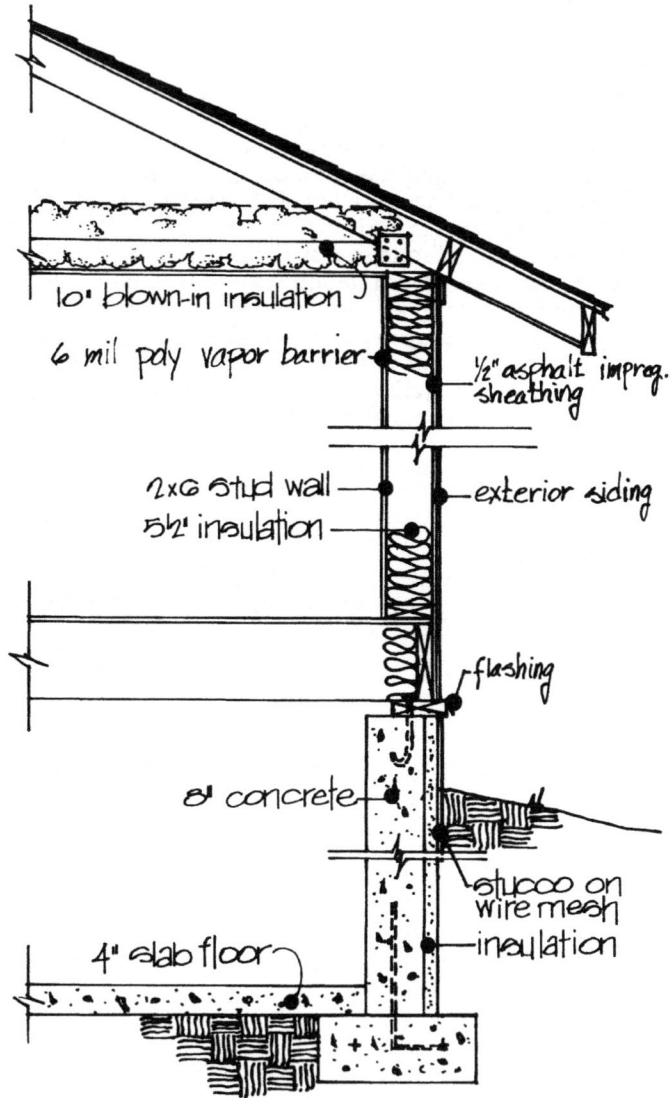

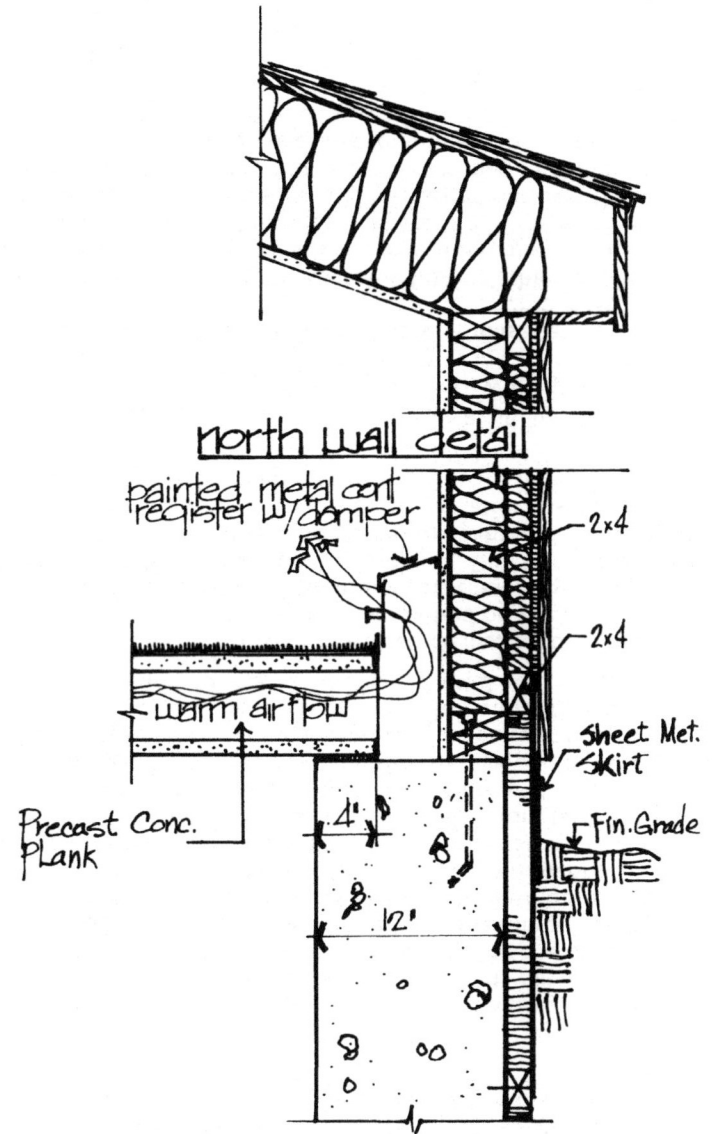

WALL SECTION

10" blown-in insulation

6 mil poly vapor barrier

½" asphalt imprg. sheathing

2×6 stud wall

exterior siding

5½" insulation

flashing

8" concrete

stucco on wire mesh

4" slab floor

insulation

WALL SECTION

WALL SECTION

north wall detail

painted metal cont. register w/damper

2×4

2×4

warm air flow

Sheet Met. Skirt

Precast Conc. Plank

Fin. Grade

4"

12"

WALL SECTION

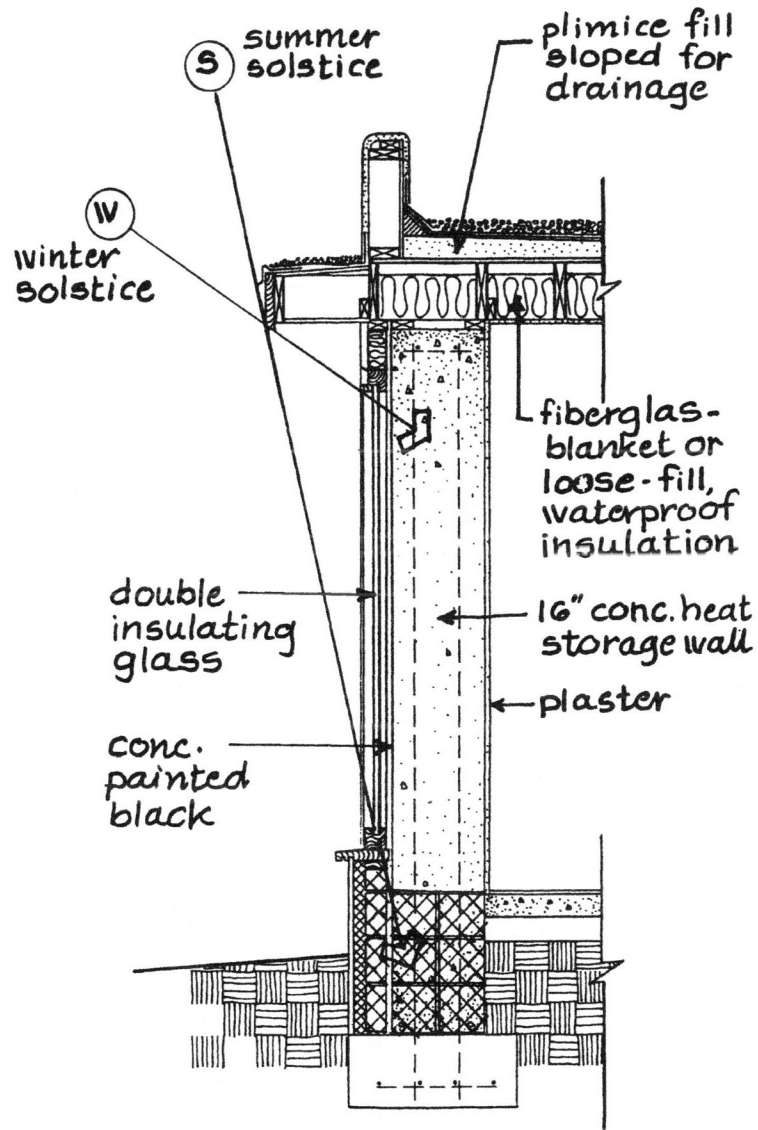

VENTLESS TROMBE WALL

S summer solstice

W winter solstice

plimice fill sloped for drainage

fiberglas-blanket or loose-fill, waterproof insulation

16" conc. heat storage wall

plaster

double insulating glass

conc. painted black

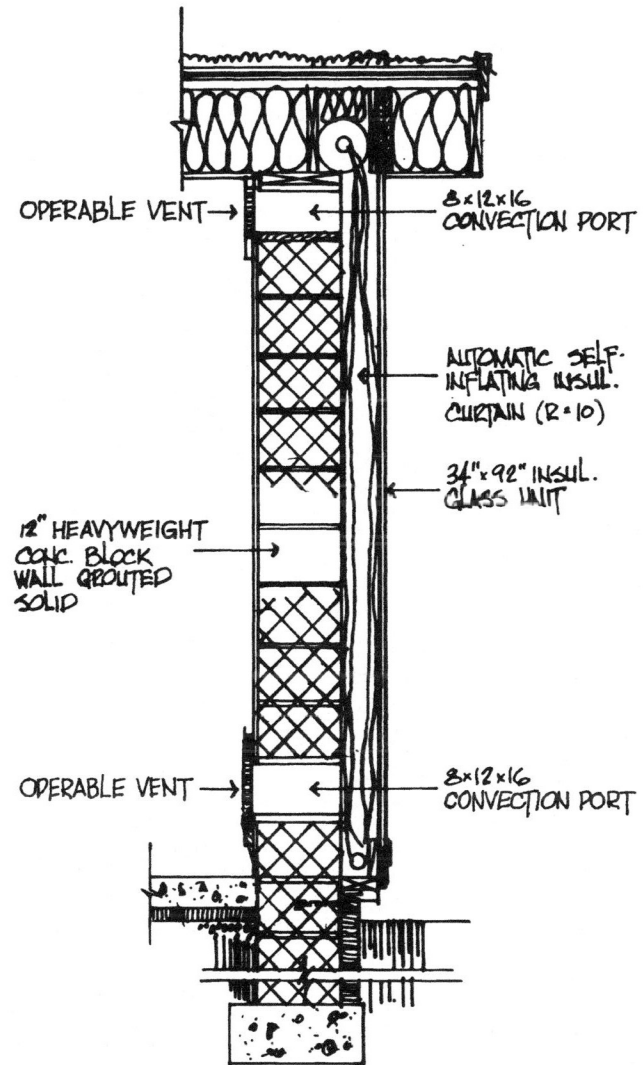

TROMBE WALL SECTION

OPERABLE VENT →

8×12×16 CONVECTION PORT

AUTOMATIC SELF-INFLATING INSUL. CURTAIN (R·10)

34"×92" INSUL. GLASS UNIT

12" HEAVYWEIGHT CONC. BLOCK WALL GROUTED SOLID

OPERABLE VENT →

8×12×16 CONVECTION PORT

TROMBE WALL DETAILS

Trombe walls are constructed of heavy masonry including brick, block, adobe, or poured concrete. Structurally, Trombe walls are no different than common masonry walls, since the thermal fluctuations of the mass are often less than exterior masonry walls. Concrete block used in a Trombe wall should be of heavy aggregate and should be filled with pea gravel, concrete grout, or a heavy mortar. Twelve inch or 16" block is commonly used, although in some situations an 8" block wall is used for a quicker transfer of heat to the interior.

Vents are generally added to Trombe walls at the top and bottom to allow for convection into the living space for heat during the day. The wall vents in the Trombe wall increase the heat gain efficiency slightly, but they do allow a faster transfer of heat to the living space. The total cross-sectional area of the bottom vent openings should be equal to one-half the plenum cross-sectional area between the masonry wall and the glazing. Deviation from this value is not critical, however the top vents must have automatic damper flaps to prevent reverse thermosiphoning at night. Frequently a simple layer of thin film plastic is taped to the interior side of the Trombe wall at each upper vent. When room air attempts to enter the Trombe wall at the top opening in the evening, the plastic is forced to close against a light-gauge screen set in the interior plane of the Trombe wall. Optional manual bottom dampers will add control, allowing the system to stagnate when desired.

The glazing on a Trombe wall should be double layer glass or plastic. Factory-sealed dual pane glass panels are preferred as they will not fog with moisture condensation. Caution should be applied to the use of acrylic plastic glazing on Trombe walls as the temperatures generated have caused warping and cracking. Gaskets or mullions which allow for sufficient thermal expansion must be used.

Trombe walls should either be vented to the outside and/or shaded to prevent them from heating the interior in the summer. It is important to insure that Trombe walls are thermally isolated from the exterior with no cold bridges at the lead or sill. The foundation wall should be insulated on the exterior from the bottom of the glazing down to the footing.

Trombe walls do not require moveable insulation, but their performance is improved by its application. The extent of the improvement will depend on the local climatic conditions. Exterior hinged insulating panels can be employed which also reflect additional sunshine onto the wall, or roll shades can be deployed between the wall and its glazing.

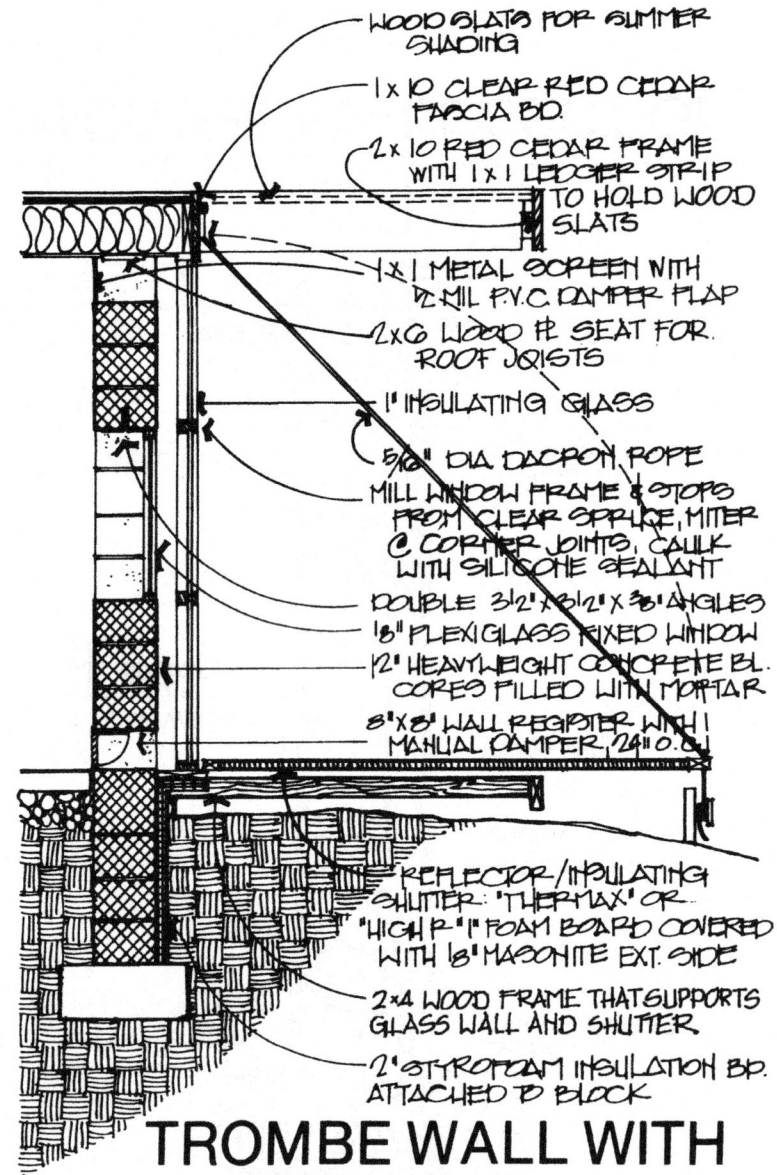

WOOD SLATS FOR SUMMER SHADING

1 x 10 CLEAR RED CEDAR FASCIA BD.

2 x 10 RED CEDAR FRAME WITH 1 x 1 LEDGER STRIP TO HOLD WOOD SLATS

1 x 1 METAL SCREEN WITH 12 MIL P.V.C. DAMPER FLAP

2 x 6 WOOD PL. SEAT FOR ROOF JOISTS

1" INSULATING GLASS

5/16" DIA. DACRON ROPE
MILL WINDOW FRAME & STOPS FROM CLEAR SPRUCE, MITER @ CORNER JOINTS, CAULK WITH SILICONE SEALANT

DOUBLE 3 1/2" X 3 1/2" X 3/8" ANGLES

1/8" PLEXIGLASS FIXED WINDOW

12" HEAVYWEIGHT CONCRETE BL. CORES FILLED WITH MORTAR

8" X 8" WALL REGISTER WITH MANUAL DAMPER, 24" O.C.

REFLECTOR/INSULATING SHUTTER: 'THERMAX' OR 'HIGH R-1' FOAM BOARD COVERED WITH 1/8" MASONITE EXT. SIDE

2 x 4 WOOD FRAME THAT SUPPORTS GLASS WALL AND SHUTTER.

2" STYROFOAM INSULATION BD. ATTACHED TO BLOCK

TROMBE WALL WITH REFLECTOR/SHUTTER

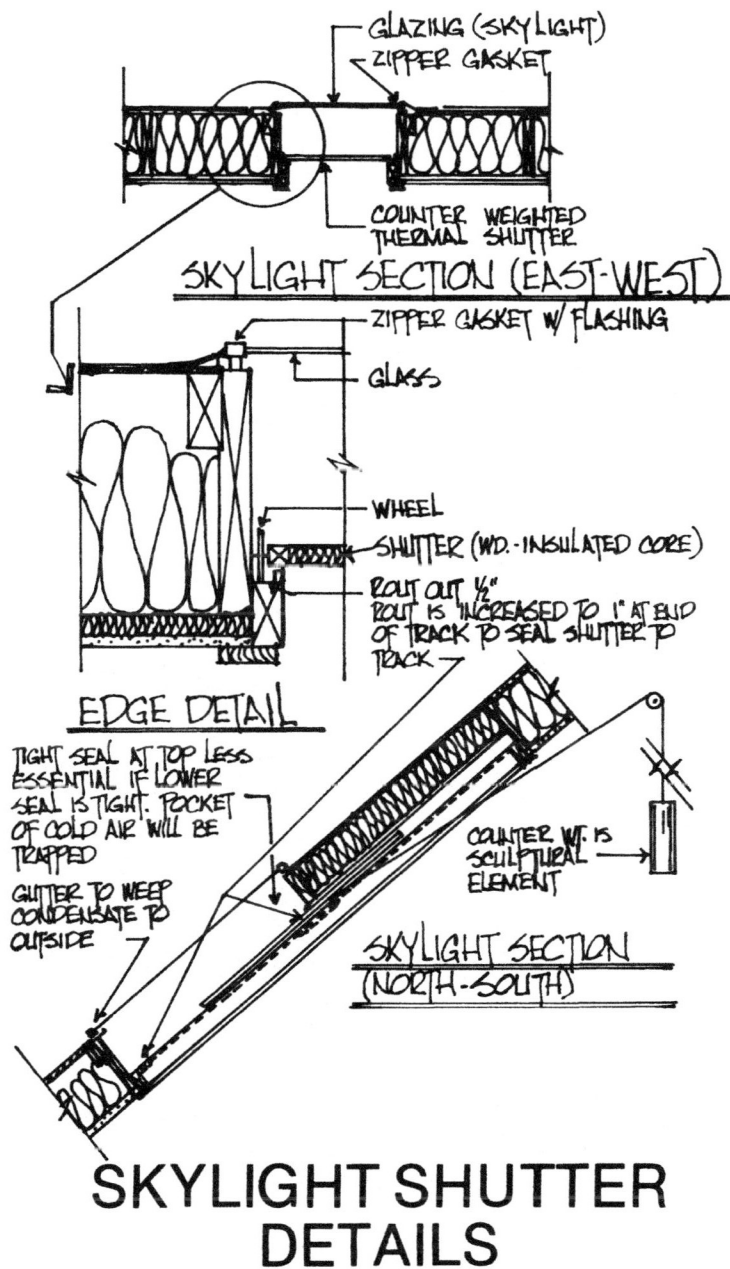

GLAZING (SKYLIGHT)
ZIPPER GASKET
COUNTER WEIGHTED THERMAL SHUTTER

SKYLIGHT SECTION (EAST-WEST)

ZIPPER GASKET W/ FLASHING
GLASS
WHEEL
SHUTTER (WD.-INSULATED CORE)
ROUT OUT ½"
ROUT IS INCREASED TO 1" AT END OF TRACK TO SEAL SHUTTER TO TRACK

EDGE DETAIL

TIGHT SEAL AT TOP LESS ESSENTIAL IF LOWER SEAL IS TIGHT. POCKET OF COLD AIR WILL BE TRAPPED

GUTTER TO WEEP CONDENSATE TO OUTSIDE

COUNTER WT. IS SCULPTURAL ELEMENT

SKYLIGHT SECTION (NORTH-SOUTH)

SKYLIGHT SHUTTER DETAILS

Other concerns in Trombe wall design include provisions for access to allow occasional cleaning, particularly at vision panels, and concerns for outgassing in certain conditions where wood members are used for framing. Care should also be taken to insure that the painted Trombe wall be able to cure before glazing is applied.

MOVEABLE INSULATION

The thermal envelope of the building can respond to changes in the outside temperatures and levels of sunlight by the use of moveable insulation. Passive solar heating system performance is usually enhanced by moveable insulation as it allows glass areas in the home to collect heat during the day and trap this heat at night. This is a new building concept for which there are as yet few off-the-shelf products, although many systems are currently being developed.

Moveable window insulation can be found in the form of shades, shutters, or curtains inside the home, slatted shades or hinged shutters on the exterior, or foam beads moved by a blower into and out of the space between dual glass panes. Shades and curtains can be used in the space between Trombe walls and glazing, and thin reflective film shades can be helpful in sloped solarium glazing. Even foam-filled overhead garage door panels can be used for moveable insulation. At an increase in cost, some systems are automatically activated, but the majority are operated manually.

The thermal design of a system must include adequate edge seals to prevent drafts behind the moveable insulation panel which will render it ineffective. The higher the R-value of the insulation panel, the tighter the air seals it must have. The thermal resistance of the panel should be as high as incremental cost savings dictate, but the difficulty of sealing the edges makes an R-factor greater than 5 impractical.

In general, moveable insulation is one of the most cost-effective components for passive solar homes, or any home for that matter, but requires care in the design selection to provide adequate thermal protection and to match the needs of the occupants.

SHADING DEVICES

Shading devices generally fall into two categories—fixed and operable. The two may be combined if conditions warrant, or if some particular aesthetic effect is desired. Fixed shading devices most often are incorporated as building elements. These include roof overhangs, floor cantilevers, stepped building masses, and wall extensions. As an integral part of the structure, these elements are strong enough to resist winds and the destructive influences of rain, snow, and ice. Decks, balconies, or exterior stairs also offer increased marketing potential and additional utilitarian value.

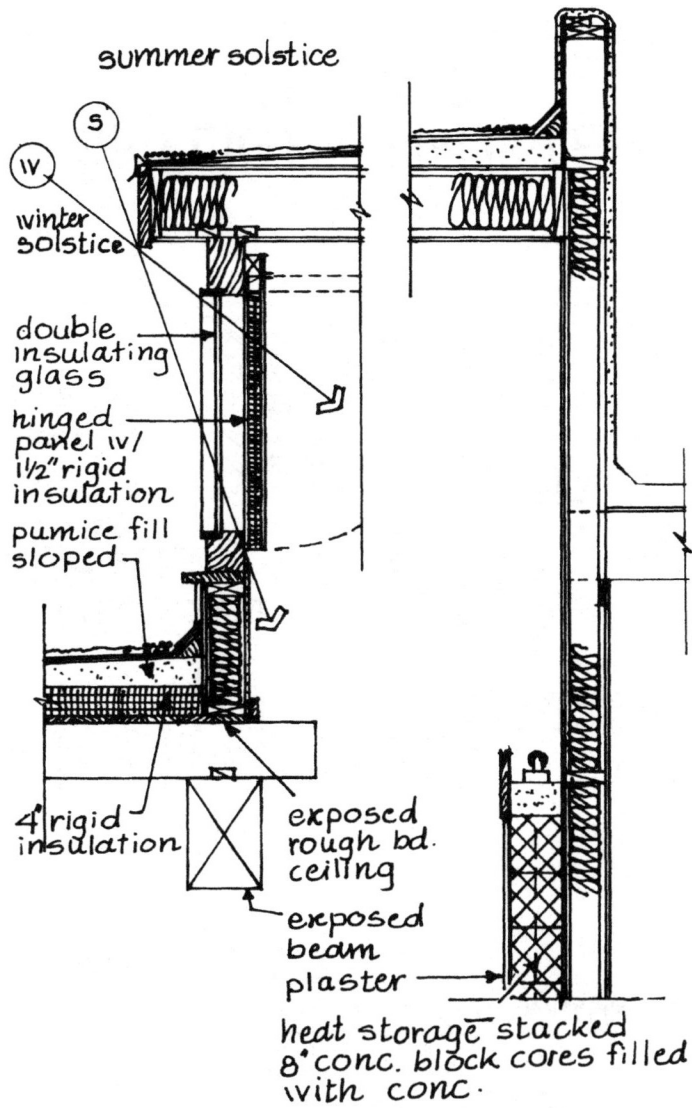

CLERESTORY WITH SHUTTER

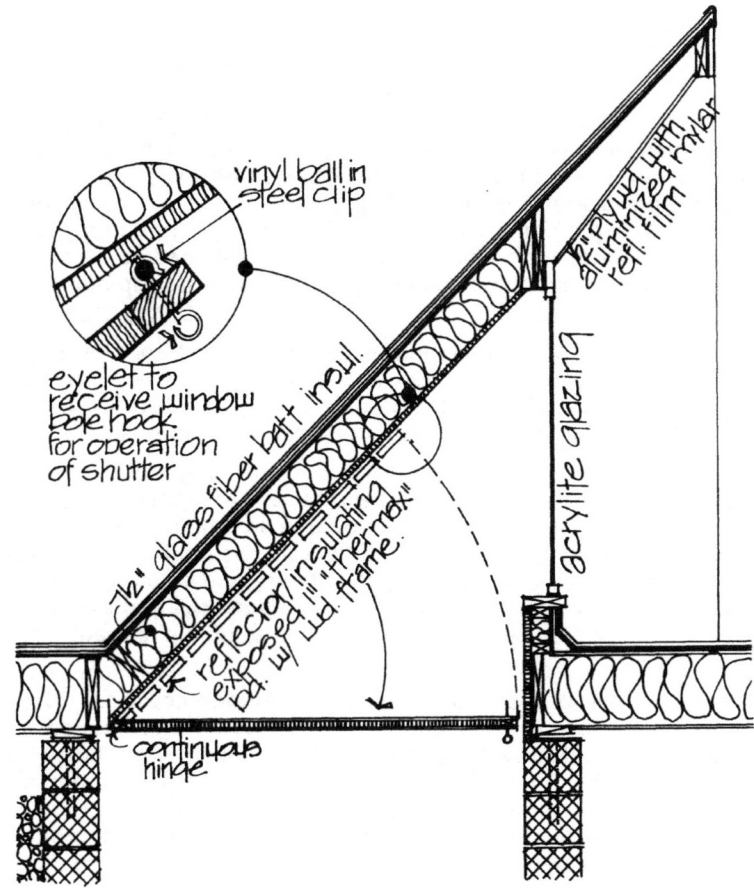

summer solstice

S

W

winter solstice

double insulating glass

hinged panel w/ 1½" rigid insulation

pumice fill sloped

4" rigid insulation

exposed rough bd. ceiling

exposed beam plaster

heat storage stacked 8" conc. block cores filled with conc.

CLERESTORY WINDOW WITH SHUTTER

vinyl ball in steel clip

eyelet to receive window pole hook for operation of shutter

½" glass fiber batt insul.

reflector insulating bd. exposed w/ wd. frame

continuous hinge

acrylite glazing

2" plywd. with aluminized mylar refl. film

The greatest limitation of fixed shading devices is their inability to adjust to the sun's continually changing position. Nor can they respond to occasional days during the heating or cooling season when unusual weather conditions cause abnormal heating or cooling demands. However, as an inherent part of the building structure, fixed shading devices are generally cost effective and easily implemented.

Cedar Slats (Removable)

Vents

Fiber glass.

gl.

Flourescent valance lighting

Triple track w/ fabric wrapped 4" foam sliding shutters.

Casement

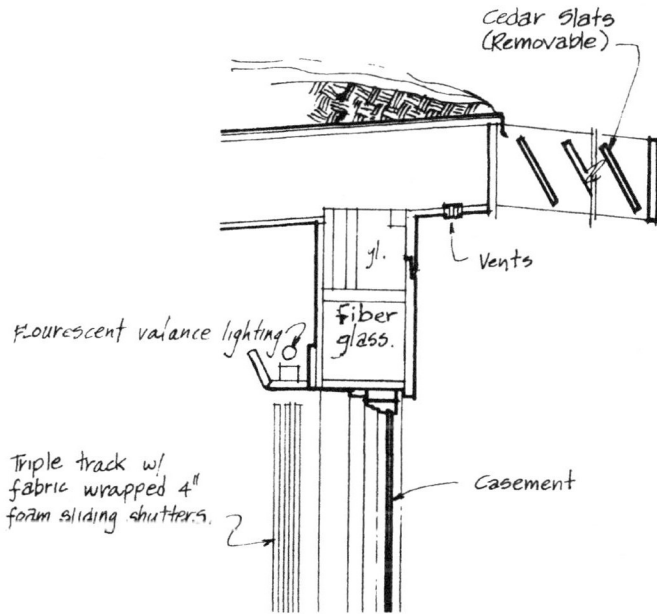

DEMOUNTABLE LOUVER DETAIL

Operable shading devices potentially offer better seasonal performance and design flexibility. These are often designed to act simultaneously as insulating or reflective panels, which enhance the overall performance and improve their relative worth. Operable shading devices are typically hinged panels or operable louvers which are connected and activated collectively. Unfortunately, the same mechanisms that make shading systems operable also make them susceptible to damage from high winds and freezing. Hinges and operating hardware used in operable shades must resist racking and the loading imposed by heavy snows and ice accumulation. Actuating mechanisms should be rigid or stable enough to prohibit flapping or other undesirable movement.

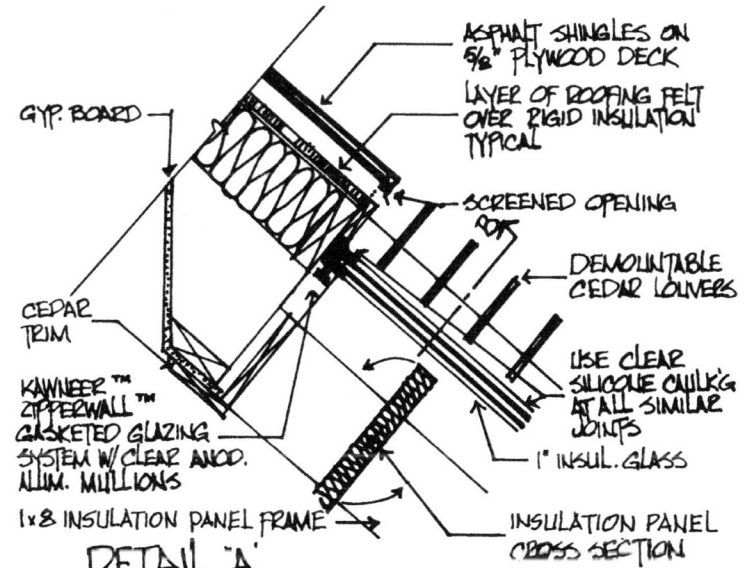

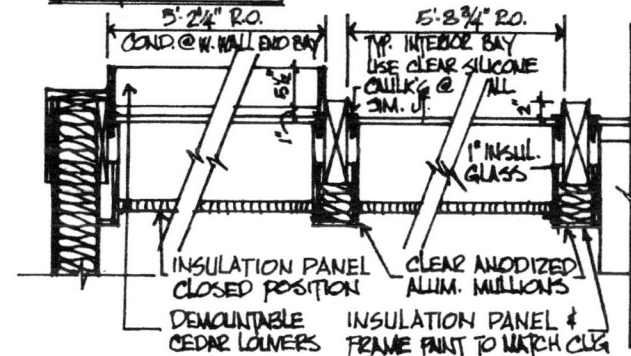

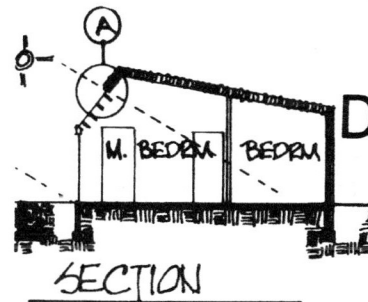

GYP. BOARD

ASPHALT SHINGLES ON 5/8" PLYWOOD DECK

LAYER OF ROOFING FELT OVER RIGID INSULATION TYPICAL

SCREENED OPENING

DEMOUNTABLE CEDAR LOUVERS

CEDAR TRIM

USE CLEAR SILICONE CAULK'G AT ALL SIMILAR JOINTS

KAWNEER ™ ZIPPERWALL ™ GASKETED GLAZING SYSTEM W/ CLEAR ANOD. ALUM. MULLIONS

1" INSUL. GLASS

1x8 INSULATION PANEL FRAME

INSULATION PANEL CROSS SECTION

DETAIL "A"

5'-2¼" R.O. COND. @ W. WALL END BAY

5'-8¾" R.O. TYP. INTERIOR BAY USE CLEAR SILICONE CAULK'G @ ALL SIM. JT.

1" INSUL. GLASS

INSULATION PANEL CLOSED POSITION

CLEAR ANODIZED ALUM. MULLIONS

DEMOUNTABLE CEDAR LOUVERS

INSULATION PANEL & FRAME PAINT TO MATCH CLG

DETAIL "B"

A

M. BEDRM   BEDRM

SECTION

DEMOUNTABLE LOUVER DETAIL

199

gromet & lacing —————————————

canvas painted bright ——————
yellow

1" conduit pipe ——————————

"balloon
curtain"

control-cord for
exterior shutters

mullion

sleeve for
balloon curtain

double-glazing

3½" steel support
frame for shutters

light-control
shutters
(exterior)

friction tube for
shutter control rope

8" concrete block

4" rigid foam insul.

footing for
shutter
frame

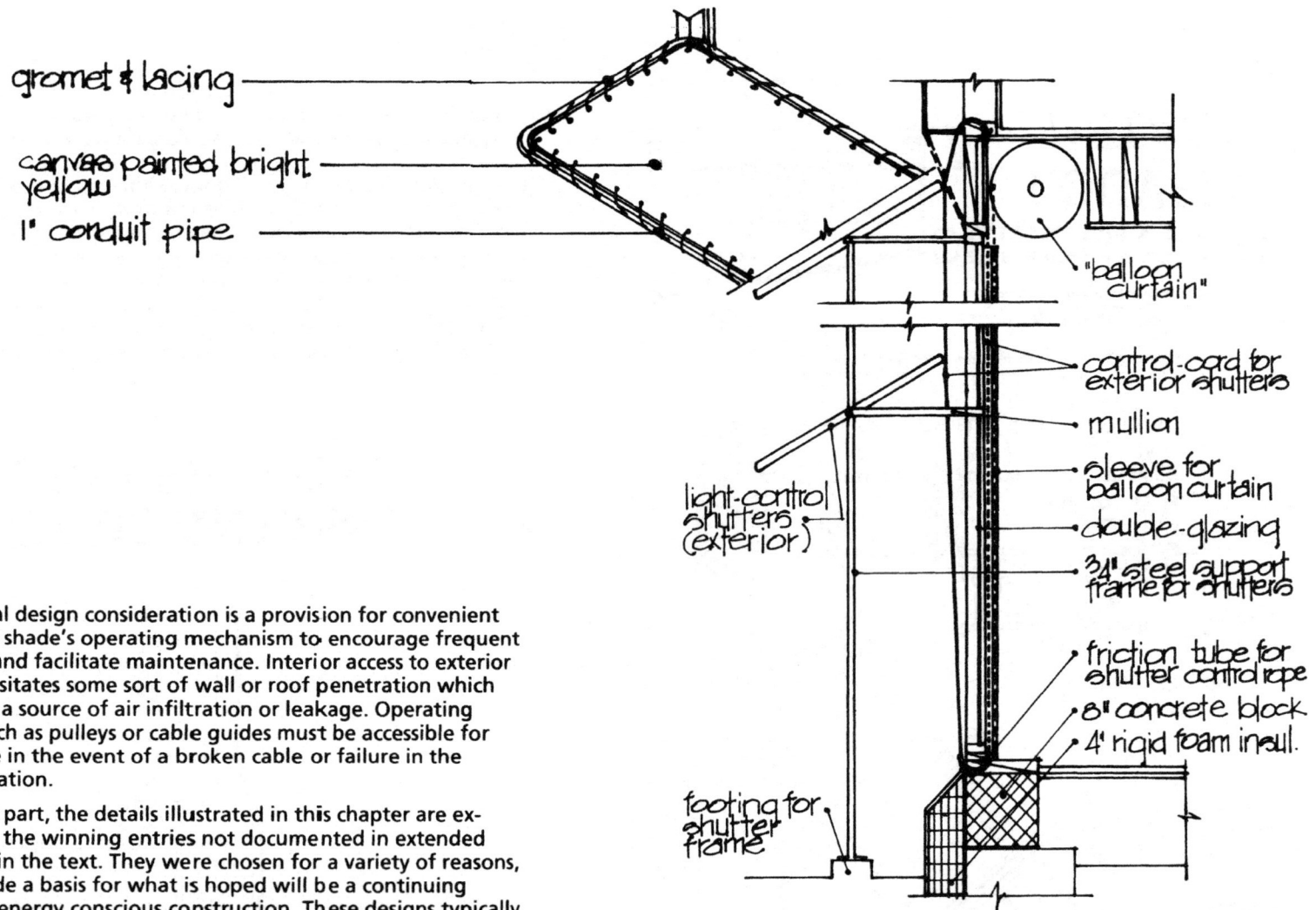

An additional design consideration is a provision for convenient access to the shade's operating mechanism to encourage frequent adjustment and facilitate maintenance. Interior access to exterior shades necessitates some sort of wall or roof penetration which may provide a source of air infiltration or leakage. Operating hardware such as pulleys or cable guides must be accessible for maintenance in the event of a broken cable or failure in the shade's operation.

For the most part, the details illustrated in this chapter are extracted from the winning entries not documented in extended form earlier in the text. They were chosen for a variety of reasons, but all provide a basis for what is hoped will be a continuing dialogue on energy conscious construction. These designs typically demonstrate techniques on detailing with which the home builder should become familiar. In many cases, improved performance might be attained with design modifications. The inclusion of these details in this text in no way implies a warranty of their accuracy, completeness, efficiency, or usefulness, or suggests that their use would not infringe patents.

EXTERIOR ADJUSTABLE
LOUVER DETAIL

FLASHING

INSULATING SHADE

AWNING

DOUBLE PANE GLASS

TILE

6" SLAB

STUCCO

CALK

FINISH GRADE

RIGID INSULATION

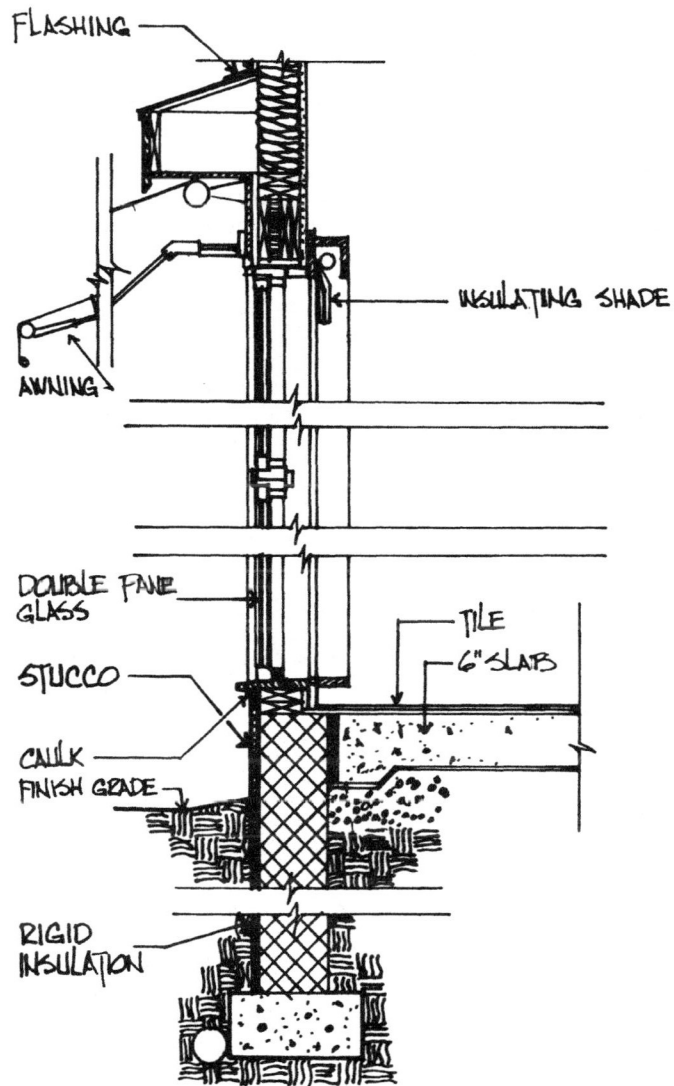

EXTERIOR AWNING & INTERIOR INSULATING SHADE

CHAPTER 9
ENERGY CALCULATIONS

All solar energy buildings, either active or passive, must first be energy conserving to make the contribution of solar energy worthwhile. Many calculation techniques are available to analyze the energy requirements of the buildings. The application form for the Passive Residential Design Competition included a step-by-step procedure for the energy requirements of a building and also estimating the solar energy contribution that a particular system would make. The following example illustrates the calculation technique for a hypothetical building. The design shown is not intended to portray any submission which was actually received at the competition but is included for illustrative purposes only.

The building under consideration is illustrated in Figures A, B, C, and D, and it has the following characteristics.

1. It is a single-family detached dwelling located in Schenectady, New York, at a latitude of 42° 50'.

2. The building is a trapezoidal shape with 1,600 square feet total heated floor area on two floors. It is slab on grade construction with the lower level tucked into a south-facing slope.

3. The passive solar system is a combination of Direct Solar Gain and Mass Trombe Wall.

The procedure, devised at the Los Alamos Scientific Laboratory (LASL), consists of two basic parts: a step-by-step procedure to complete the Building Thermal Load Profile in 10⁶BTU (Table A), and the Auxiliary Load Profile in 10⁶BTU (Table B).

TABLE A provides a Building Thermal Load Profile for each month of the year. This load profile is based on a simple degree-day analysis, assuming a single Building Loss Coefficient in BTU/DD from the building's heated space to the outside, in the absence of solar gains.

TABLE B provides an Auxiliary Load Profile in 10⁶BTU based on the LASL Solar Load Ratio Method. The total monthly solar energy transmitted through the solar collection surface is determined, and this is divided by the total monthly thermal load listed in Table A to determine the monthly Solar Load Ratio (SLR). The monthly Solar Heating Fraction (SHF) is then determined from a selected curve and the monthly auxiliary energy (backup heat required) is calculated.

The key performance estimate obtained is the total annual heat energy (BTU year) required to maintain the home at 65° F.* This number, when divided by the house floor area (in sq. ft.) and divided by the annual heating degree days (DD) specific to that location, provides a common-basis figure of merit for comparing different buildings. Through use of solar gains, this number can usually be reduced to well below 5 BTU/FT²/DD.

To complete Table A, Building Thermal Load Profile, use the following steps:

STEP 1: Determine the Modified Building Loss Coefficient

METHOD 1:
Two alternate methods may be used. If the hourly design load has been determined, based on an assumed design temperature, then the Modified Building Loss Coefficient may be calculated directly as shown below. This short technique is used if heat loss calcula-

*A typical constant design temperature; however, 70°F is used for this example.

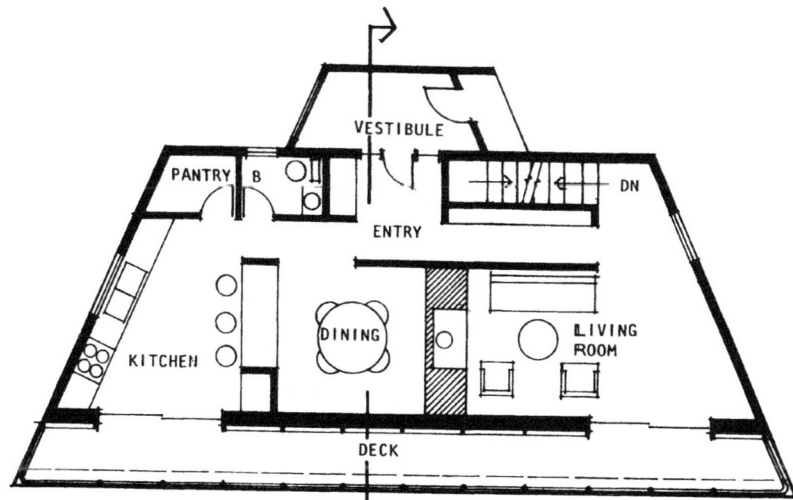

FIGURE A   UPPER LEVEL

VESTIBULE

PANTRY   B

DN

ENTRY

DINING

KITCHEN

LIVING ROOM

DECK

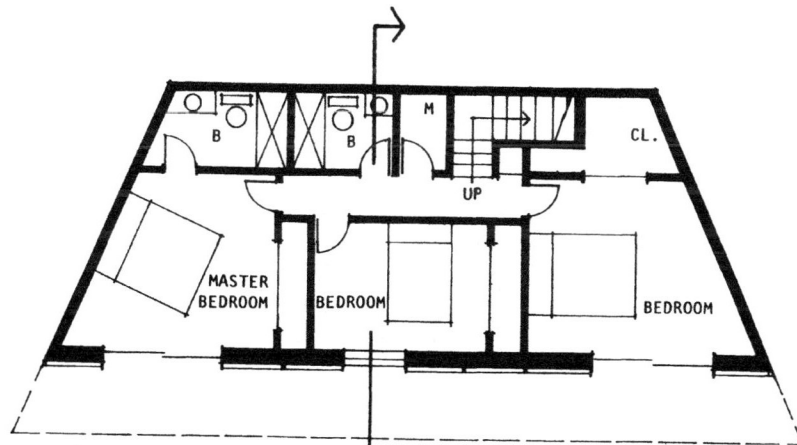

FIGURE C   SOUTH ELEVATION

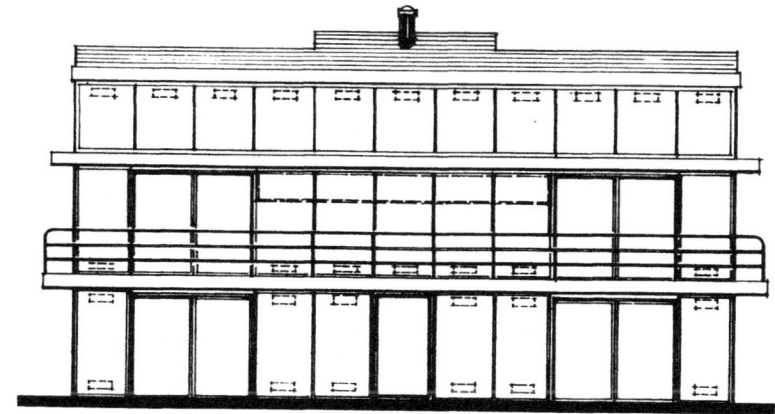

FIGURE D   LOWER LEVEL

B

B   M

CL.

UP

MASTER BEDROOM

BEDROOM

BEDROOM

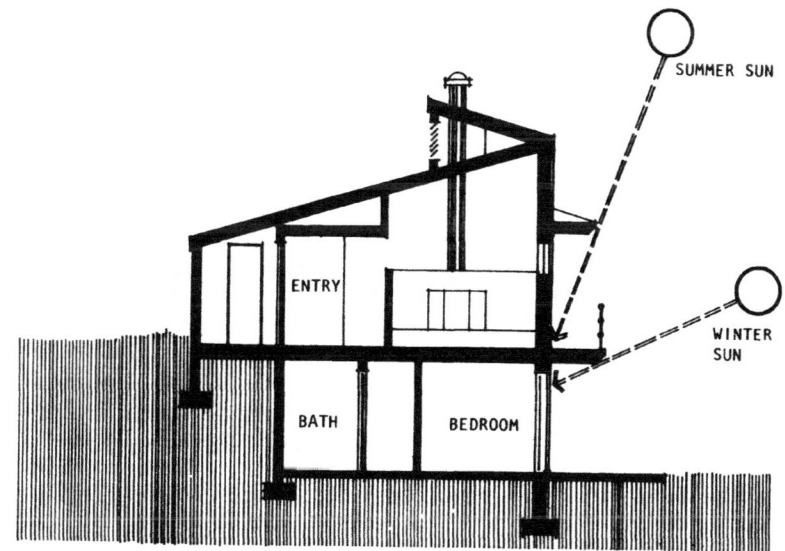

FIGURE D   SECTION

SUMMER SUN

ENTRY

WINTER SUN

BATH

BEDROOM

203

tions have already been done based upon the peak, or design heating load, with both internal load and solar gains being excluded. The Design Heating Load is taken from a <u>heat loss</u> calculation.

Surface Type	Area (ft²)	U-Value (BTU/hr/°F/ft²)	U x A (BTU/hr/°F)
CEILING	840	0.03	25.2
FRAME WALL	640	0.05	32
CONCRETE WALL (BELOW GRADE)	432	0.129	56
MASS WALL	400	0.22	132
GLASS (DOUBLE GLAZED EMITTANCE COATING)	280	0.35	122.5
CONCRETE SLAB	800	0.10	80
DOOR	21	0.10	2.1

Total (Building Skin Conductance): $\boxed{450}$ BTU/hr °F

Modified Building Loss Coefficient BTU/Degree-day

$$= \frac{24 \text{ hr} \times \left(\text{Design Heating Load,}\right) \text{BTU/hr}}{\left(\text{Inside Temperature Assumed,}\right) \text{°F} - \left(\text{Outside Design Temperature,}\right) \text{°F}}$$

$$(15,980) \text{ BTU/DD} = \frac{24 \times (\quad 49,950 \quad) \text{ BTU/hr}}{(\quad 70 \quad) \text{°F} - (\quad -5 \quad) \text{°F}}$$

METHOD 2:

If the design load has not been calculated previously, then a heat loss calculation must be done first to arrive at the Modified Building Loss Coefficient. In this method, the area of each window, wall section, door, roof, etc. is calculated and listed. Then the appropriate U-Value (BTU/hr./°F/FT²) is listed and multiplied by the respective area to tabulate the U x A products (BTU/hr./°F). The sum of the U x A products is the Building Skin Conductance (BTU/hr./°F)

In performing these calculations, it is important that the characteristics of the materials are accurately listed. When such devices as moveable insulation are used, which result in a lower U-value, an estimate must be made of the amount of time during which the insulation will be in place and the amount of time during which the insulation will not have an effect on heat loss. For example, it might be reasonable to assume that moveable insulation panels covering the windows of a home would be open for twelve hours each day and closed for twelve hours each day. In this case, the U-value which should be listed would be the average of the two conditions. A typical value for the total Building Skin Conductance on an average sized house would be 300-400 BTU's/hr./°F. The following table lists the values for the example problem.

One of the most important elements contributing to energy requirements in a home is the infiltration of outside air through cracks which occur around doors and windows and through the joints between different building materials. Even though it seems that a house is tightly constructed, there is still a surprising amount of outside air which will leak in and result in a higher heating requirement. Since all of the air which enters the house from the outside must be heated to the desired room temperature to maintain comfort, it is important to account for this in the heat loss calculation.

The type of construction has a very significant impact on the actual amount of infiltration which occurs. New houses, which are carefully built to minimize infiltration, will frequently experience a quantity of infiltration that is equal to approximately 1/2 to 3/4 air changes (of the internal volume of the house) per hour. Older houses, which are not as tightly constructed, could easily experience infiltration as high as 1-1/2 or even 2 air changes per hour. It can easily be seen what an important impact this has on the total heating requirements for the home. It is unlikely that a rate much lower than 1/2 - 3/4 of an air change will be experienced in any building which is constructed above grade. Buildings which are sheltered by the earth may experience slightly lower values for infiltration.

To calculate the heat required to make up for infiltration, the internal volume of the house is multiplied by the specific heat of air (0.018 BTU/FT³/°F at sea level) and by the number of air changes which occur per hour (ACH). A typical value for this type of calculation in a modestly sized home might be 150 to 250 BTU/hr./°F. The following calculation is again from the example problem.

$$\text{Infiltration Load} = \begin{bmatrix} \text{Volume} \\ 16,000 \end{bmatrix} \times \begin{bmatrix} \text{Specific Heat} \\ 0.018 \end{bmatrix} \times \begin{bmatrix} \text{ACH} \\ 0.75 \end{bmatrix} = \boxed{216} \text{ BTU/°F/hr}$$

Based upon the building skin conductance and the infiltration load, the Modified Building Loss Coefficient from the example can be calculated by multiplying the sum of those two numbers by 24 hours per day.

$$\text{Finally:} \quad 24 \text{ hr} \times \left[\begin{pmatrix} \text{Building Skin} \\ \text{Conductance} \\ \text{in BTU/°F/hr} \end{pmatrix} + \begin{pmatrix} \text{Infiltration} \\ \text{Load} \\ \text{in BTU/°F/hr} \end{pmatrix} \right]$$

$$\begin{matrix} \text{Modified} \\ \text{Building Loss} \\ \text{Coefficient} \\ \text{BTU/DD} \end{matrix} = 24 \text{ hr} \times \left[\begin{pmatrix} 450 \end{pmatrix} + \begin{pmatrix} 216 \end{pmatrix} \right] = \boxed{15,980} \text{ BTU/°F}$$

With an assumed inside temperature 15,980 BTU/°F = 15,980 BTU/DD

In careful building construction, the Modified Building Loss Coefficient will probably be approximately 7-8 BTU/Degree Day for each square foot of building floor area. In this example, 15,980 ÷ 1,600 – 9.9 BTU/DD/FT'.

When Step 1 has been completed, the Modified Building Loss Coefficient obtained can be used to complete Columns (1) and (2) in Table A: Building Thermal Load Profile.

Identify Monthly Heating Degree Days in Column (1).

STEP 2: Multiply Column (1) by Modified Building Loss Coefficient to obtain Gross Monthly Load (10⁶BTU/month). List in Column (2).

TABLE A. BUILDING THERMAL LOAD PROFILE

	(1) Monthly Degree-Days °F	Modified Building Loss Coefficient	(2) Losses Gross Monthly Load 10⁶BTU/mo.	(3) Non-Solar Gains Internal Sources MBTU/mo.	(4) Needs Net Thermal Load MBTU/mo
Aug.					
Sept.	123	15,980	1.97	1.5	.47
Oct.	422	15,980	6.75	1.5	5.25
Nov.	756	15,980	12.0	1.5	10.5
Dec.	1159	15,980	18.5	1.5	17.0
Jan.	1283	15,980	20.5	1.5	19.0
Feb.	1131	15,980	18.0	1.5	16.5
Mar.	970	15,980	15.5	1.5	14.0
Apr.	543	15,980	8.7	1.5	7.2
May	211	15,980	3.4	1.5	1.9
June					
July					
TOTAL	6650		105.3	13.5	91.8

*List data only for months over 100 degree days.

205

STEP 3: List monthly heating from internal sources other than auxiliary heating systems such as lights, stove, water heater energy retained in the house, dryers, people, etc. (10⁶BTU/month) in Column (3).

In order to complete Column (3), a bit of background information on internal loads is required. The amount of internal heat which is generated by people and appliances varies significantly depending upon the size of the family and their living habits. A recent investigation of the significance of internal loads was conducted and published in the Minimum Energy Dwelling Workbook, sponsored by the Department of Energy and the Southern California Gas Company.[1]

This investigation indicated that a family of four has a typical hourly internal heat production which averages approximately 2,000-2,250 BTU's/hour, as shown in Figure E.

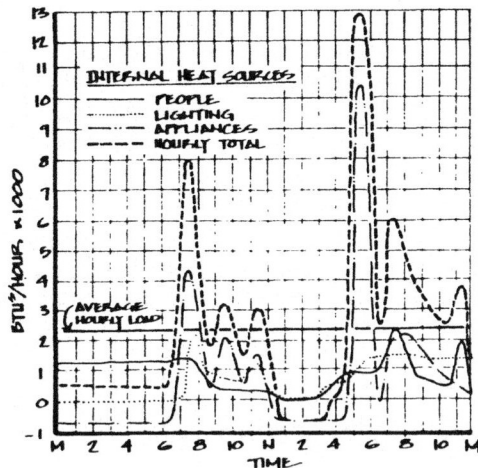

FIGURE E

Internal Heat Production For A Typical Weekday

1. Prepared by Burt Hill Kosar Rittelmann Associates, available from National Technical Information Service, U.S. Department of Commerce, Springfield, VA 22161 (No: SAN/1198-1)

Based upon this estimate, a reasonable value for the figures in column (3) is approximately 1.5 10⁶BTU/month (2,000 BTU/hour x 24 hours/day x number of days in the month). It should be noted that a smaller family would have a lower internal load, as would a family which spends less time in the home. It can be seen from Figure E that there are distinct peaks which occur in the internal generation of heat, particularly around the dinner hour. At times, this may not be useful energy, since it may tend to overheat the house. Although the internal generation of heat can contribute a significant amount to the heating requirements, it should not be assumed that all of the internal heat that is generated is a useful contribution to the heating load, and it is recommended that a conservative approach be used in generating this number in order to achieve realistic results.

Using the information tabulated thus far, and listed in the appropriate columns in Table A, the Building Thermal Load Profile can be completed.

STEP 4: Subtract Column (3) from Column (2) and list in Column (4). This is the Net Thermal Load. Add all columns to show yearly totals.

Once Table A has been completed, the final values are used in Table B: Auxiliary Load Profile. If you remember from the introduction, Table B, when completed, will give you the auxiliary energy needed to maintain the building's established interior temperature of 65°F.

To complete Table B, follow Steps 1 through 8:

STEP 1: List in Column (1) the Net Thermal Load (10⁶BTU/month) from Table A, Column (4) or Column (5).

STEP 2: List in Column (2) the monthly solar radiation incident on a horizontal surface for the proposed site (BTU/month-ft.²).

NOTE: Do not use ASHRAE clear-day tables or cloud cover factors. Values listed must be based on National Weather Service measured data or other reliable measured data.

STEP 3: Enter Latitude of Site (L): _42° 50'_

STEP 4: Subtract Solar Declination at Mid-Month (given in Column (3) from Latitude and list in Column (4). This is L-D.

TABLE B. AUXILIARY LOAD PROFILE

Month*	(1) Net Thermal Heating Load, (from Table A)	(2) Solar Radiation on a Horizontal Surface	(3) Solar Declination at Mid-Month (D)	(4) L-D	(5) Solar Radiation Absorbed	(6) Solar Load Ratio, (SLR)	(7) Solar Heating Fraction (SHF)	(8) Auxiliary Energy
	MBTU/mo.	BTU/mo. ft²	degrees	degrees	MBTU/mo.			MBTU/mo.
Aug.			14.0					
Sept.	0.47	33,120	2.8	40.03	12.9	27.45	1.0	
Oct.	5.25	24,986	−9.1	51.93	14.5	2.76	0.865	0.71
Nov.	10.5	14,250	−18.6	61.43	11.1	1.06	0.52	5.04
Dec.	17.0	11,842	−23.1	65.93	10.3	0.61	0.34	11.22
Jan.	19.0	14,818	−21.4	64.23	12.5	0.66	0.36	12.16
Feb.	16.5	20,720	−14.0	56.83	14.0	0.85	0.44	9.24
Mar.	14.0	31,248	−2.8	45.63	14.9	1.07	0.52	6.72
Apr.	7.2	37,530	9.1	33.73	11.6	1.61	0.68	2.3
May	1.9	47,244	18.6	24.23	10.0	5.26	0.99	0.02
Jun.			23.1					
Jul.			21.4					
Totals	91.8							47.4

*List data only for months over 100 heating degree-days.

STEP 5: List in Column (5) the Solar Radiation Absorbed. For vertical, south-facing double glazing (normal transmittance 0.74), this can be determined directly from the following equation:

$$\begin{bmatrix} \text{Solar Radiation} \\ \text{Absorbed,} \\ \text{BTU/month} \end{bmatrix} = \begin{bmatrix} \text{Horizontal} \\ \text{Solar Radiation} \\ \text{(Column (2))} \\ \text{BTU/month-ft}^2 \end{bmatrix} \times \begin{bmatrix} \text{Net Solar} \\ \text{Collection} \\ \text{Area, ft}^2 \end{bmatrix} \times \begin{bmatrix} \text{Solar} \\ \text{Absorptance} \end{bmatrix} \times \begin{bmatrix} (0.226) - (.002512)(L\text{-}D) + (.0003075)(L\text{-}D)^2 \end{bmatrix}^*$$

*Note: Instead of solving this equation for latitude correction. the value may be obtained from the attached plots for each month.

Procedures for determining the Solar Radiation Absorbed for non-south facing or non-vertical surfaces can be found in Chapter 26, 1977 ASHRAE Handbook of Fundamentals.

The Solar Absorptance factor is the absorptance of the surface inside the glazing. For a direct gain building, a value of .90 (light interior) to .95 (dark interior) is recommended.

The last factor in the above equation can be calculated or taken directly from the Latitude Correction Table shown in Figure F.

STEP 6: To obtain the Solar/Load Ratio (SLR), divide column (5) by Column (1) and list this as Column (6).

STEP 7: Look to Figure G for SHF. Using the Solar Load Ratio from Step 5, obtain the value of monthly Solar Heating Fraction (SHF) from Figure G. Enter these values in Column (7). Also plot these points on Figure G and label each point with the month.

For Direct Gain and Roof Pond buildings, it is recommended that the curves labeled "Water Trombe" on Figure G be used.

For Solarium buildings, it is recommended that the curves labeled "Mass Trombe" on Figure G be used.

For Thermosiphon buildings, determine the Solar Load Ratio based on the type and location of the collector(s).

STEP 8: The Monthly Auxiliary Energy (backup heating required), calculated from the following formula, is listed in Column (8):

$$\begin{bmatrix} \text{Monthly} \\ \text{Auxiliary} \\ \text{Energy} \\ \text{BTU/Mo.} \end{bmatrix} = (1\text{---}SHF)\begin{bmatrix} \text{Net Monthly} \\ \text{Load (Column 1),} \\ \text{BTU/Mo.} \end{bmatrix}$$

We could summarize these charts in a final comparison number; Heating Energy number: Add the total of Column (3) of Table A (internal sources) to the total of Column (8) of Table B (auxiliary heating) and divide this sum by the product of the heating degree days (total Column (1), Table A) and building floor area:

$$\begin{bmatrix} \text{Heating Energy} \\ \text{BTU/DD-ft}^2 \end{bmatrix} = \frac{\text{(Internal sources)} + \text{(Auxiliary heating)}}{\text{(Degree-days)(Floor area)}}$$

$$\text{Heating Energy} = \frac{\begin{bmatrix} 13.5 \times 10^6 \end{bmatrix} + \begin{bmatrix} 47.4 \times 10^6 \end{bmatrix}}{\begin{bmatrix} 6650 \text{ DD} \end{bmatrix} \times \begin{bmatrix} 1600 \text{ FT}^2 \end{bmatrix}} = \boxed{5.72}$$

In homes which have a significant percentage of their total heating requirements met by solar energy, this final number should probably be less than 5 BTU/DD/FT², and may be as low as approximately 3 BTU/DD/FT².
This table can also give you solar percent contribution as:

$$1 - \frac{\text{Auxiliary Requirements}}{\text{Net Thermal Load}} = \%$$

It should be noted here that the discrepancy between the expected values and what was actually computed is due to a slightly undersized Trombe wall in the sample problem.

LATITUDE CORRECTION
VALUE OF $0.226 - (0.002512)(L - D) + (0.0003075)(L - D)^2$

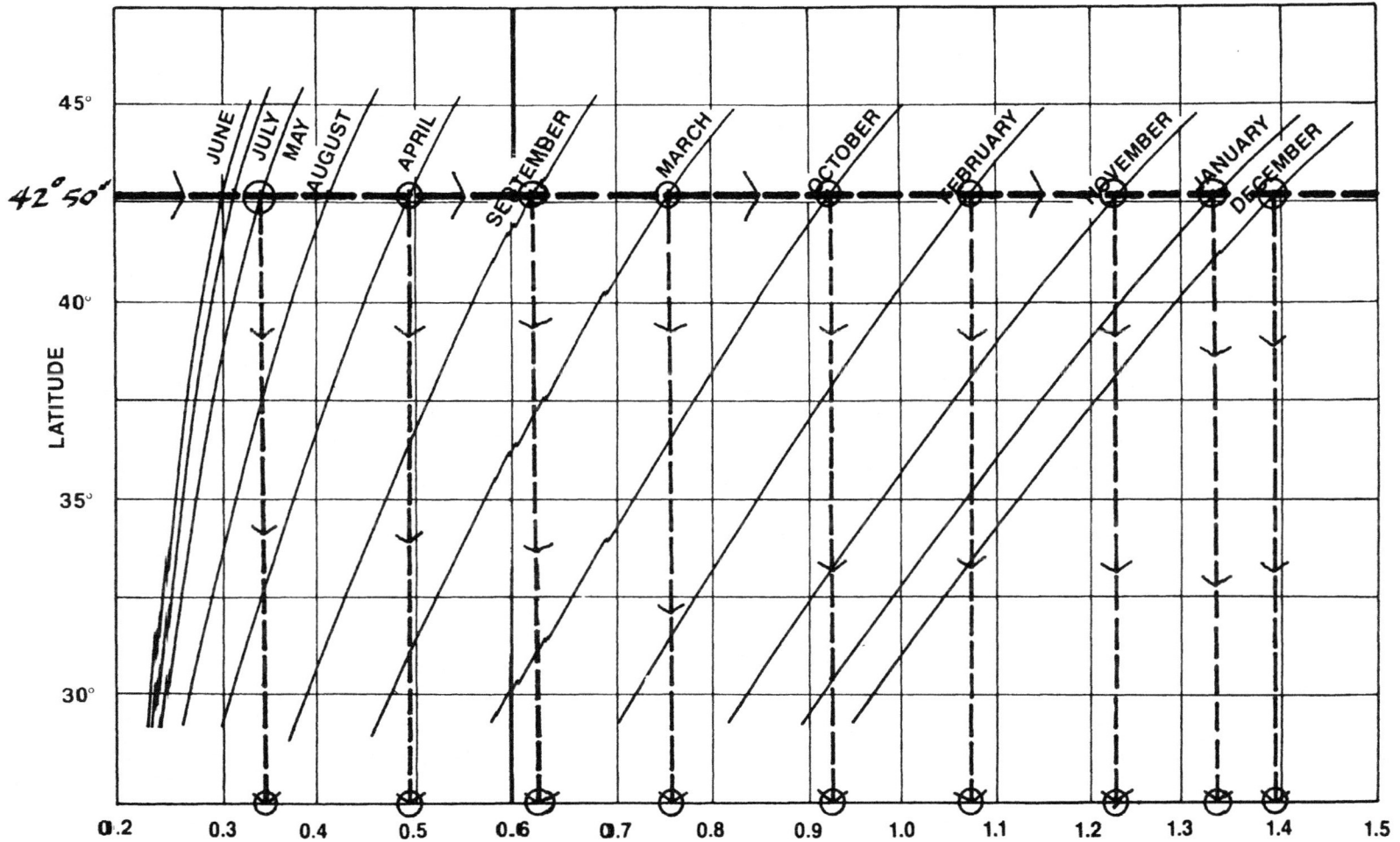

FIGURE F

209

ASSUMPTION

This shorthand calculation technique, which was used in the application form for the Passive Residential Design Competition and Demonstration, is a simple technique, the accuracy of which is dependent upon several assumptions. One of the most important assumptions is that adequate storage mass has been provided to store the solar energy for use in the building. Almost all solar energy systems, either active or passive, make use of a storage element, since the amount of solar energy which can be collected during the day usually exceeds the immediate heating requirements. Therefore, it is important to provide storage, so that heating requirements can be met when solar energy is not available.

In passive solar systems, it is important (1) to provide sufficient thermal storage mass to prevent large changes in internal air temperature during the day, and (2) that the storage mass be placed so that it may be correctly struck by the incoming solar energy; then the storage capacity should be approximately 30-45 BTU's/°F per square foot of solar aperture. This range in storage capacity was assumed for the shorthand calculation procedure given above To check whether the actual storage capacity falls within this range, complete the following table and divide by the collector area.

c. Thermal Storage

(1) Storage Element	(2) Material	(3) Density #/ft³	(4) Thickness ft.	(5) Area ft²	(6) % of Surface Exposed	(7) Exposed Mass (M) #	(8) Specific Heat (CP) BTU/#/°F	(9) Storage Capacity MCp BTU/°F
WALL	CONCRETE	144	× 12"	× 400	× 100	= 57,600	× 0.2	= 11,520
FLOOR	CONCRETE	144	× 6"	× 400	× 0.75	= 21,600	× 0.2	= 4,320
BED	WATER	62.5	× 12"	× 100	× 100	= 6,250	× 1.0	= 6,250

TOTAL STORAGE CAPACITY = 22,090

(1) Floor, wall, ceiling, rockbed, etc.
(6) In direct sun, December 21 at solar noon, if applicable
(7) product of columns (3), (4), (5), and (6).
(9) Product of Columns (7) and (8). The total storage capacity should be approximately 30 BTU/°F per square foot of aperture.

$$\frac{11,520 + 4,320 + 6,250 \; (BTU/°F)}{680 \; (FT^2 \; of \; aperture)} = 32.49 \; BTU/°F/FT^2$$

MONTHLY SOLAR HEATING ESTIMATOR

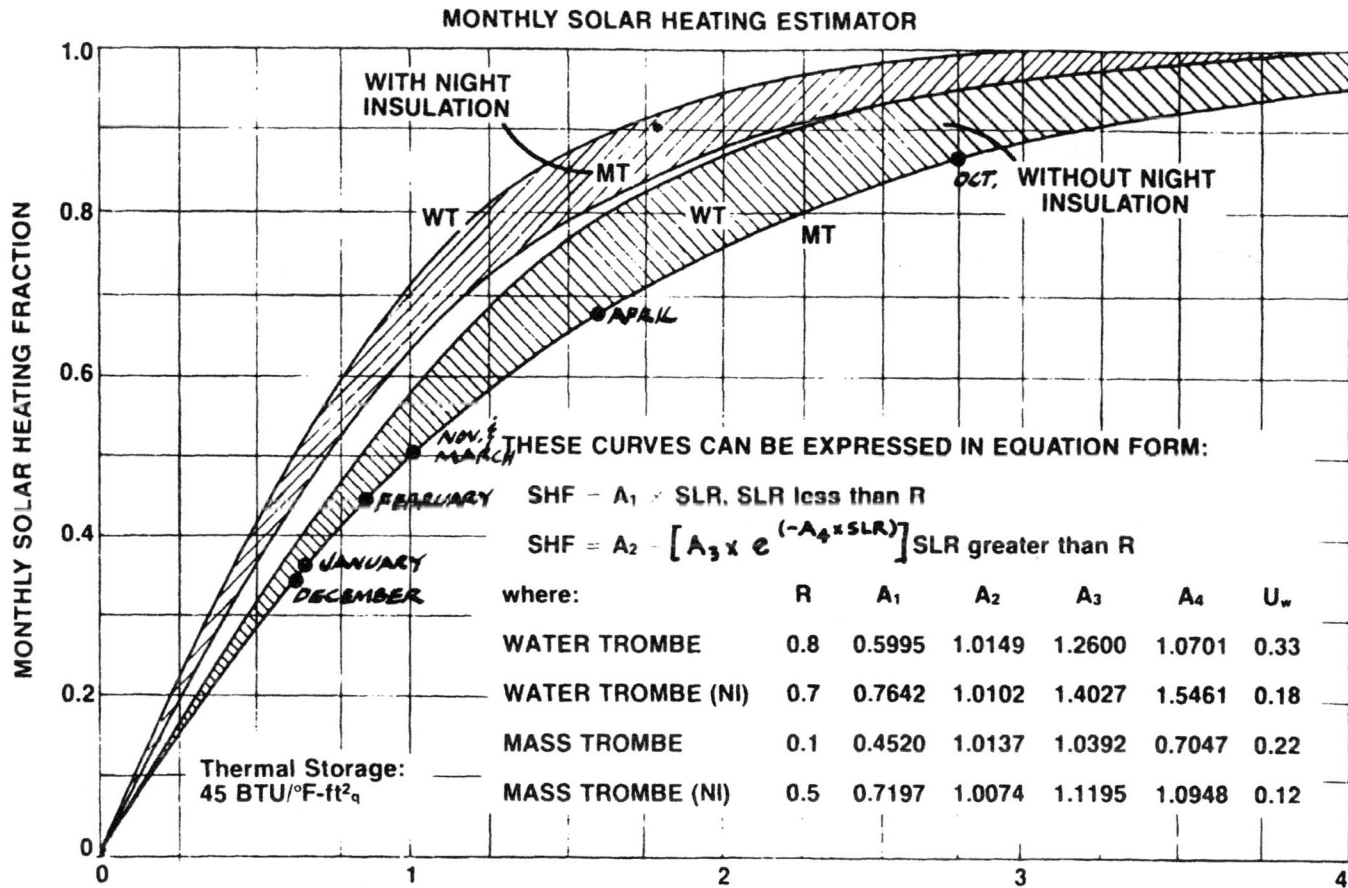

WITH NIGHT INSULATION

MT

WT

WT

MT

OCT. WITHOUT NIGHT INSULATION

APRIL

NOV. & MARCH

FEBRUARY

JANUARY

DECEMBER

Thermal Storage: 45 BTU/°F-ft²q

THESE CURVES CAN BE EXPRESSED IN EQUATION FORM:

$$SHF = A_1 \times SLR, \quad SLR \text{ less than } R$$

$$SHF = A_2 - \left[A_3 \times e^{(-A_4 \times SLR)} \right] \quad SLR \text{ greater than } R$$

where:

	R	A_1	A_2	A_3	A_4	U_w
WATER TROMBE	0.8	0.5995	1.0149	1.2600	1.0701	0.33
WATER TROMBE (NI)	0.7	0.7642	1.0102	1.4027	1.5461	0.18
MASS TROMBE	0.1	0.4520	1.0137	1.0392	0.7047	0.22
MASS TROMBE (NI)	0.5	0.7197	1.0074	1.1195	1.0948	0.12

FIG. 1 MONTHLY SOLAR LOAD RATIO (SLR) = $\dfrac{\text{MONTHLY SOLAR ENERGY ABSORBED}}{\text{NET MONTHLY THERMAL LOAD}}$
(including the static conduction through the solar wall, $A_w \cdot U_w \cdot DD$)

NOTE: Use Water Trombe (WT) Curves For Direct Gain & Roof Pond
Use Mass Trombe (MT) Curves For Sunspace

FIGURE G

Since concrete has a density of approximately 150 pounds per cubic foot, sufficient storage mass will be provided by a concrete wall approximately 1 to 1-1/2 feet in thickness. Other materials can be used to provide an equivalent amount of storage.

When the thermal storage materials are placed in a different configuration, such as in a concrete floor slab, only a limited thickness of material will actually provide useful heat storage. In floor slabs, for example, it would be reasonable to assume that the top 6" or so of concrete provides useful storage. Additional material placed below this level will have some benefit, but it will be quite difficult to extract the heat from this part of the storage mass to provide useful energy to the home. Therefore, in the calculations, the designer should be aware of the relationship between the storage mass and the occupied living spaces.

Since solid materials have a lower specific heat than water (which has a specific heat of 1.0), this must be accounted for in the calculations. For example, concrete has a specific heat of approximately 0.2. This is the amount of heat in BTU's which is required to raise one pound of material one degree F.

If water is used as the storage element, then capacity can be obtained by providing approximately four to six gallons of water per square foot of solar aperture. If a solid material, such as concrete, is used for heat storage, then an equivalent amount of mass must be provided.

It is most important to place the primary living spaces close to the major solar collection and storage elements so that they may obtain the most direct benefit from the solar energy which is collected. Since passive systems rely upon a non-mechanical circulation of heat, less collected solar energy will reach portions of the building which are remote from the solar collection element. Sometimes, it is not possible to place sufficient storage mass in the immediate area of the solar aperture. In such cases, massive elements of the building which are not close to the solar aperture will provide some storage benefit. However, if the storage mass is disconnected from the solar aperture, considerably more quantity of mass will be required to maintain a comfortable swing of internal air temperature. As a general rule of thumb, storage mass which is placed in a remote portion of the building must be approximately four times as large as storage mass which is placed in the vicinity of the solar aperture.

To summarize the analysis of a building's energy requirements, it is useful to graph the seasonal Heating Load Profile. This graph should show the total heating load together with the contribution from Internal Sources, Auxiliary Energy and Solar

Energy. The load profile for the hypothetical building used to illustrate the calculation technique is shown in FIGURE H. On the following pages are load profiles for each of the projects selected for in-depth discussion and analysis.

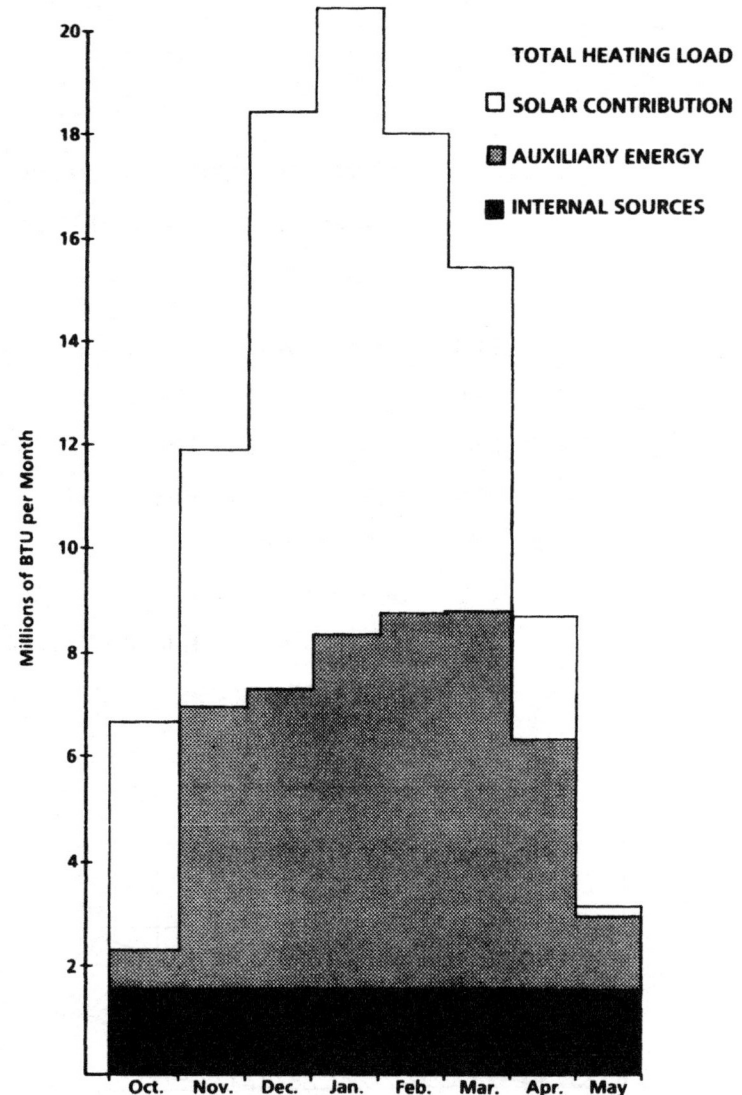

Figure H

212

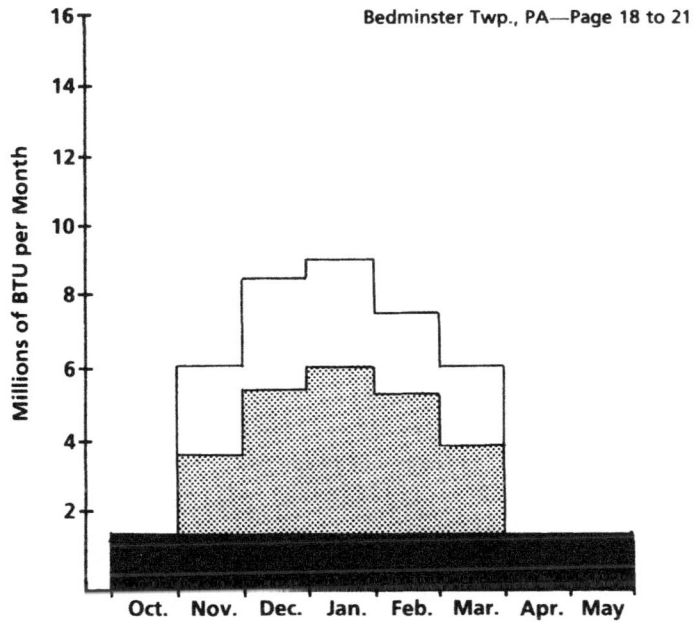

Bedminster Twp., PA—Page 18 to 21

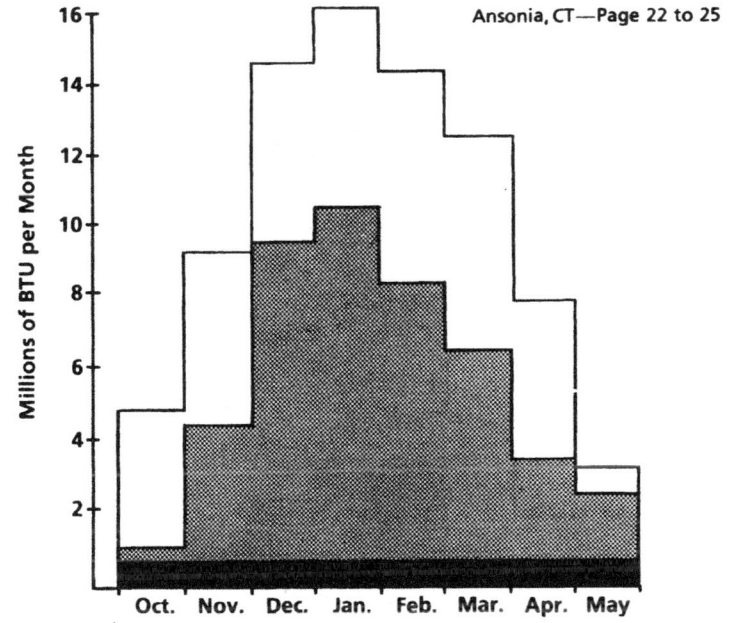

Ansonia, CT—Page 22 to 25

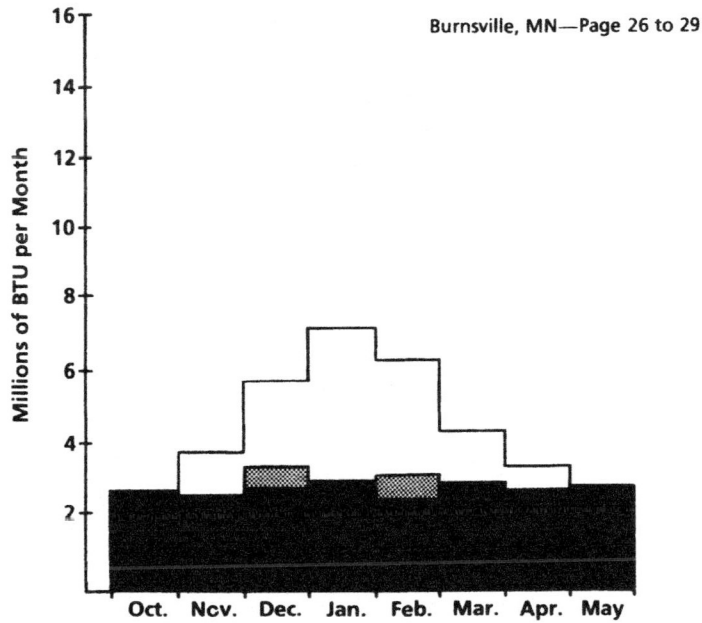

Burnsville, MN—Page 26 to 29

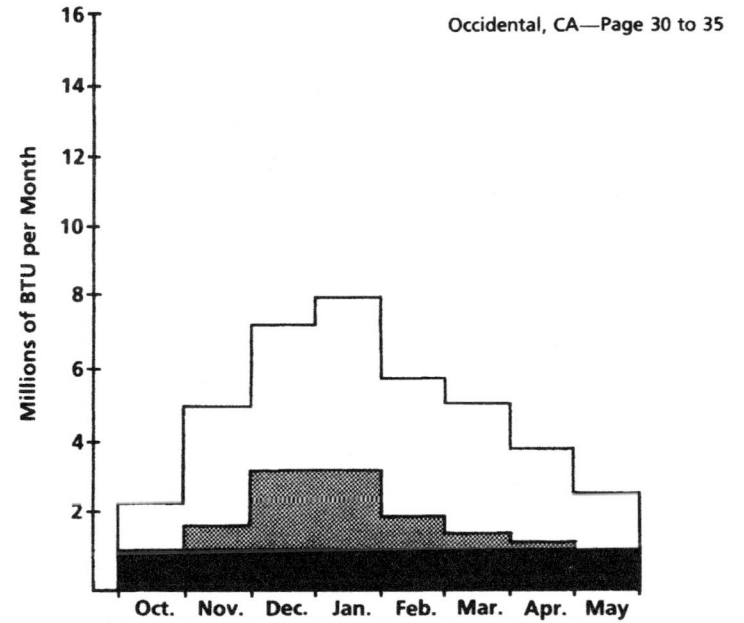

Occidental, CA—Page 30 to 35

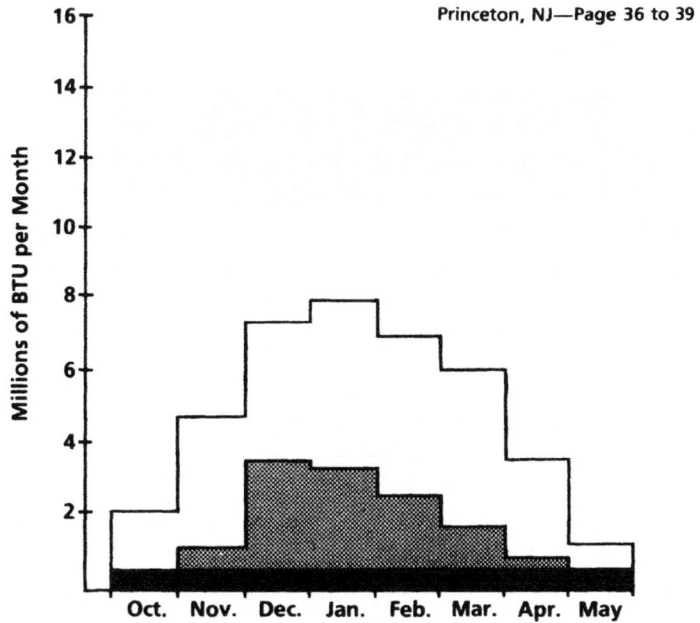

Princeton, NJ—Page 36 to 39

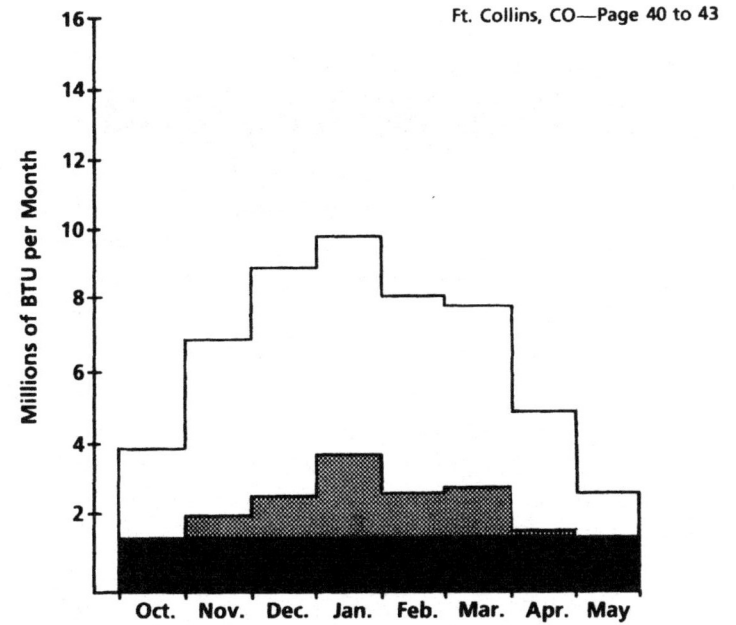

Ft. Collins, CO—Page 40 to 43

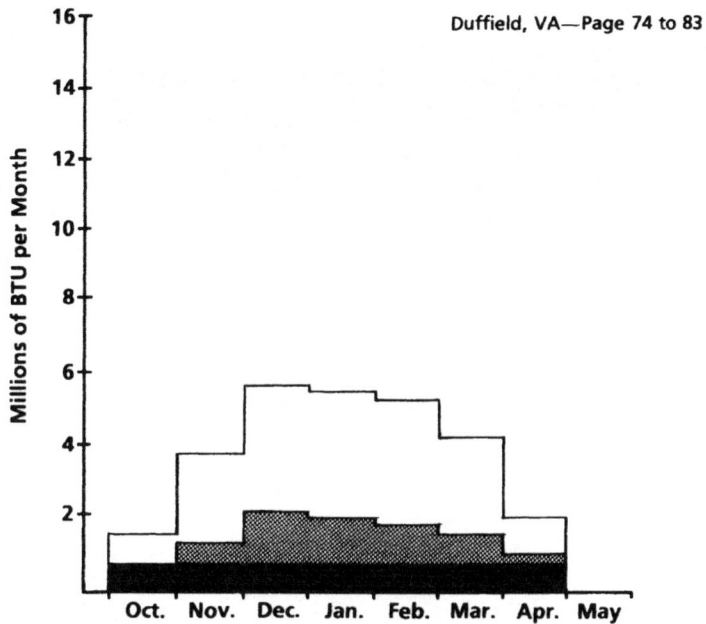

Duffield, VA—Page 74 to 83

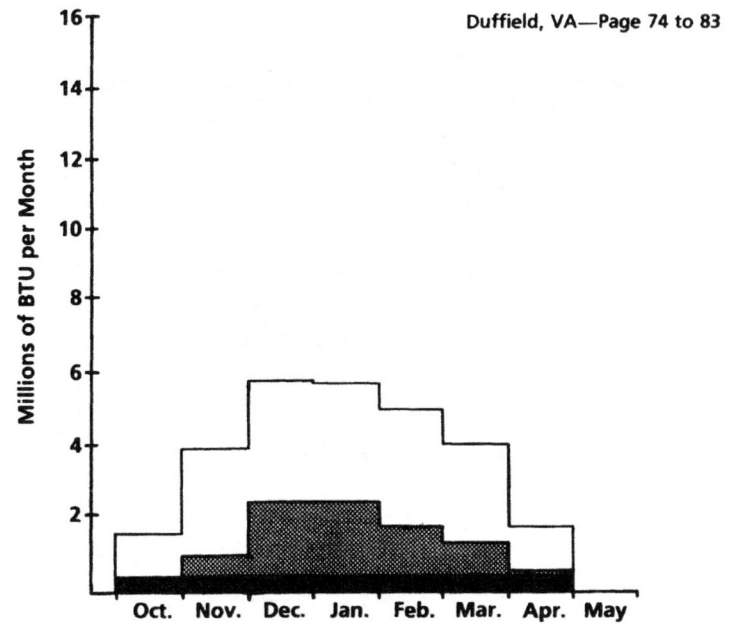

Duffield, VA—Page 74 to 83

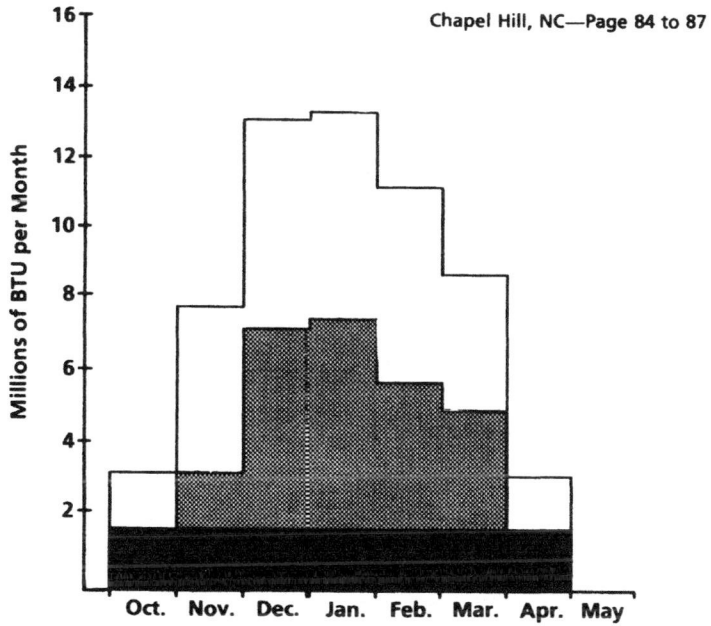

Chapel Hill, NC—Page 84 to 87

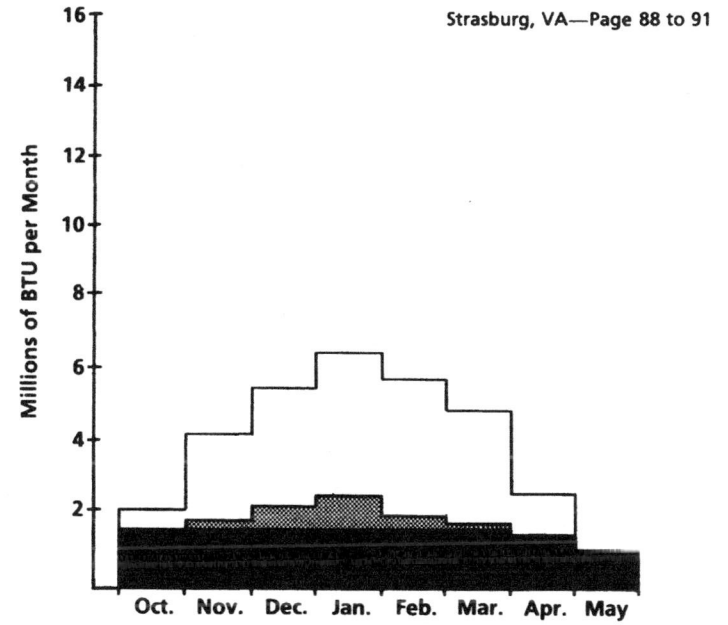

Strasburg, VA—Page 88 to 91

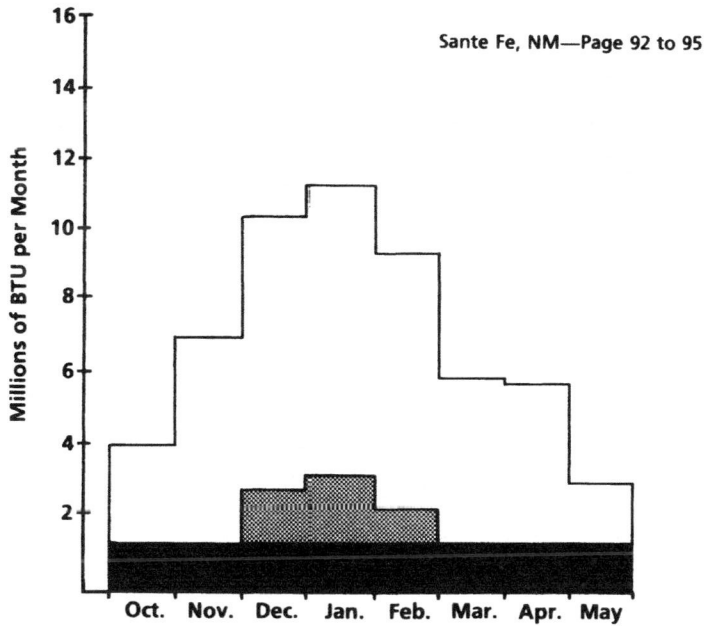

Sante Fe, NM—Page 92 to 95

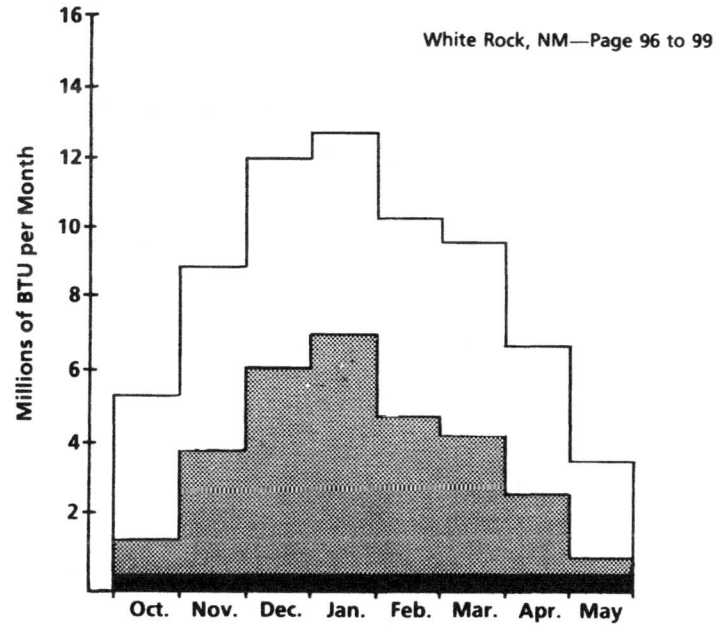

White Rock, NM—Page 96 to 99

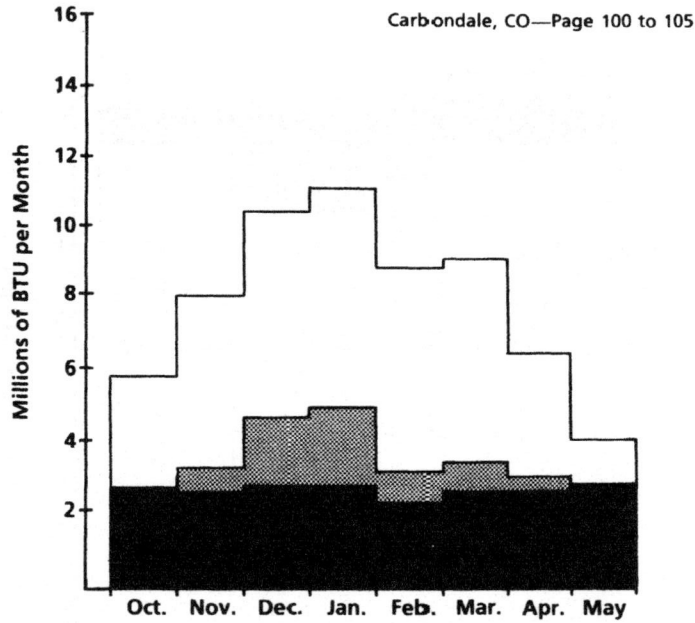

Carbondale, CO—Page 100 to 105

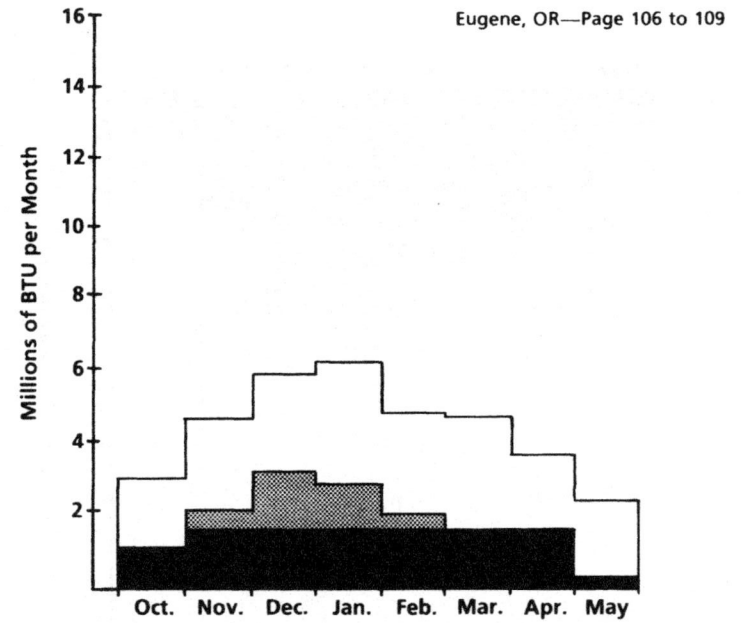

Eugene, OR—Page 106 to 109

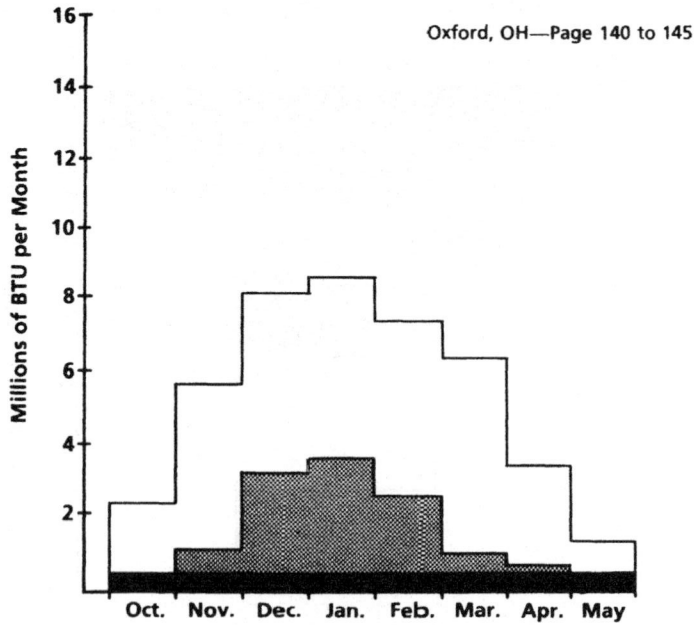

Oxford, OH—Page 140 to 145

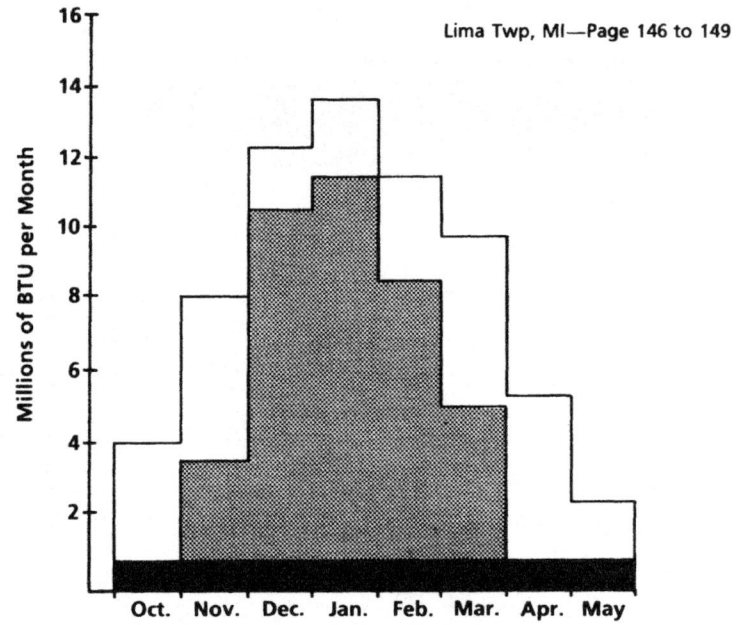

Lima Twp, MI—Page 146 to 149

216

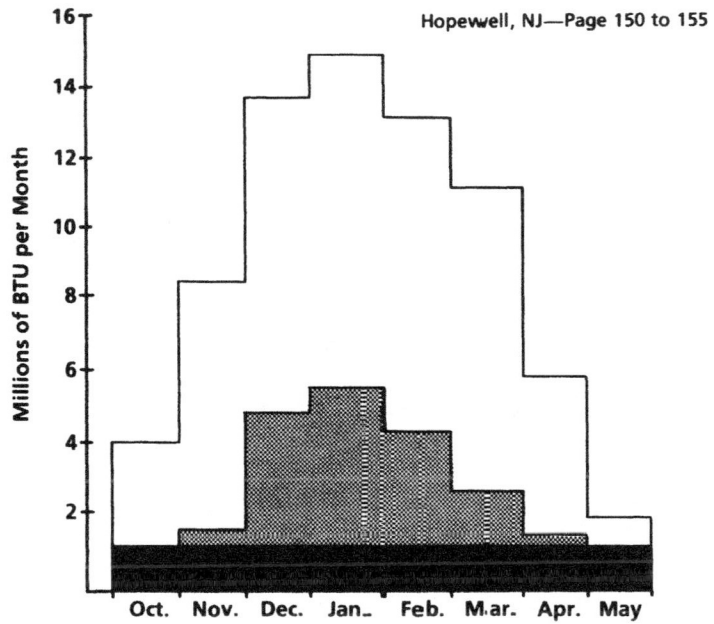

Hopewell, NJ—Page 150 to 155

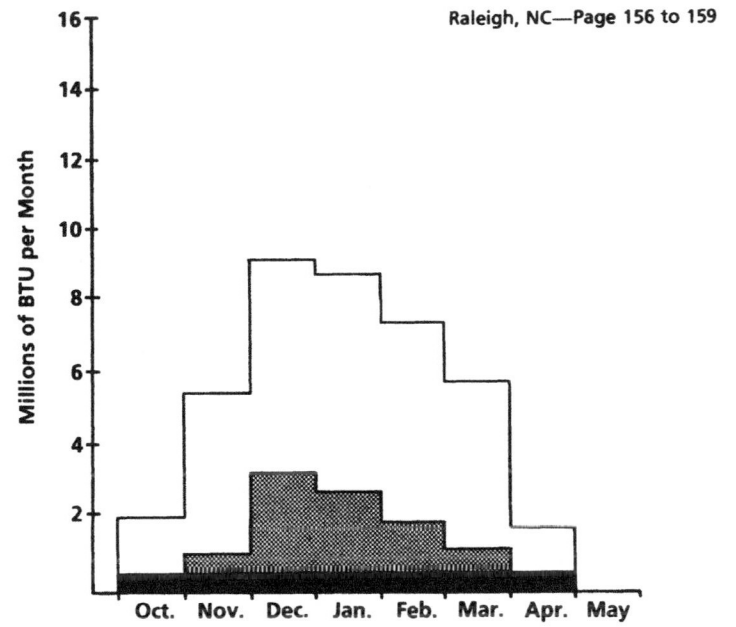

Raleigh, NC—Page 156 to 159

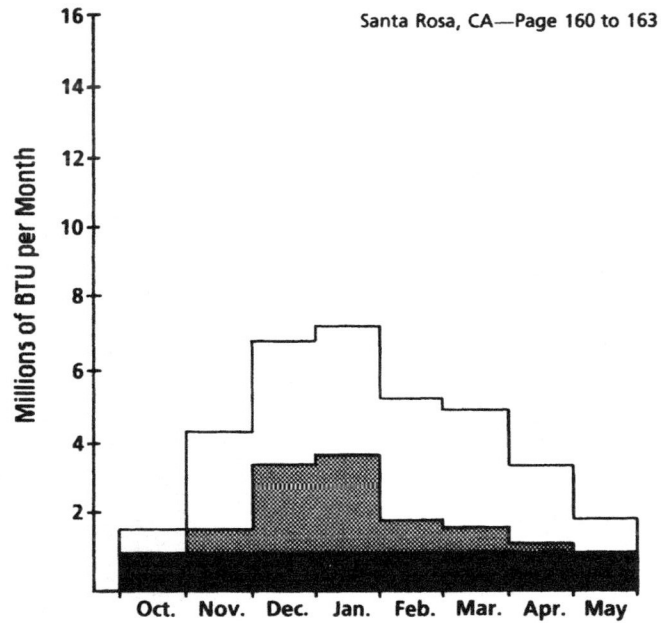

Santa Rosa, CA—Page 160 to 163

EPILOGUE

SOLAR DOGHOUSE

NOTE

The passive solar residential competition implied, but did not specify, that these residences were to be for humans. Don Finkell of Knoxville, Tennessee, has a sense of humor and had some time on his hands, and put together a special application, in consultation with his dog/friend/client, "Sampson". While we cannot make a formal design award for this design, it was so well done and so imaginative that a special Honorable Mention category was created for **SAMPSON'S SOLAR DOGHOUSE.**

Knoxville, TN

Don Finkell
Knoxville, TN

HEATED AREA: 5 FT²

NUMBER OF DEGREE DAYS: 3,478

NET THERMAL LOAD: .150 10^6BTU/YR

AUXILIARY ENERGY: .000 BTU/DD/FT²

YEARLY SOLAR FRACTION: 77%

SAMPSON'S SOLAR DOGHOUSE

CONTEXT

A new passive solar design is soon to be unleashed on the American consumer. This scheme for Sam's doghouse is a mass-producible home for man's best friend, complete with thermal mass, solar reflector and summer shading. The designer of this house can certainly be said to have collared this market.

CONSERVATION

This modest shed for shedding is a beautiful expression of energy-conscious design. With modestly sized openings, compact floor plan and reduced perimeter area, Sam's house should be the cat's meow. As Sam put it himself, "My owner put this thing together in a couple of weekends with 2 sheets of 1/2" marine grade plywood, one sheet of 1" rigid insulation board with reflective surface, some 2 x 4's, a quart of white paint and primer, 2"-wide duct tape, some caulk, nails, old bricks and glass. He scrounged up most of it around the house and got the rest at a builder's supply store. They thought he was a nut!"

HEATING

The low winter sun penetrates the south-facing glass and heats the bricks in the floor. A possible improvement to the design would be a sleeping loft which would permit Sam to move to the northern part of the home during the daytime so that the light can reach the thermal mass in the floor. Pavlov has done some excellent work on this variable. Since there is essentially no heating load and Sam's own metabolism would sufficiently heat this well-insulated home, the solar contribution of this scheme has taken a back seat to the warmth the owner will feel, knowing that Sam is comfortable.

COOLING

Both the dog and the direct gain system are unregistered, which could cause some overheating and control problems. Dogs in heat, as we all know, are a control problem.

CONCLUSION

This is a hard-nosed solution to a cold-nosed problem. An entirely new meaning has been developed for the phrase "pet rock". We are proud to be able to bestow on this design a special "Best of the Breed Award".

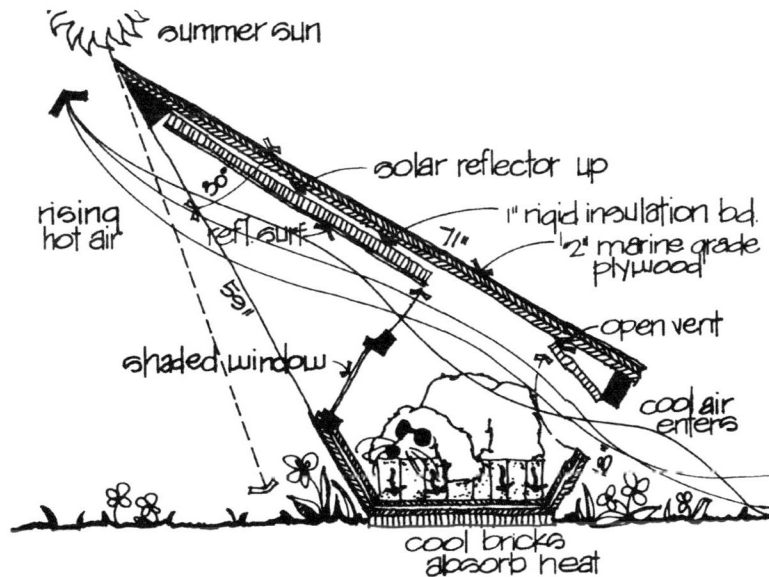

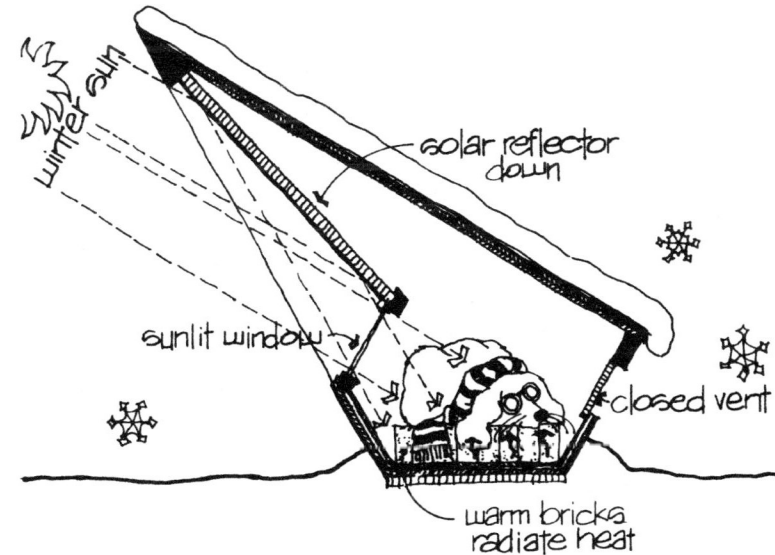

GLOSSARY

Absorption—ratio of solar radiation absorbed by a surface to the amount that strikes it (an important aspect of collector efficiency).

Active Solar Energy System—a solar heating and/or cooling system using mechanical methods of heat distribution.

Airlock Entry—a vestibule enclosed with two air-tight doors for permitting entrance without tremendous air or heat exchange.

Atrium—a closed interior court to which other rooms open, often used for sitting and plants.

Auxiliary Heating (see Backup Heating System)

Backup Heating System—A constantly available source of heat energy which is brought into operation when the solar system storage has been exhausted and the need for heat exists.

Building Skin (see Surface-to-volume ratio)

BTU—British Thermal Unit—basic heat measurement, equivalent to amount of heat needed to raise 1 pound of water 1° Fahrenheit.

Charge—putting heat into storage through radiant absorption or convective heat transfer (blown in). (Also see Discharge and Supercharge.)

Clerestory—vertical window placed high in wall near eaves, used for light, heat-gain, and ventilation.

Collection—the act of trapping solar radiation and converting it to heat (also see Distribution and Storage).

Collector Aperture—the glazed opening being used for admitting solar radiation.

Conduction (or Conductivity)—the transmission of heat from molecule to molecule.

Convection—heat transfer through a fluid (such as air or liquid) by currents resulting from the natural fall of heavier, cool fluid and rise of lighter, warm fluid.

Deciduous Trees—trees which shed their leaves each winter at the end of the growing season.

Degree Day (DD)—the degree day is a unit of heat measurement equal to one degree variation from a standard temperature in the average temperature of one day. If the standard is 65°F and the average outside temperature is 50°F for two days, then the number of degree days is 30.

Direct Solar Gain (see Chapter 3).

Discharge—removing heat from storage by radiation or convective heat transfer (blown out). (Also see Emission, Charge, and Supercharge.)

Distribution—the act of moving collected heat to needed areas. (Also see Collection and Storage).

Diurnal (or Day-Night)—Diurnal means a daily event as opposed to nocturnal happenings. When used in solar energy work, it refers to the effect of moisture in the atmosphere upon the collection of incident radiation by the collector. Because of diurnal cycling, periods of high humidity generally occur in the early morning hours near sunrise. The diffuse component of solar energy is therefore a reduced factor in the early morning hours of operation. In the late afternoon the humidity is usually lower on the diurnal cycle and collector efficiency stays higher than it does in early morning.

Earth Berms (or Berming)—a mound of earth either abutting a house wall to help stabilize temperature inside house, or positioned to deflect wind from house.

Emission (or Emissivity)—the ability to radiate heat in the form of long-wave radiation.

Evergreen/Coniferous Trees—trees which do not shed their leaves at the end of the growing season.

Glazed Area (or Glazing)—for solar collection, glazing refers to all materials which are translucent or transparent to short-wave solar radiation, including glass, plexiglass, Kalwall™, etc.

Greenhouse (see Chapter 5)

Heating Load—the term refers to the amount of BTU's required to perform the task of water and/or space heating.

Heat Sink—a massive body which can serve to absorb and store solar heat.

Hybrid Solar Heating System—solar heating system that combines active and passive techniques.

Indirect Solar Gain (see Chapter 4)

Infiltration—the unwanted admittance of air through cracks and pores which increases heat transfer.

Internal Mass—massive materials with heat storage potential contained within the building as walls, floors, or free-standing elements.

Isolated Gain (see Chapter 5)

Moveable Insulation—insulation placed over windows when needed to prevent heat loss or gain, and removed for light, view, venting, or heat.

Passive Solar Energy System—a solar heating and/or cooling system using natural means of heat distribution—generally building's structure itself forms solar system.

Peak Load—the design heating and cooling load used in mechanical system sizing. Usually set to meet human comfort requirements 93%-97% of the time.

Plenum—a cavity of air space through which air is moved. In some passive solar designs a plenum may be used to evenly distribute heat which otherwise would collect at a single point.

Radiation (or Radiant)—the process in which energy in the form of rays of light and heat is transferred from body to body without heating the intermediate air.

Retrofit—to add a solar heating or cooling system to an existing home, previously conventionally heated and/or cooled.

Rock Bed (Remote)—a heat storage container filled with rocks, pebbles, or crushed stone.

R-value—capability of a substance to impede the flow of heat. The term is used to describe insulative properties of construction materials. (Also see Thermal Resistance.)

Selective Surface Coating—specially adapted coating with high solar radiation absorptance and low thermal emittance, used on surface of an absorber plate to increase collector efficiency.

Solar Fraction—the percentage of a building's net heating load met by solar gain.

Solar Gain—the absorption of heat from the sun. The amount of solar radiation (BTU's) received on an identified surface.

Solarium (or Greenhouse, Sunspace, and Isolated Gain) (see Chapter 5).

Solar Mass Wall (see Trombe Wall)

Stack Effect—The ability to set up a large enough temperature difference to effect the displacement of warm air by cooler air in a thermal chimney, such that the lighter warm air rises through a distribution space. (Also see Thermal Chimney and Stratification.)

Stagnation—trapped heat, no air movement.

Storage—Storage refers to storing heat collected during the day in excess of immediate requirements for use overnight or on cloudy days. Unless infinite storage is provided, a system cannot be fully utilized at all times except during periods of high demand. (Also see Collection and Distribution.)

Stratification—the temperature gradient distribution of a material or substance. A water-filled container used in solar heat storage usually will be warmer at the top than at the bottom. This difference is expressed as stratification or layering. (Also see Stack Effect and Thermal Chimney.)

Supercharge—process of heating a storage material or room air beyond its heat capacity. Used in solarium or isolated gain.

designs as a means of increasing storage temperature. (Also see Charge and Discharge.)

Surface-to-volume Ratio (or Building Skin)—the ratio of exposed surface of a building to occupied volume. A measure of exposure to harsh climate conditions causing unwanted heat loss and heat gain. (Smaller numbers are desirable.)

Thermal Capacitance—heat storage potential of a substance. Mass x specific heat of the substance = thermal capacitance.

Thermal Chimney—a vertical cavity through which heated air moves as a result of the stack effect. Used as a means of passive solar heat distribution or induced ventilation. (Also see Stack Effect and Stratification.)

Thermal Envelope—the enclosure of a building which exhibits thermal resistance.

Thermal Resistance—the ability of a substance to impede the flow of heat. (Also see R-value)

Thermosiphon (see Convection)

Time-lag Heating—a process of heating a building's interior by using the heat loss properties of massive materials to delay the movement of solar heat.

Trombe Wall (or Solar Mass Wall)—a wall that absorbs collected solar heat and holds it until it is needed to heat house interior.

Trombe Wall Cavity—the space between a solar mass wall and its exterior glazing in which air is heated. This air will rise and may be vented into the building's interior for distribution.

U-value—the capability of a substance to transfer the flow of heat. Used to describe the **conductance** of a material or composite of materials used in construction. The reciprocal of resistance:

$$\frac{1}{U \cdot A} = R$$

Ventilation, Induced—the thermally assisted movement of fresh air through a building. (Also see Stack Effect and Thermal Chimney.)

Ventilation, Natural—the unassisted movement of fresh air through a building.

Water Storage Wall (see Trombe Wall)

BIBLIOGRAPHY

ALTERNATIVE NATURAL ENERGY SOURCES IN BUILDING DESIGN
. . . A. J. Davis and R. P. Shubert; Passive Energy Systems, P.O.
Box 499, Blacksburg, VA 24060, 1977, $7.00.

ARCHITECTURE AS ENERGY . . . M. Villecco; DESIGN QUARTERLY
(103):1-36, 1977.

DESIGN CRITERIA FOR SOLAR HEATING BUILDINGS . . . Everett M.
Barber and Donald Watson, AIA, Sunworks Inc., P.O. Box 1004,
New Haven, CT 06508, 1975, $10.00

DESIGN WITH CLIMATE . . . V. Olgyay; Princeton University Press,
Princeton, NJ 08540, 1963, $28.50.

DESIGN FOR A LIMITED PLANET . . . Norma Skurka and Jon Naar;
Ballantine Books, New York, 1976, $5.95.

DESIGNING AND BUILDING A SOLAR HOUSE, YOUR PLACE IN THE
SUN . . . Donald Watson, Garden Way Publishing, Charlotte, VT
05445, $8.95.

EARTH COVERED BUILDING FOR ENERGY CONSERVATION CON-
FERENCE PROCEEDINGS . . . Frank L. Moreland, Director, Center
for Energy Policy Studies, Institute of Urban Affairs, University of
Texas at Arlington, Arlington, TX 76019.

EARTH INTEGRATED ARCHITECTURE: AN ALTERNATIVE METHOD
FOR CREATING LIVABLE ENVIRONMENTS WITH AN EMPHASIS
ON ARID REGIONS . . . J. W. Scalise (ed); Architecture Founda-
tion, College of Architecture, Arizona State University, Tempe,
AZ 85281, 286 pp, $10.00.

ENERGY CONSERVATION IN BUILDING DESIGN, 1974 . . . The
American Institute of Architects, 1735 New York Avenue NW,
Washington, DC 20006, $5.00.

ENERGY, ENVIRONMENT AND BUILDING . . . Philip Steadman;
Cambridge University Press, 32 East 57th Street, New York, NY
10022, $5.95.

THE FUEL SAVERS . . . Dan Scully, Don Prowler and Bruce Anderson; Total Environmental Action, Inc., Church Hill, Harrisville, NH
03450, 1978, $3.75.

LET'S REACH FOR THE SUN—30 ORIGINAL SOLAR HOME
DESIGNS . . . George Reynoldson and the Space/Time gang;
Space/Time Designs, Inc., P.O. Box 4229, Bellevue, WA 98009,
July 1978, $9.95.

NATURAL AIR CONDITIONING WITH ROOF PONDS AND MOVABLE
INSULATION . . . H. R. Hay and J. I. Yellott; ASHRAE TRANS.
75(1):105-177, 1969.

NEW ENERGY TECHNOLOGIES FOR BUILDINGS . . . Schoen, Hirsh-
berg & Weingart; Jane Stein, (ed); Ballinger Publishing, 17 Dun-
ster St., Cambridge, MA 02138, $5.95.

OTHER HOMES AND GARBAGE: DESIGN FOR SELF-SUFFICIENT LIV-
ING . . . J. Leckie et. al.; Charles Scribner's Sons, New York, NY
10017, 1975, 302 pp, $9.95.

THE PASSIVE SOLAR ENERGY BOOK . . . E. Mazria; Rodale Press,
Inc., Emmaus, PA 18049, (available in January, 1979).

PASSIVE SOLAR HEATING AND COOLING, CONFERENCE AND
WORKSHOP PROCEEDINGS: Albuquerque NM, May 18-19, 1976,
Report No. IA-6637-C, 1977, 355 pp. Available from National
Technical Information Service, 5285 Port Royal Road, Springfield,
VA 22151, $10.50.

PASSIVE SOLAR HOME FOR NORTHERN CLIMATES . . . D. Marier
and A. Marier; ALTERNATIVE SOURCES OF ENERGY (25):5-11,
1977. (Complete set of blueprints available from: Don and Abby
Marier, Route 2, Box 74, Milaca MN 56353, $15.00).

PASSIVE SOLAR: STATE OF THE ART . . . Proceedings of the 2nd
National Passive Solar Conference, March 16-18, 1978, University
of Pennsylvania, Philadelphia, PA, 1978. Three volumes. Available from Mid-Atlantic Solar Energy Association, 2233 Gray's
Ferry Avenue, Philadelphia, PA 19146, $20.00.

REGIONAL GUIDELINES FOR BUILDING PASSIVE ENERGY CONSERV-
ING HOMES . . . AIA Research Corporation. Available in January,
1979 from the US Government Printing Office, Washington, DC
20402.

RESEARCH EVALUATION OF A SYSTEM OF NATURAL ARCHITEC-
TURE . . . K. Haggart; Report No. PB 243 498. Available from
National Technical Information Service, 5285 Port Royal Road,
Springfield, VA 22151, $10.50.

SHARING THE SUN: SOLAR TECHNOLOGY IN THE SEVENTIES, Joint
Conference of the American Section, International Solar Energy
Society and the Solar Energy Society of Canada, Winnipeg,
Manitoba, August 15-20, Volume 4, Solar Systems, Simulation,
Design. Complete Proceedings, Individual Volumes of Papers
available from American Section, International Solar Energy So-
ciety, 300 State Road 401, Cape Canaveral, FL 32920.

SIMULATION ANALYSIS OF PASSIVE SOLAR-HEATED BUILDINGS . . .
D. Balcomb and J. Hedstrom; Report No. LA-UR-76-1719, Los
Alamos Scientific Laboratories, Solar Energy Lab., Mail Stop 571,
Los Alamos, NM.

SOLAR DWELLING DESIGN CONCEPTS . . . AIA Research Corporation; Stock No. 023-000-00334-1, May 1976, 136 pp., $2.30. Avail-
able from Superintendent of Documents, Government Printing
Office, Washington, DC 20402.

SOLAR ENERGY HOME DESIGN IN FOUR CLIMATES . . . Total Environmental Action (TEA), Church Hill, Harrisville, NH 03450,
$12.75.

SOLAR GREENHOUSE: DESIGN CONSTRUCTION AND OPERATION
. . . R. Fisher and B. Yanda; John Muir Publication, Santa Fe, NM,
1976, $6.00.

SOLAR HEATED BUILDINGS: A BRIEF SURVEY (13th edition) . . .
William A. Shurcliff; 19 Appleton Street, Cambridge, MA 02138,
$12.00.

THE SOLAR HOME BOOK, HEATING, COOLING AND DESIGNING
WITH THE SUN . . . Bruce Anderson, Brick House Publishing
Company, Church Hill, Harrisville, NH 03450, $7.50.

SOLAR HOUSES, 48 ENERGY-SAVING DESIGNS . . . Louis Gropp; A
House and Garden Book, Pantheon Books, New York, NY, 1978,
$8.95.

SUNSET HOMEOWNER'S GUIDE TO SOLAR HEATING . . . Sunset
Books, Lane Publishing Co., Menlo Park, CA 94025, $2.95

THERMAL ANALYSIS OF A BUILDING WITH NATURAL AIR CONDI-
TIONING . . . J. I. Yellott and H. R. Hay; ASHRAE TRANSACTIONS
75(1):178-188, 1969.

THERMAL MODEL FOR A SOLAR HEATING BUILDING . . . M. G.
Davies; BUILDING SCIENCE SUPPLEMENT (ENERGY AND HOUS-
ING): 67-76, 1975.

30 ENERGY EFFICIENT HOUSES—YOU CAN BUILD . . . Alex Wade
and Neal Ewenstein, Rodale Press, 33 East Minor Street, Emmaus, PA 18049, 1977, $8.95.

UNDERGROUND HOUSING . . . R. F. Dempewolff; SCIENCE DIGEST
78(5):40-53, November, 1975.

THE USE OF EARTH COVERED BUILDINGS . . . National Science
Foundation; Stock No. 038-00000-286-4, 1976, $3.25. Available
from Superintendent of Documents, Government Printing Office, Washington, DC 20402.

WHY NOT BUILD THE HOUSE RIGHT IN THE FIRST PLACE? . . .
Raymond Bliss; BULLETIN OF THE ATOMIC SCIENTISTS: 32-40,
March 1976.

SURVEY OF PASSIVE SOLAR BUILDINGS . . . AIA Research Corpora-
tion; Stock No. 023-000-00437-2, February 1978, 177 pp., $3.75.
Available from Superintendent of Documents, Government
Printing Office, Washington, DC 20402.

THE THERMAL ADMITTANCE OF LAYERED WALLS . . . M. G. Davies
BUILDING SCIENCE 8:207-220, 1973.

The National Solar Heating and Cooling Information Center has
compiled several bibliographies on passive systems and related
solar subjects. Some of these are:

Passive Solar Energy Designs and Systems
Underground Houses
Solar Greenhouse Bibliography and List of Plans
Energy Conservation in Buildings.

The Center also maintains lists such as:

Solar Equipment Manufacturers (Description Code SYP: Passive)
Passive Design Tools.

To receive the above bibliographies and lists and other information on passive solar systems write to the National Solar Heating
and Cooling Information Center, P.O. Box 1607, Rockville, Mary-
land 20850.

CREDITS

The information presented in this report is the result of the efforts of many people—the applicants in the design competition, the individuals who participated in the evaluation of the applications, and the writers and illustrators who prepared the material for publication. In addition to the HUD staff, these include:

Application Evaluation
A.I.A. Research Corporation
Brian H. Ford
Greg Gibson
Steve Kleinrock
Vivian Loftness

Dennis A. Andrejko (D. Wright, A.I.A.)

Dave Bainbridge (California State Energy Commission)

Herman Barkman (Barkman Engineering)

Boeing Aerospace Co.
Jim Nenninger

Burt-Hill, Kosar & Rittelman
Steve Denbow
Harry T. Gordon
Bob Kobet

Douglas Coonley (Chi Housing, Inc.)

Dubin-Bloome Associates
Jules Alcorn
Sel Bloome
Fred Dubin
Gerald Falbel
William Klingsborn
Bernard Levine
Robert Martin
Glen Tucker
Bill Maxfield

Franklin Research Center
Tom Giller
Tom Lent
Gene Nieri
Molly Sayvetz
James Schardt
Frank Weinstein

Drew Gillett

William Glennie

Marshall Hunt (California State Energy Commission)

Tim Johnson (M.I.T.)
Ralph Jones

Denny Long (California State Energy Commission)

Los Alamos Scientific Laboratories
Doug Balcomb
Bob McFarland
Bill Wray

Richard Larry Medlin

Bill Mingenbach (The Architects Taos)

Dan Nall

225

National Bureau of Standards
Ed Arens
Bill Ducas
S. Robert Hastings
Dennis Jones
Gene Metz
Kalen Ruberg

Bill Otwell

Peter Powell (Mark-Beck Associates, Inc.)

Gordon Preiss (Solar Processes Inc.)

Travis Price (Sun Harvester Corporation)

Real Estate Research Corporation
Len Carnevale
J. Norwell Coquillard
Peter Gennuso
Linda Hardway
Maxine V. Mitchell
Ann Morrow
Iveana Shook
Stephen Spigel
Margaret Winter
Randy Zeldin

Steve Robinson

Solar Energy Research Institute
Gregory Franta
Keith Haggard
Steve Hogg
Michael Holtz
Ron Judcoff

Total Environmental Action
Winslow Fuller
Joe Koehler
Charles Michael
Paul Pietz
Dan Scully
Paul Sullivan

Peter Van Dresser (Sundwellings Program)

George Way, A.I.A. (Tackett, Way, Lodholz & Associates)

David Wright, A.I.A.

John Yellot (Arizona State University)

Publication Preparation
AIA/RC
Vivian Loftness
Peter Whitehead

Consultants
Peter Calthorpe
Steve Denbow
Harry Gordon
Doug Kelbaugh
Bob Kobet
Ed Mazria
Susan Nichols
Don Prowler

Ballinger
Martin Conrad
Rita Eng
William J. Fisher
Steve Katz
Arvind Padwall
Richard J. Palmer
Robert H. Rand
Judi Repp
Alberto Salvatore
David Wang

Chilton
John Cantwell
Sally Collins
Ed Dahl
Gail Krueger
Tom Stavrakis
F. Phillip Verrechia

Franklin Research Center
Dan Shimberg, Project Manager
George Royal, Technical Editor
Marsha Biderman, Designer
Rich Barrett
Anne Peters
Rita Wilson

Consultant
William Langdon

CPSIA information can be obtained
at www.ICGtesting.com
Printed in the USA
LVOW02s0416130317
526972LV00002BA/412/P